DIGITAL SPIRIT

Minding the Future

Jan Amkreutz

ISBN: 1-4107-5640-8 (e-book)
ISBN: 1-4107-5639-4 (Paperback)
ISBN: 1-4107-5638-6 (Dust Jacket)

Library of Congress Control Number: 2003094481

This book is printed on acid free paper.

Printed in the United States of America
Bloomington, IN

1stBooks - rev. 07/03/03

For

Marlies
Michael & Aileen
Suzanne & Scott
Elisa & Otis
Izabella
Emma
Andrew
Maxwell

Table of Contents

Confessions to the Reader

As a young man, I read Teilhard de Chardin's *Phenomenon of Man*, which is arguably the most prominent work about the evolution of all that exists: the evolution of the universe, the evolution of life, and the evolution of the human mind. Teilhard presents a vision in which a 'layer of human thought' will emerge and lead to a point of convergence in the future. Teilhard called this layer the *Noosphere* and the point of convergence *Point Omega*, which he envisions to be God. This view fascinated and mystified me. Was the world entering its final leg of existence; was the universe on its final approach to a spiritual landing strip?

The question followed me throughout my life, until finally, after working for three decades in the application of computer technology, as a programmer, teacher, researcher and entrepreneur, I read Ray Kurzweil's *The Age of Spiritual Machines*. Kurzweil envisions intelligent machines that will surpass human intelligence in the *coming decades*, when the robots will take over the management and control of the world. Incredibly, Kurzweil's very different vision of the future left me with the same question. Is humanity facing the beginning of an end? Are the robots the 'flesh and blood' of Teilhard's layer of thought?

I asked myself, again, 'what in the world is going on?' My search for an answer is the motivation for writing this book, and the journey of that search is the book's content. Part of the answer lies behind the glossy, and not so glossy, surfaces of the gadgets produced by the profession I had pursued with such passion. Beyond cell phones and PC's, and beyond spreadsheets and the Internet, a new reality is engulfing the globe. This reality is digital and will change our minds in ways we can hardly fathom. It changes the question 'What in the world is going on?' into a broader one: 'what in the universe is going on?' This is the question I try to answer in this book.

My *first confession* to you, dear reader, is this: my view of the future is a heartfelt positive one. Does this mean that I see the 'glass' representing the present to be 'half-full' instead of 'half empty'? The answer might surprise you, because the answer is *no*. The true answer is a more interesting and a more challenging one. My positive view is about the glass itself. For 15 billion years, from the bubble that burst into the universe at the time of the Big Bang, to the digital bubble that burst in the stock markets at the end of the 20th century, nature has continuously and relentlessly expanded the glass *itself*. I call it the glass of opportunity. That glass represents the expanding repertoire of the choices we have, and the new choices that we create. Our frame of mind will determine the choices we make, and the choices we make will change our frame of mind. This interaction puts us on the spiraling road that is our future. Today, we are about to cross a new bridge in evolutionary history. Crossing that bridge accelerates the expansion of our glass of opportunity to a size beyond belief. Its opportunities cover the full range, from a self-inflicted Armageddon to a Garden of Eden. To find a sustainable path in between the two is the task ahead of us. The bridge we are about to cross is built in bits and bytes, the 'atoms' of the maturing digital world. This bridge will expand, and change, our human minds. That change is the bedrock of my passionate belief in human nature, and its ability to find and define the winding path of human *being*.

My *second confession* to you concerns language. Having been born and raised in The Netherlands, I miss the well-worn clothing of my native language in writing this book. On top of that, my life-long passion was the language of engineering, and the language of computing. When that shows, dear reader, I will ask for your smile.

My *third confession* concerns language also. Just as the peoples, cultures, religions, and the villages of our planet have their own language, every science has its own language as well. Languages evolve in harmony with the view of the world of the people that use the language. Since the publications of renowned scientists like Teilhard de Chardin and Julian Huxley, the perspective of one coherent evolutionary story of our existence has just begun to emerge, and hence, we still do not have a vocabulary suitable to describe the continuity of our own evolutionary history. I consciously try my best to stay within scientific boundaries, simply because the current state of the scientifically knowable is a miracle of human comprehensibility in itself. Having said that, I ask for the benefit of the doubt when I stray in my use of words, as the lack of vocabulary that can describe the universe, the diversity of life, and the creations of the human mind, as the one brilliant symphony that it is, restrict me.

If, dear reader, you view the book as a cocktail of scientific insights and metaphorical fiction, with sprinkles of contention, assertion and conjecture - a cocktail not shaken, but just slightly stirred, we can tune in on the same frequency of thought as we embark on a journey that changed my mind; a journey that will change your mind as well, one way or another. This is my promise to you. It is why I wrote this book.

Jan Amkreutz
Somers, Montana, USA, March 24, 2003

DIGITAL SPIRIT

Minding the Future

"Now is the watershed of Cosmic history. We stand at the threshold of the New Millennium. Behind us yawn the chasms of the primordial past, when the universe was a dead and silent place; before us rise the broad sunlit uplands of a living cosmos. In the next few galactic seconds, the fate of the universe will be decided.
Life –the ultimate experiment- will either explode into space and engulf the star-clouds in a fire storm of children, trees, and butterflies wings; or Life will fail, fizzle, and gutter out, leaving the universe shrouded forever in impenetrable blankness, devoid of hope. Teetering here on the fulcrum of destiny stands our own bemused species. The future of the universe hinges on what we do next."

Marshall T. Savage, The Millennial Project.[1]

Prologue: Minding the Future

"Most people, even though they don't know it, are asleep. They are born asleep, they marry in their sleep, they breed children in their sleep, they die in their sleep without ever waking up. They never understand the loveliness and the beauty of this thing that we call human existence."

Anthony de Mello[2]

Her face is radiant, bright, yet stern. The frown between her eyes tells her mood. She is angry.

"I am not angry, Mr. J. But I am concerned. Very concerned. I wonder what in the world is going on. It is nothing short from astonishing."

I will introduce her to you in a little while, dear reader. I'm curious to hear what she has to say. She is quite formal, most of the time, hence the 'Mr. J'.

What is so astonishing, Novare?

"We, the people of the planet Earth, are asleep. It is an amazing scene, if you look at it from a place beyond time and space. Amongst the two hundred billion galaxies in the universe, only one of them has a name that sounds like candy: The Milky Way. In a rural area of this galaxy, a little globe is spiraling around one of the galaxy's billions of stars; a star that we call our Sun. This star keeps all six billion people on the tiny globe alive, and keeps each of them on the move through the universe at a breathtaking speed. And most of the people are asleep."

"At first, I thought it was a dream, JJ."

When Novare appears, Alvas is always nearby. He is the informal one. Just calls me 'JJ'. A dream, Alvas?

"At the speed of light, we play the game of pointing the weapons of words and bombs and spy-satellites. The astonishing thing is that we point them inward, thus constricting space and time. To me, it feels like a scary phobic dream: people locked up in dark narrow tunnel with no place to go, ignorant of the universe of fresh air and new adventures that surrounds them."

"Imagine, Mr. J, the phobia is self-imposed! Of course, it is easy for me to feel this way after helping you write this book. I know about the reality that unfolds beyond the ring of satellites, because I know about evolution and the potential of the human mind. For the time being, however, we, the people of the planet Earth, are asleep: we made the bad dream our reality."

"We keep on dreaming while the wakeup calls thunder around the planet, JJ. ' War', 'Terrorism', 'Corporate fraud', 'Aids', 'Ethnic cleansing', 'Poverty' and 'Hunger' are the names of the dissonant tunes that try to wake us up. Digital technology carries the tunes around the world, eliminating space and time, obliterating the excuse for the arrogance of ignorance. Yet, we easily find the snooze button of our daily routine, and just keep hitting it, day after day. Every time we hit that button, we extend the dream, because, after all, we have to 'make' a living. The sounds get louder with each passing day, but we are learning to live with this gnawing feeling of unease about the world around us."

"We want to keep the peace of mind inside of us, and the physical pain of our wars outside of us, Mr. J. The war on terror, the war on hunger, the war on poverty: they are somewhere "out there." We would like to keep it that way. We hear that *times have changed* and that the *only constant is change*, as if it is a new trend, a fashion of some sorts. But then, what is it that changed? Many of us feel a growing disparity of the world inside our mind, the one we painstakingly learned since our birth, and the world "out there," the one that gives us a paycheck, social security, and wars. Why is the consumption of alcohol, mind-altering drugs, legal and illegal ones alike, at an all-time high? What is going on? What kind of a storm is brewing? Where are we heading? Where is *the world* heading, Mr. J?"

That is the central question of this book, Novare. Thanks for helping out. My objective is to shed a new light on the long-term implications of that question by proposing a new framework for answering it.

The feeling of an impending tsunami comes at a time that has no precedence in the history of the universe. In just one century, human ingenuity delivered the technologies that provide us with an array of tools to shape our own future. At one end of the spectrum, we have the choice to cure diseases, and to produce new forms of life through genetic engineering. At the other end of the spectrum is our choice to destroy all of human life at the push of a button. Never before did a species have that spectrum of choices between these boundaries of life and death.

It took fifteen billion years of cosmic evolution, and four billion years of biological evolution before the first human mind started thinking. This took place just thousands of years ago. Did evolution stop at that moment? Or did, as some believe, technology take over the task of nature and continued the process of evolution of the world? The answers are no, and no.

We, the people of the planet earth, are about to wake up to the incredible depth of the creative powers of our own human mind. As we shed the Darwinian mindset of defenseless randomness, we will rediscover where we are in the history of the universe. We will realize that we are at the frontier of evolution, and that *we are* that frontier. Now, with the human mind given to us by time, we will see that we *are* creating our own future, whether we realize it or not. Evolution is how the world came to be, and evolution is how the future will develop. With the emergence of human knowledge, evolution continues its journey of expanding space and time, and shaping every event and every experience in its unfolding history: the history of the future.

A tsunami is about to hit the shores of our existence. However, it does not signal The End, or the coming of Judgment Day. Rather, it signals a new beginning, the beginning of the era of humanity, ruled by what is exclusively human: a mind that has the creative powers to shape new futures. We arrived at a new milestone in the evolution of human knowledge, and the emergence of new opportunities that this knowledge reveals. This milestone marks the beginning of acceleration, rather than the approach to a looming singularity. This is not some utopian new world; it is already happening.

The technology came quietly, nobody planned for it, and nobody controls it. Over the last fifty years, big chunks of it developed in isolation, and for different purposes. Now, it is starting to surface, patches of it, like the first bubbles rising to the surface in boiling water. We know those bubbles as 'websites', their invisible connections as 'the Internet'; its totality we call 'cyberspace'.

That is it, *the Internet?* some might ask. They might call the Internet 'A useful tool', or 'hype', or 'the future of commerce'. Many people, if not most, see the Internet, and cyberspace, as a phenomenon that is secondary to commerce, on line shopping, or at best a handy tool for communication or playing computer games. They see its applications as playful waves on the ocean's surface on a peaceful summer night. They fail to perceive the depth and the power of the underlying currents that are heading for the shores of humanity: a tidal wave that will expand human thinking. The wave is about to engulf us, permeating our lives, and changes us, forever. It is the wave of a new reality: the digital reality.

We are weaving this new reality, piece-by-piece, and layer-by-layer. You and I, all of us. We are shaping the colorful patches of its four-dimensional quilt. They are the products of our individual minds. Thoughts and emotions, inventions and scientific discoveries, rules and laws, movies, music and poetry are the unique components, each carrying individual meaning, put in the fabric of this new reality, built with bits and bytes. We dress them as images, movies, diagrams, spreadsheets and music. In this new reality, we already conceive, design and engineer, advertise and sell the products of our imagination, before they ever hit the store-shelves or the UPS delivery van.

We are weaving this new reality from patches of bits and bytes, and the yarns that we call 'the world wide web'. It is the reality in which we already engineer and test bridges, highways, and airline schedules, before they ever become reality. From it, we can launch, manage and control space probes, marketing campaigns, and missiles. Digeality is the design lab, workbench, factory, and corporate meeting room for the future, all arranged and connected into one digital landscape. In the near future, we will test our new cars in digital crashes. The "dummies" of the future consist of bits, not atoms. Weather will manifest itself in digital reality before hitting the shores on the planet. All of us are going to have our presence in this new reality. The shape and structure of our bodies, the sequence of its genes and its medical history will move around in digital form, tracking and guiding our movements in space and time of the real world. Our digital image will monitor and maintain the vital signs of our real bodies through the digital world in our clothing. Our digital twin will know our wishes, desires and preferences, and will understand the meaning of our expressions.

In digital reality, we will recreate the soil and the atmosphere of Mars before we get there ourselves. In it, we will design and test new forms of life on distant planets, and eventually, create dwellings and new life forms, remotely, using nanotechnology and biotechnology, thus transforming digital creations into physical reality. Our expectations of intelligent alien life forms might prove to be right after all. Only, we might create them, not discover them by surprise.

This extension of our physical world is coded in digital languages, its renditions are becoming as real as the 'real world' itself. Yet it is different. And then, there are new realities. Worlds we could hardly imagine before. It is the tapestry of connected human thought, originating from logic, imagination or intuition. To give this new reality a name, *digeality* seems as good as any. Digeality then, expands the horizons of our real world, adding to it, and modifying it. It surrounds us, permeates our lives, and connects us. We talk to it, and in it. It eliminates the barriers of space and time, or more accurate, it has a time-space continuum of its own, changing our own perspective on time and space. We create knowledge in it ourselves, and discover the knowledge of others, we combine domains of knowledge, create new perspectives, hence new knowledge. We learn how our fellow human beings *think*, and in doing so, we will change our own, or their, mindsets. Digeality is becoming an integral and intimate part of our every day life, our politics, our sciences, and doing business.

The enormous power of computers leads some to the prediction that robots will take over from us. The world, as we know it, is going to end, leaving us behind in a doomed land of terror, or in a blessed land of human zombies. Such views will prove to be shortsighted. The impact of this new reality reaches far beyond 'surfing the Internet', 'extended enterprise', 'virtual nations', or 'Robo sapiens'. It will change the way we think about our daily life, be it family life, professional life, healthcare, education, government, or the use of genetic engineering. It will change our expectations. A new mindset will emerge that determines the future of human evolution.

Reality did get ahead of our way of thinking. Billions of us are ignorant and even fight the reality of evolution. How can we face the reality of change if we ignore the change of reality itself? How can we mind about our future if we keep fighting the evolution of the past? Our mindset is still stuck in the static worldview that prevailed for millennia, and still prevails today.

It took the human mind only thousands of years to bring us to where we are today. We begin to understand the nature of the universe, we can design molecular machines that make nuclear weaponry seem like child's play, but we barely scratched the surface, quite literally, of the mysteries of the human mind. No wonder, we have yet to learn how to use it. Knowledge, and all of its manifestations, shapes the reality of our world, and our mindset determines the way we deal with that reality. We can now identify the three intertwined spirals of thought that run through the story of this book.

The first spiral is the evolution of the realities we live in: the universe, life on the planet earth, and the products of human thought. I will look at the past to shine a new light on the future.

The second spiral is the power of the human mind. I will suggest that we are just crossing the threshold of its true potential.

The third spiral is the emerging reality of the digital world. I will expose digeality as the drawing board, the laboratory, and the factory, for the creation of the new realities of the future.

"JJ?"

Yes, Alvas.

"Strong language, JJ. Aren't you over-extending yourself?"

I can read it from her face, Alvas. Novare has an answer for you.

"Indeed, I do! Listen to me, Alvas! Has your dream ever been shattered by a single remark, in one tiny fraction of a second? *Nice idea Jones, but you do not understand this organization. Our company will not go out on a limb!* This might be your boss speaking. You *know* that your idea would catapult your company to the halls of corporate fame, but your dream is in ruins, shot to pieces by a single remark. Or a 'friend' shattering your dream after a passionate presentation of your new business-plan: *Quit dreaming. Get real!* You got the message: your idea is beyond your, or their (?), comprehension of reality. You want to play the violin; they tell you to stick to your chainsaw. The message is plain and simple: You are reaching beyond your grasp."

Alvas looks puzzled. "What are you trying to tell me, Novare?"

"How would children ever learn, if not for reaching out to fill their minds with the unknown, way beyond their grasp? How could art, and science, exist without reaching into uncharted territory? Of course Mr. J. has to reach out, sometimes beyond his grasp. Without that, he would just be like another hamster that keeps running around in the cage of the well-known and only perceives the surface of reality as it rolls by."

"That would be no fun at all."

"Remember, Alvas. You and I are going to help!"

"That *will* be fun."

The universe reached out in a perpetual wave of energy, expanding its chaotic distribution of particles and energy, so that stars, galaxies, and solar systems could form. Living nature reached out in similar ways, continuously experimenting with new combinations of DNA to form millions of species, build new organs, trying out new life forms, and finally, producing a species that was capable of reaching out to discover America, produce penicillin and invent the light bulb.

Reaching out into no-man's-lands is the beginning of any change and of true innovation. Intuition and imagination, creativity and inventiveness, love and passion, are but a few of its powerful manifestations. Reaching out is in our mind, it is in our genes, and it is in every particle of the universe. It defines humanity.

During our journey, we will visit with the understanding of many great minds. Their explanations of reality are the landmarks on the road to the emergent present. Sometimes I reach beyond them and venture into the unknowable territory of the future, where we can no longer count on biological evolution to take the decisions away from us, and where we cannot count on the machines either: how would they know the best future for our *human* nature? We are facing the challenge of co-creating the future. It seems that we still have to wake up to that. Which we are about to do.

"Let's have some fun, JJ."

"Serious fun, Alvas. Not your kind of fun."

"I know Novare. When you are around, there is no fun at all. But then, I'll be there too."

PART I: GENERATION OF CHANGE

June 26, 1945

WE THE PEOPLES OF THE UNITED NATIONS DETERMINED:
To save succeeding generations from the scourge of war, which twice in our lifetime has brought untold sorrow to mankind, and
To reaffirm faith in fundamental human rights, in the dignity and worth of the human person, in the equal rights of men and women and of nations large and small, and
To establish conditions under which justice and respect for the obligations arising from treaties and other sources of international law can be maintained, and
To promote social progress and better standards of life in larger freedom.

Preamble to the Charter of the United Nations, 26 June 1945.[1]

"But the spread, both in width and depth, of the multifarious branches of knowledge during the last hundred odd years has confronted us with a queer dilemma. We feel clearly that we are only now beginning to acquire reliable material for welding together the sum total of all that is known into a whole; but, on the other hand, it has become next to impossible for a single mind fully to command more than a small specialized portion of it. I can see no other escape from this dilemma (lest our true aim be lost forever) than that some of us should venture to embark on a synthesis of facts and theories, albeit with second-hand and incomplete knowledge of some of them – and at the risk of making fools of ourselves."

Erwin Schrödinger, Dublin, September 1944.[2]

"Change is hard, but stagnation is fatal..."

Peter Bishop[3]

The declaration of the United Nations expresses the new hope in the middle of the 20th century, after half a century of unprecedented physical violence and a century of unprecedented scientific discoveries and technological advancements, that no single human being could no longer comprehend as one totality, as Erwin Schrödinger's words from 1944 describe.

In Part I of our journey, I will start at exactly the time Schrödinger wrote down his words: September 1944. From the perspective of that moment in time, we will visit with some of the minds that brought us there. I will suggest that at that moment, the gap between our scientific view of the world and the 'common mindset' around the globe has already started: a gap that would grow in the second half of the 20th century, and become a deep chasm of knowledge, that I describe as the 'accelerating friction in the axle of evolutionary change'. While the gap is growing, human minds invented a way to describe reality in a radically new way: digital technology became a reality. In the second half of the 20th century, change would accelerate. The generations that were alive in it, are the generations of change. The product of this era is the generation of change. Our physical reality changed; the change of our mindset did not keep the pace of that change.

Chapter 1: Mindset

Games of War

It was on December 10, 1945. Albert Einstein addressed the audience on the occasion of the fifth Nobel Anniversary Dinner at the hotel Astor in New York. World War II was over. A new organization had announced its charter to the world. The United Nations were determined: the world would be a place of hope and promise. The charter agreed upon just five months earlier, in June of 1945, left no doubt, the nations of the world would stand shoulder to shoulder now. Together, they would take on the fight for social justice, freedom and better standards for living for everyone. Together, they could, and they would, win that fight. The title of Einstein's speech was *The war is won, but the peace is not.* Einstein was skeptical about the expectations held by so many:

> *"The war is won, but the peace is not. The great powers, united in fighting, are now divided over the peace settlements. The world was promised freedom from war, but in fact, fear has increased tremendously since the termination of the war. The world was promised freedom from want, but large parts of the world are faced with starvation while others are living in abundance. The nations were promised liberation and justice…*
> *…The picture of our postwar world is not bright. So far as we, the physicists, are concerned, we are no politicians and it has never been our wish to meddle in politics. But we know a few things that the politicians do not know. And we feel the duty to speak up and to remind those responsible that there is no escape into easy comforts, there is no distance ahead for proceeding little by little and delaying the necessary changes into an indefinite future, there is no time left for petty bargaining. The situation calls for a courageous effort, for a radical change in our whole attitude, in the entire political concept. May the spirit that prompted Alfred Nobel to create his great institution, the spirit of trust and confidence, of generosity and brotherhood among men, prevail in the minds of those upon whose decisions our destiny rests. Otherwise, human civilization will be doomed."*[1]

"The situation calls for a courageous effort, for a radical change...." Einstein said fifty-seven years ago. What words would he add, if he could address the world today?

It is December 10, 2002. The window in my office frames the view of the snow-covered mountains of rural Montana, when the TV-set wakes me from my reverie. "And now, live from Oslo... where President Jimmy Carter will give his acceptance speech for the 2002 Peace Nobel prize. As always, CNN will bring it to you live."

"Your majesties, members of the Nobel Committee in Norway, your Excellencies, distinguished guests; it is with a deep sense of gratitude that I accept this prize..."

Exactly 57 years, to the day, after Albert Einstein delivered his speech, ex U.S. president Jimmy Carter addresses the world:

"...In Washington and in Moscow, we knew that we would have less than one-half hour to respond after we learned that intercontinental missiles had been launched against us. There had to be a constant and delicate balancing of our great military strength with aggressive diplomacy, always seeking to build friendships with other nations, large and small, that shared a common cause..."
"...12 years ago [Soviet] President Mikhail Gorbachev received recognition for ending the Cold War that had lasted 50 years. But instead of entering a millennium of peace, the world is now in many ways a more dangerous place. The greater ease of travel and communication has not been matched by equal understanding and mutual respect. There is a plethora of civil wars, unrestrained by rules of the Geneva Convention, within which an overwhelming portion of the casualties are unarmed civilians who have no ability to defend themselves..."

Two wars, bridged by the rainbow of peace we call the Cold War. What changed?

"...The world has changed greatly since I left the White House. Now, there is only one superpower, with unprecedented military and economic strength. The coming budget for American armaments will be greater than those of the next 15 nations combined. And there are troops from the United States in many countries throughout the world. Our gross national economy exceeds that of the three countries that follow us..."

Reality changed.

"...In order for us human beings to commit ourselves personally to the inhumanity of war, we find it necessary first to dehumanize our opponents, which is in itself a violation of the beliefs of all religions. Once we characterize our adversaries as beyond the scope of God's mercy and grace, their lives lose all value. We deny personal responsibility when we plant land mines, and days or years later, a stranger to us, often a child, is crippled or killed. From a great distance, we launch bombs or missiles with almost total impunity, and never want to know the number or the identity of the victims."

Did our mindset change?

"We have not yet made the commitment to share with others an appreciable part of our excessive wealth. This is a necessary and potentially rewarding burden that we should all be willing to assume. Ladies and gentlemen, war may sometimes be a necessary evil. But no matter how necessary, it is always evil, never a good. We will not learn how to live together in peace by killing each other's children. The bond of our common humanity is stronger than the divisiveness of our fears and prejudices. God gives us a capacity for choice. We can choose to alleviate suffering. We can choose to work together for peace. We can make these changes. And we must."[5]

A few months later, when I read Einstein's and Carter's words again, I wonder about the changes in the half century that separate them. Half a century of changes in the physical reality that we call 'our world'. Half a century of human thinking. Did our thinking change? Did the *way we think* change?

My thoughts spiral around, bouncing off the edges of my mind. My thoughts bounce forth and back between conscious attempts to find answers and the feelings that seem to bubble up from my unconscious. What happened? What went wrong after the emerging hopes of the 1940's? Vaguely and subconsciously, I notice an eerily familiar image on CNN. What was that? Where did I see that before? As my mind wanders and wonders, Novare appears out of nowhere. She always does. I never need to call her. She just appears. As always, Alvas is with her, wearing a big grin.

"You want to go back, JJ?"

"Everything is prepared, Mr. J. The team is ready, the bugs worked out. Are *you* ready?"

I am confused. "Sure" is all I remember saying...

Today is September 27, 1944.

The Netherlands are chilly this time of year, but even without a coat, I felt warm. A small team of people had made it possible for me to visit the little town, at exactly the right time. In many parts of Europe, the war has reached a turning point. You can feel it. The war is ending. The beginning of the end had started on the beaches of Normandy, a little over three months ago. Now, you can hear it.

The different timbre of the English language sounds liberating and melts away the icy reality of German occupation. You can smell it. The smoldering remains of cities hide the canvas for a new future underneath. A new hope is born. The seeds of expectation for a brighter future are being sown in the fertile soil of that hope. Visions of new opportunities fuel the desire for better times.

Passierschein

Der deutsche Soldat, der diesen Passierschein vorzeigt, benutzt ihn als Zeichen seines ehrlichen Willens, sich zu ergeben. Er ist zu entwaffnen. Er muß gut behandelt werden. Er hat Anspruch auf Verpflegung und, wenn nötig, ärztliche Behandlung. Er wird so bald wie möglich aus der Gefahrenzone entfernt.

OBERBEFEHLSHABER
der alliierten Expeditions-Armeen

Englische Übersetzung nachstehend. Sie dient als Anweisung an die alliierten Vorposten.

SAFE CONDUCT

The German soldier who carries this safe conduct is using it as a sign of his genuine wish to give himself up. He is to be disarmed, to be well looked after, to receive food and medical attention as required, and to be removed from the danger zone as soon as possible.

SUPREME COMMANDER,
Allied Expeditionary Force

Illustration 1: Pamphlet distributed at the end of World War II to persuade German soldiers to surrender. Courtesy: Paul Weelen[2]

I am back in the little town, just a stone's throw from the German border, longing to find my parents. I look at the old church; the new one has not been built yet. The chestnut trees to the left of the church are as magnificent as they would be later, just not quite as tall. The leaves are coloring and the first chestnuts have fallen, scattered everywhere. Instinctively I start to pick up some of them, when I notice the pamphlets. Nobody else was around to pick them up, and no children to collect the chestnuts. The children had left, together with their parents.

Yesterday, the Americans had urged, no, commanded everybody to leave town; the town I have called home for so long. "*De Amerikane sund heij!*" ('*The Americans are here!*'), everyone told me with the excitement of reborn expectations. "The Americans are only two kilometers away," the shopkeeper said. "I'm staying; I want to be in my shop when they arrive. Just head west and you'll find your folks; the last ones just left." With the German border only a few hundred meters away to the east, that advice makes sense. He offers me his bicycle, of German make, "as long as you bring it back and visit my store!" I rearrange my gear, especially the gloves, mount the bike and head west. After not more than half a kilometer

I witness the scene that was etched in my memory by stories about the evacuation that now had just begun.

A long stream of people winds its way through hills and meadows and valleys, into the unknown, hoping to find a friendly face that will offer shelter. From my vantage point, the September colored landscape seems to slide under their feet as if it wants to make their travel effortless. I make my way alongside the people with bikes and charts, careful not to get too close to them. Then, not more than fifty meters away, I recognize them. The urge to talk to them is unbearable, but I know what will happen if I do. Walking, guiding the bike with my right hand, I nudge forward as far as I can, without touching or talking. Even with all my new gadgets, changing history is impossible. I stop, and wait, and listen.

The tall man walks in front of his wife. Their coats are warm, long and dark, just like I remember. I know what will follow, but still, I hold my breath when a somber looking Dutch official with restless eyes stops the man. The Americans had asked him to select 'trustworthy' individuals that could identify 'Nazi-collaborators' among the fleeing people. The man decides that my father belonged to the 'trustworthy', and starts to tie a white scarf around his arm that would identify him as an 'official'. "Don't let them put one of those on you!" my mother whispers. Quickly, she rips off the scarf, and pushes him forward, gently, yet insistently. "You never know what comes from that, just keep going, we have to find food for the baby," she whispers again. My father stumbles forward. The Dutch official understands. He smiles. My mother smiles back warmly, and so they go on, part of a new future that had barely begun. My father turns around to face his young wife: "Thanks." He too smiles, as he finds his bearings and continues to push the baby carriage that he had smuggled across the German border with the help of a friendly Zollbeambte, just a few weeks ago.

"Fate, " he thinks, "determined by such a simple action." Accepting the white scarf could have changed their future, as he would be forced to stay behind, leaving his family to its fate. Their son was only twelve days old, and wide-awake but quiet, maybe enjoying the ride. Wondering what could be going on in the mind of his son, and what would determine his fate, he continues his walk towards the hope that is the war's end. I ran forward to face him, not able to control the emotion controlling my motion. As my glove touched his arm, I almost fainted, but before I did, I fade, back into reality...

...My heart still pumping in my throat, I take off the gear. First the gloves. They are ultra-sensitive when touching holographic images, have forced pressure feedback, and exercise pressure on the image, so that the software can perform the necessary transformations. When I tried to touch my father's arm, the emergency exit procedure had been activated. It had been an unwritten rule in current virtual reality not to interfere with your own history. Second, I removed the contact lenses, containing the best high-resolution image projectors available today. Lastly, my shirt, pants, shoes and socks with the computer equipment woven into their fabric. The inner lining was an ultra-thin temperature controlling mesh that doubled as a monitoring device for my vital signs, as part of an experiment that studied the impact of virtual reality experiences on the human mind. The shoes had served me well. The new integrated positioning-and-motion-control module had performed beautifully, even close to so many people.

Although relieved to be back in my home office, one thing bothered me enormously. The white scarves were supposed to be black! That was my memory, and that is what I had told my team of programmers. What happened? Did my senses play a trick on me? Or was my memory distorted? Did I just dream that the scarves were white? Or black? Could it be that virtual reality had made up its own mind, just as Ray Kurzweil[3] predicted it would? As I look out the window, I decide to check out the software later. Now I have to prepare for tonight's lecture, the first one in a series on the use of digital technology in the design and construction of buildings.

It is the summer of 1978. I had arrived in Canada just a few weeks earlier, and we had been lucky to find a marvelous apartment on Nun's Island, in the middle of the majestic Saint Laurent River. My family loves Montreal, and I am very proud of my invitation as visiting professor in "Computer Aided Design" at Concordia University. With all the hype about Computer Aided Design, my first lecture focuses on what computers cannot do. These days, some predict that computers will take over the role of the human designer. "Computers are going to be as smart as humans, it is just a matter of time" was their mantra. Proponents argue that computers will be able to solve the world's pains better than humans will, or they foresee a seamless symbiosis of "man and machine." Some opponents want to stop computers from becoming smart, 'to prevent an Orwellian scenario'; after all, we are just six years away from 1984. Others find 'just the usual hype' in all the promises, or view computers as handy gadgets, or clumsy tools.

Influenced by my twenty years as a computer programmer, I am confused about the issue. On the one hand, I clearly see digital technology as the most powerful human invention in the history of human thinking. On the other hand, I feel that the depth of the human mind is vastly underestimated. However that might be, in tonight's lecture I hope to bring across some of the differences in the way the human mind produces 'thought' and how computers produce 'results' following algorithms. Edward de Bono's[4] explanations of human thinking, as described in his book '*Lateral thinking*', will be of great help. The bottom-line? Thinking about 'how computers think' is easy, because it is sequential, no matter which way you look at it.

Human thinking, in essence, is not sequential, although most schools educate our children as if it were. I hope that my students will bear with me. They will have the future, like me, to find out where digital technology will go...

...I wake up with a jump out of my chair, right here in my office. It is March, and the year is 2003. Spring is in the air. What had happened? A virtual meeting with my parents in 1944, while working in Montreal, in 1978?

I must have dozed off. I remember that my mind was wandering between the consciousness and its intimate unconscious companion. Well, I guess my unconscious mind simply took control in an apparent effort to prove its own value. While mixing up time and space, and everything else, my mind weaves its own pattern of history. Time is a profoundly strange thing. I had jumped through fifty years in just a few seconds, as if our mind weaves time from the yarn of our experiences, rather consuming time. Amazing. I even brought the pamphlet back from my virtual reality visit. As if I needed printed proof that I was there, back in 1944.

While I still wonder about the black-or-white-scarf, CNN establishes, once again, the final link with the current reality, because suddenly, I do remember the image that was so eerily familiar, when I started my little journey through the space and time of the last fifty years. Different leaflets, millions of them, have been dropped!

A new war had started as 'the war on terror' in Afghanistan, and now, expands into Iraq. "This is a different kind of war"; "this is a global war. It will last a long time," we are told.

Two pamphlets, and two wars. One was called a 'World War', this one a 'Global War'. 'Globalization' must have many meanings. Back to reality. Whatever 'reality' is.

Illustration 2: Millions of leaflets dropped over Iraq in March 2003.

"Excuse me Mr. J?"
Go ahead Novare.
"That issue about the scarf? I thought you wanted to know what happened. It is really easy."

You know what happened?

"Sure I do. You heard the evacuation story many times over the years. In those stories, the scarf always had a black color, so Alvas stored that in your memory. Then you read that book recently, remember? It told the true story about the end of the war. The scarf turned out to be white and that is the information you gave to the software people. When you were out there, the software showed you a white scarf, while I kept insisting on black, because that was the information that Alvas handed to me. Of course, I knew better, but I had to be consistent."

Ah, that explains it. Thank you Novare.

"There is one more thing you might not be aware of."

What is that?

"Remember when you heard people shouting: *The Americans are close*?"

Yes I do.

"That was the first human expression you ever heard. Those words were shouted during the first seconds of your life, out there, in the streets of your hometown. The moment you heard them in virtual reality, was the moment of your birth. We did not find a better way to model that in the software. Anyway, those words must have made a deep impression on you. No surprise you now live in the United States."

The Old War: a World War, and it was ending. The ending was a new beginning, the beginning of a new hope. A New War: a Global war, is just beginning. And a Cold War, the rainbow that connects the two. Like an umbrella under which a generation kept an image of peace. The rainbow began with the terror of war, and ended with the war of terror. What did we gain?

Most of us feel that something has changed. And that that 'something' has a fundamental nature. It is not just the scientific breakthroughs and the invention of new technologies, or the new products that saturate the world. It feels more like the tectonic shifts of continents. As if our way of thinking does not match the realities around us. We are unable to keep our peace of mind in the midst of a changing reality that we cannot grasp. "Information society," "knowledge worker," and "globalization" are the words used to label the change. Biotechnology and nanotechnology: more labels for new promises, and new threats. Something is cooking. Something is bubbling. Are we walking on a stage that is about to collapse? Or is the world pregnant of something new altogether? Something that will change the way we think?

Nature created a reality. Human 'nurture' changes that reality with every passing day. Nature created the human mind. Human nurture changes the mindset of that mind with every passing day. Are the two out of sync? What is going on?

"Now, that's a mental pretzel, if I ever saw one, JJ."

A mental pretzel, Alvas?

"Black scarf, white scarf, dreams, virtual reality, *real* reality, wars, mindsets, Einstein, Carter. Which is which? I'm pretty darn good at sorting out all of your concoctions, but that one beats me. All kind of events tangled up in one big mess. That's what I call a mental pretzel."

I do want to focus on the mind, Alvas, and on reality. And you are right; reality and our mental picture of that reality are tangled up in mysterious ways. What is reality? Is it what we think it is? Or is it something different altogether? If someone's mindset creates false images of reality, we call it illusions.

"But if we all agree on such illusions, we call it 'common sense'?"

There you go. In this chapter, I want to draw attention to the relationship between reality and mentality, because I assume a growing friction between the two. I like your 'mental pretzel' idea, because it does point out the mysterious entanglement of mentality and reality.

"What does that have to do with war and peace JJ?"

"For once, Alvas, *think*." Novare is frustrated. "Both wars, the old and the new, started in the human mind. The tools of warfare are inventions of the mind, and the tools to end them as well. It all starts, continues, or ends in the mind."

"Is peace a state of mind then?"

"You got it."

"Then why can I have perfect peace of mind, while there is a war going on in reality?"

"Because, Alvas, you are ignorant."

"Or you, Novare, are arrogant."

OK guys, just calm down. I apologize, dear reader. Sometimes I think I can control them, but after all, they are my mind.

Peace of Mind

The instruments of war are inventions of the mind. Starting and ending wars begins in the human mind, arguably the most powerful instrument in the known universe. Peace is a state of mind, although peace of mind and a state of war have been incredible allies throughout human history in the struggle to transcend the paradox. Our minds perceive reality and generate perspectives, as we learn the rules of law, the laws of physics, the physics of living, and the life of emotion and thought. These perspectives help us to sort out reality. Dreams, feelings, emotions, mathematical formulas, and the languages we speak, help us shape the things we think and the way we think them. Impressions and expressions, perceptions and conceptions: they are the products of our mind, and in turn, they determine our mind-set. That mind-set is our view of the world, determines the way we think, and produces our actions in that world.

Hitler's mindset started the Old War. His mindset was based on the belief in the supremacy of the Aryan race; a twisted interpretation of the continuation of biological evolution, concocted in an evil mind. This mindset allowed the treatment of members of other races as *Untermensch*, like animals, causing the destruction, and the mutilation of millions of living human beings. It represented the biggest, the ugliest and most methodical physical destruction in the history of the human mind. The Old War applied the products of the industrial revolution to perform its physical brutality. The trains, planes, ships, automobiles and weapons, and the phones, radios and radar were the instruments that made the war possible.

Bin Laden's mindset started the New War. His mindset forged by a twisted interpretation of the teachings of the prophet Mohamed, he uses the instrument of terror to force change. Terror does not seek the invasion of new territory. Terror is an invasion of the human mind. The attacks in the United States on September 11, 2001 constitute the biggest single act of terror in modern American history. Now however, the horrendous physical destruction was just the means to an end. The attack was only possible by applying the products of the information revolution of the previous fifty years. Flight simulators, online flight information, digital satellites, cell phones, email, websites, and digital money were the instruments that made the attack possible. A human mindset is the source of the terror. Digital technology provided the tools to enable physical destruction. The destruction caused the intended trauma to the mindsets of many in the United States, and around the world. Now, a different mindset is emerging; one that should eradicate terror from the planet. What will the new mindset be? What tools will it decide to use? Will the mind dictate the physical tools of warfare? Or will the war be fought with the tools of the mind?

I believe it will be both, the smart bombs and the smart minds. The first might be the necessary evil that leads the way to the second.

Reality is a funny thing. Your reality is yours alone. And mine is mine alone. My mind managed to travel from one war to the next, in one dream, which might have taken all of a few seconds. Our personal reality is in our mind; inside, not outside. It is what gives us identity. Part of it came 'with the package'. It came with our genes. Part of it is what we accumulated throughout our life. We learned it. That is how each of us developed our own mindset. With every experience we learn, and with every experience our mindset changes. Our mindset *is* our view of the world. Mindsets live a ghostly existence. As Sir Charles Sherrington puts it:

> *"Mind, the anything perception can compass, goes therefore in our spatial world more ghostly than a ghost. Invisible, intangible, it is a thing not even of outline; it is not a "thing." It remains without sensual confirmation and remains without it forever."*[6]

Yet, our mind determines how we see the world. We cannot trade a mindset for somebody else's; it is uniquely ours. Our minds are confused by new realities. What is the right way of thinking? How do we view a world that is changing so rapidly? Alcohol, sugar, caffeine, and other legal or illegal 'anti-depressants keep us running, while we try to 'make sense' of the world. Many are splitting their lives into two different realms: 'making a living' on the one hand, and 'getting a life' on the other, as if two different mind-sets rule those different realms. It seems as if two sets of values emerged, one for conducting business, and one for conducting 'life'. 'Business is business'; life seems to be something different altogether. We separate the two to stay 'whole'.

Let us return to mind changing business of terror. The attack on some of the core symbols of US culture was enormously successful. People live in fear. The threats are real. They change people's view of the world. The shape of this new emerging perspective will depend on the portrait that is painted by the leaders of society, and the leaders of the world. When they 'speak their mind', they access and influence the minds of hundreds of millions. That is a responsibility of enormous proportions, because even if the war on terror eliminates the lives of those that carry a twisted mindset, it will not eliminate the mindset of millions of people that have no need for newly imported western values. As the American futurist, Joseph Coates puts it:

> *"We must understand Islam, we must understand Islamic countries, and we must understand the actions that need to be taken to extinguish anti-Western beliefs and attitudes. My own observation is that over the last five decades American policy, both in government and business, introduced Western ways, Western tools, and Western developments ... often if not always with the endorsement and encouragement of the local governments. This Westernization has had the effect of making a small number of individuals wealthy, and delivering virtually nothing of significance to the mass population. That is the underlying basis of the hate of the West and toward the United States, throughout Islam. If the assembly of scholars confirms that in detail – perhaps taking Iran as the model example – that should lead to different government policies, different advice to business, and perhaps even to different constraints on international businesses in terms of plans,*

policies, programs and behavior... Don't demonize the enemy. It's easy to do and far more difficult to retract, recant and heal the breaches that demonization creates when the trouble is over."[7]

Both wars have one horrible thing in common. It is easier to change the physical reality of the world around us, than to change the human mindset. To erase the life of humans is easy. Changing their mind is the more difficult task. Understanding different mindsets seems to be a necessity to devise better solutions that transcend the differences, and might present the biggest challenge of the 21st century.

War and Peace, they begin and end with our mindset. How we treat our neighbors and people half way around the world, how we treat the planet we borrow from nature, how we handle our job and our careers, how we choose the politicians that represent us, and how we treat our children and our spouses, it all depends on our individual mindset. Nature, in a fifteen billion-year-long evolution, shaped a foundation for our mindsets: a genetic foundation, that all of us share, around the globe.

Our mind is part of our 'human nature'. It is the most amazing part, because it is the only part of our humanity that is changeable. With every single experience, everything we learn, every encounter with the world around us, this mindset changes, whether we like it or not and whether we are conscious of it or not. All of the scientific breakthroughs and all of the inventions in the last century originated in the minds of a very small number of people. Large numbers of people translated the new knowledge into products of practical use. Together, they changed the world, especially in the last fifty years.

The new reality is the connected world of digital 'thoughts', real time access to everything that happens anywhere on the planet, and access to all of the knowledge the world has accumulated. Over the course of the last fifty years, huddled under the umbrella of the cold war that kept the world in check, we quietly doubled the number of us that live on the earth and from the earth. We may start to realize that we are, now tightly packed, sitting in the same boat. We are drifting on the currents of drastically changing realities. At the helm, a mindset that has served us well for the last centuries is frantically trying to steer the boat. The currents are treacherous, the boat is leaking, the navigational landmarks have changed, and the captain is asleep. That captain is our individual mindset that still sees the same navigational checkpoints in the same landscape as a century ago.

No longer can one part of the world have peace of mind, while another part wages the wars of reality. The emerging digital reality takes away the boundaries between human mindsets, eliminating the isolation that time and space used to provide. The world is connected, and so are human minds. Thoughts, ideas, and emotions will blend, and result in new ones; in every corner of the planet.

Minding this new reality requires a 'new mind', a mind that is at peace with itself, and with the reality around it. The power of the human mind, combined with the power of digital technology makes that possible. For we are at a turning point in history. Not just a turning point in our recorded human history. We have arrived at a cusp in the history of our universe. The events of the last century are witness to this turning point in the evolution of human thinking. We are at the beginning of an amazing acceleration of human creativity that will change reality, again. We are in the process of waking up to this fact. Our mindset stayed behind in a rapidly changing reality. The world outside is ringing the alarm bells. The sounds are frightening, disturbing and confusing. They are the sounds of a wakeup call, and therefore painful, especially on the Monday morning of our current reality. When we wake up, we will realize the enormity of our potential, and the enormity of our responsibilities. A new mindset will emerge, and that mindset will create new futures.

The world is burning. The fire we see is the fire of change. It has been burning for fifteen billion years, when our existence began. It is the fire of evolution. Only recently did we start to discover its fire, because it intensified to the explosive threats of nuclear, chemical, or biological warfare, as hunger, poverty, coercion and ignorance of human dignity fuelled the flames. Evolution continues, while many still do not see the creative potential of its latest accomplishment: the human mind. Indeed, billions of people do not even acknowledge its existence. Evolution is the fire of change that created the reality of today.

Somewhere in that fire of change, we might find the clues that can serve as our guides for the future. Maybe, nature reveals some fundamental secrets for its continuous innovation, and for the direction of it's unfolding. When it does, we might develop a different perspective on our destiny. We might discover the qualities of nature that can guide us when we design new futures, turning destiny into possible destinations.

This, dear reader, is my mindset for writing this book. The book then, is an attempt to share that mindset with you. To do a better job at that, or maybe just to convey my own confusion, I will from time to time call in the help of my own two mind companions that you've already met briefly. Their names are Novare and Alvas. They have always been with me, and you might recognize them, too. They are the voices that argue with me, and while they do, they 'make up' my mind. Who knows what they are, or where they come from? Are they logic and intuition, brain and heart, reason and emotion? Or do they represent 'holding on' and 'letting go', the past and the future, prose versus poetry? They are confusing characters, yet they give me the clues and the evidence when *I* make up my mind. Sometimes one morphs into the other, sometimes they fight, sometimes they dance, and sometimes one of them doesn't show up at all. Allow me to introduce them as best as I can. The introduction can only be cursory, because if I knew their exact nature, I probably would no longer need them.

On the left side of my mind is Novare, the smarter of the two, although Alvas will tell you differently. A smart woman, articulate, logical, and precise. At the same time, she is the one with an incredible imagination. She is always alert, and usually assertive, but that can differ. You never know where she is, until she appears, seemingly out of nowhere. The *nowhere* sealed the fate of her name, because that is what Alvas called here: '*Nowhere*'. To my right then, is Alvas; he was the first to appear as my companion, so I agreed with Novare's name. 'First birthright', if you will. In an attempt to be 'creative', he applied a poetic sound to the name, and called her *Novare.*

"It had a Spanish sound to it JJ, and I like Spanish. Anyway, not only am I older than her, I am the justification for her existence. Without me, there would be nothing for her to do. I choose that name, because she is *nowhere* to be found when I need her. You should learn to control her JJ. Also, that remark about Novare being smarter? That hurts my feelings, *and* my intelligence. I'm sure readers will decide for themselves who the smarter one is."

Clearly, Novare doesn't like that.

"Alvas, you can tell the difference between '2' and '3', but you cannot put 'm together. Show some modesty for a change. But modesty is not your strongest claim to fame. About that appearance out of nowhere? I am always available; if only you would be more disciplined and stop your constant dreaming. As you well know, the true meaning of my name is *now – here.* I am always here, when I am needed."

"Wow! Then tell me, Novare, why are you not there when I need you most? You only appear when you feel like it. In an emergency, I have to decide for myself, because you are *nowhere* to be seen. Both, JJ and you would have been dead, if it weren't for my constant care."

It is only fair to say that Alvas *is* always active, no matter what happens. In that sense, he is the more reliable one. In fact, it is the origin of his name. Since I can always count on him, I simply called him *Always.* Naturally, he worked on it, and changed it to *Alvas.* "Sounds better," he said, "sounds Spanish." And that was fine with me.

"To me, his name is just a symbol for his slowness. He needs to consider *All – the - Ways* that can possibly lead to a solution for a problem I have. He changes his mind constantly, and he gives me things he does not even *understand.* You should really clarify his job description Mr. J. At least *my* role is clear. Everybody knows *what* I do, although nobody knows *how*. I determine what Alvas thinks. I am the future, while he maintains the past. Granted, the past grows, as the future gets smaller, so he will keep his job. And for the record, I agreed to that Spanish sounding name, because it is the most widely spoken single language in the world, although to me it sounds Italian. I *know* what goes on outside, you see, unlike Alvas."

Is there anything at all that you and Alvas agree on?

"Sure JJ. All there is is change. Both of us live of change. The flow of experiences. *Go with the flow, or take Prozac.* That's my motto, as you know."

What about you Novare?

"I agree, Mr. J. Change is how my awareness works. If there is no change, I go to sleep. For me change is opportunity, and opportunity is change. That is the difference with Alvas, you see. He just processes the change."

"And without me, Novare wouldn't even be able to notice change. I give her the evidence. Period."

Do we agree on the prime importance of change?

They both nod in agreement. That is promising. And unusual.

Novare is the *Now-Here* and the *Nowhere* of my mind. I know when she is 'here', and when she is, I know it is 'now'. When I search for her however, and try to locate her, she is nowhere to be found, no matter how hard I try. Alvas is always around, I can just feel it. Alvas knows *all-the-ways* of interpreting the things I know, and gives the results of his interpretations back to Novare. They use my nervous system and my brain as the canvas to paint the picture that is I, and at the same time make that canvas come alive, and make it interact with the worlds inside and outside of my body. That canvas, and Alvas and Novare, blend into one, to form that powerful, yet mysterious mechanism we call mind. Sometimes I am tempted to call Novare my consciousness, and hence, Alvas my unconsciousness. At other times, I think that they have nothing at all to do with consciousness. In everyday thinking, it seems to be more complicated than that, and such distinctions seem inadequate and too simplistic. I wish I could give a better description of their roles, but unfortunately, or maybe fortunately, I cannot.

However, they are the keepers of many gates, at least so it seems. They keep the gate between past and future, as they are the observers and interpreters of change, which marks the floating present. They are the gate between the inside and the outside worlds, the gate between emotion and reason, the gate between the personal and society, and the gate between objective and subjective.

Together, they continuously create and change our mindset. That mindset determines what we do and how we do it. Individual mindsets rule our individual lives, and, in many cases, the lives of others. As individual mindsets interact, or do not interact, collective mindsets emerge. They are the mindsets of communities, societies, religions, cultures and nations. These collective mindsets rule the world we live in. They shape the lives of families and the culture of companies. They are the source of freedom and oppression. They control the choice for war and peace. The constitutions of nations, the charter of the United Nations, the Geneva Convention, and the declaration of Human Rights, the death penalty, and the call for amnesty, all of them are expressions of such collective mindsets. The emerging 'mentality' that results from the interactions of all individual and collective mindsets, shapes global mindset on the planet Earth. Our mindset shapes our individual lives, and our global mindset. Our collective global mindset shapes our global destiny, modifies our individual mindset, and hence, modifies our individual lives. To some, this might seem a long road to state the screamingly obvious. Why is it then that we seem to be unable to replace the reality of war by the mentality of peace? Why do we seem to be unable to transform destiny into destination? Are we still living with a mindset that remained unchanged for the last ten-thousand years of human existence? Maybe the reality of the universe, the reality of this planet, the reality of life on this planet, and the reality of the creative powers of the human mind are way ahead of our mentality that is stuck in that ten-thousand year-old mindset?

"When did our mindset start to stay behind reality, JJ?

"Alvas, didn't you listen? That started right in the middle of the twentieth century."

Actually, Novare, the gap started to widen much earlier, maybe as early as four centuries ago. However, the most fundamental discoveries took place during the last century, especially the first half of the twentieth century. By the middle of the twentieth century, the world had changed because of the technologies based on the new discoveries, but only a few people – the scientists and the engineers - realized the true nature of those discoveries: the widening of the gap started to accelerate.

"From that moment on, the discrepancy started to grow exponentially, correct?"

"And the discoveries of the *evolutionary* nature of the world is the reason for the discrepancy?"

Exactly.

"Are you going to explain, what this evolutionary nature is?"

The discoveries of the last century are fundamental to the explanation of our existence, and to the ways in which our existence changes, in the past, the present and the future. The frontiers of scientific knowledge mark the boundaries of what we are capable of doing. In the next chapter, we will look at those frontiers. We might not fully understand all of them, but even a glimpse will give us a 'feel' for their significance as the mental tools that will shape our future.

Chapter 2: Changing Minds

"Well do I know that I am mortal, a creature of one day. But if my mind follows the winding paths of the stars Then my feet no longer rest on earth, but standing by Zeus himself I take my fill of ambrosia, the divine dish."

Ptolemy, book 1, Almagest, 2nd century AD

"An important scientific innovation rarely makes its way by gradually winning over and converting its opponents: it rarely happens that Saul becomes Paul. What does happen is that its opponents gradually die out, and that the growing generation is familiarized with the ideas from the beginning."

Max Planck, New York 1949

Focus on the Universe

Alexandria, Egypt, 2nd Century AD

His name was Claudius Ptolemy, a mixture of the Greek Egyptian 'Ptolemy' and the Roman 'Claudius'. This indicated that he was a descendant from a Greek family living in Egypt and that he was a citizen of Rome. During the period from 127 to 141 AD, Ptolemy carried out his astronomical observations in Alexandria in ancient Egypt, and published his theory in a thirteen-volume treatise called the Almagest. The earth was the center of the universe. Around the earth was a sphere that contained all of the heavenly bodies, including the Sun and the planets. All of these heavenly bodies had fixed positions in this sphere, and Ptolemy developed the mathematical formulas to describe the motions of these bodies. He based his view on the earth-centric system already developed by Aristotle six centuries earlier.[1] Ptolemy's model of the universe would prevail for the next 1400 years. The view of the universe was simple enough. Plato had created the view of the existence of a world of universal truths and Aristotle added the view of a static Earth as the center of the universe. Ptolemy added a mathematical description to the newly defined satellites of the earth.

Before Ptolemy, there had been a 'descending' view amongst the Greek philosophers. In fact, both Plato and Aristotle expended a lot of thought refuting the views of that particular man…

Greece, 5th Century BC

This man was the Greek philosopher Heraclites, who lived from 536 - 470 BC. From his writings, he does not appear the most pleasant amongst men. He considered war to be "the father of all and the king of all; and some he has made Gods and some men, some bond and some free." However, his famous 'Panta Rei' (everything flows) survived through the ages, and regains new significance in light of the scientific discoveries of the last 150 years that revealed the evolutionary nature of the universe and of biological life. Heraclites was a mystic, but then, there was not observable evidence to go on in those days. No microscopes, no telescopes, and no sensitive instruments at all that would extend the reach of the human senses. Imagination was then, and is now, the most important creative force. However, imagination, if rendered into a 'reality' that escapes experimental verification, is called metaphysics. That was true then, as it is now.

Heraclites perceived everything as originating from fire: "[This world] … was ever, is now, and ever shall be an ever-living Fire, with measures kindling and measures going out." Everything was in an eternal flux. Change is what characterizes everything in existence. "You can not step twice into the same river; for fresh waters are ever flowing in upon you," and "The sun is new everyday," are some expressions of Heraclites' disciples to describe his vision of constant change. Along with the perpetual flow of things, he believed in the constant tension between opposites, and the constant cycle of opposites meeting, creating tension, and giving birth to a dynamic harmony between the opposites that produces a transcending unity that changes constantly. Later, Plato and Aristotle would refute Heraclites' vision that 'nothing ever is, everything is becoming'.[2]

Since Heraclites, philosophy as well as science, have been on an everlasting search for constancy; an ultimate truth, expressed in a set of rules, that provides a never changing framework for security, providing the 'peace of mind' that comes with the knowledge that one is safely navigating the unknown waters of life. Religion provides this permanency in the form of the eternal never changing truth, which is God, and in the form of immortality of the human soul. However, science would only find more and more evidence that supported Heraclites' view of Panta Rei. It seems that the constancy of nature as we experience it can only be found in the source of change itself. Heraclites' fire now becomes the symbol for the source of permanency: the eternal process of transformation that is at the core of everything in the universe. Today, we may apply Heraclites' perpetual flow to the process we now know as 'evolution'. It is in that fire of change, where we will look for the clues that can serve as guides for the future.

Frombork, Poland, 1543 AD

In 1543, the world changed, although recognition of that change had to wait until a century afterwards. Written with the same clarity and method as Ptolemy's Almagest, De Revolutionibus Orbium Coelestium ("On the Revolutions of the Celestial Orbs"), was published in Nürnberg, Germany. The writer was Nicolaus Copernicus, a Polish physician, lawyer and church administrator, who spent his spare time in astronomical research. At the time of publication, in 1543, he lived in Frombork (Frauenburg), Poland. Copernicus is said to have received a copy of the printed book for the first time on his deathbed. (He died of a cerebral hemorrhage.) He would never experience the waves of protests against his heliocentric theory.

Copernicus came from a middle class background and received a standard humanist education, studying first at the University of Krakow (then the capital of Poland) and then traveling to Italy where he studied at the universities of Bologna and Padua. He eventually took a degree in Canon Law at the University of Ferrara. At Krakow, Bologna and Padua he studied the mathematical sciences, which at the time were considered relevant to medicine (since physicians made use of astrology). Padua was famous for its medical school and while he was there Copernicus studied both medicine and Greek. When he returned to his native land, Copernicus practiced medicine, though his official employment was as an administrator in the cathedral chapter, working under a maternal uncle who was Bishop of Olsztyn (Allenstein) and then of Frombork (Frauenburg).[2]

While in Italy, Copernicus visited Rome, and it seems to have been for friends there that in about 1513 he wrote a short account of what has since become known as the Copernican theory, namely that the Sun (not the Earth) is at rest in the center of the Universe. *The Earth revolved around the sun, and around its own axis.* As with many discoveries in the next centuries, the world changed - our understanding of it that is - but ignorance will suppress that understanding for generations to come. Moreover, people will die because of that.

One supporter of Copernicus' theory of a moving earth was Italian scientist Giordano Bruno, born in 1548 near Naples, Italy. After receiving an education as being part of the Catholic Church, (He was a Dominican for that period), he taught at Oxford for a while. In 1592, he was arrested by the Inquisition, and put on trial for heresy for promoting the ideas of Copernicus. After a seven-year trial he was burned at the stake in Rome in 1600.[3]

The mindset of the time did not agree with the changing view of reality. God had created man, and everything that surrounds him. The human being was the real center of a universe that was at his disposal. With the security of the Earth as a permanent stage, the universe is the décor surrounding it; man was the supreme being, created in God's image, and only second to Him. He could not possibly be just a tiny part in a world that moves around in a vast universe. God had created the universe as a service to men, certainly not the other way around. Prophecies were taken as the infallible descriptions of reality, astronomical theories as mathematical exercises that lived as closed systems within themselves, not linked to reality. When we look at some of the religious fundamentalist positions today, it seems that this tradition is still being continued. In the United States, some propose to introduce (in some cases with success) creationism (the religious belief that human beings have been created 'as is' by God) as a science-like course, taught in parallel to the science of evolution.

Eppur Si Muove, Part 1

Cape Canaveral, USA, October 18, 1989

A roar shakes the ground as Space Shuttle Atlantis climbs into the sky.

Illustration 3: When crewmember Shannon Lucid released the Galileo spacecraft from the Shuttle, it began a six-year journey to that giant, colorful planet: Jupiter. The spacecraft is named in honor of the first modern astronomer --- Galileo Galilei, who made the first observations of the heavens using a telescope in 1610. Courtesy NASA[5]

The Galileo spacecraft rides in the payload bay, ready to begin a long journey to the realm of the outer planets. Its mission is to study Jupiter and its moons in more detail than any previous spacecraft.

Four hundred and eighty years before the launch of the Galileo spacecraft, in the summer of 1609, Galileo Galilei heard about a spyglass that a Dutchman had shown in Venice. Galileo was born in 1564 near Pisa, Italy; the same year in which Shakespeare was born and the year in which Michelangelo and Calvin died.

Galileo applied his own technical skills as a mathematician and as a worker, and used this idea to build a series of telescopes with an optical performance much better than that of the Dutch instrument. He described his astronomical discoveries using the new instrument in a short book called Message from the stars (Sidereus Nuncius) published in Venice in May 1610. It caused a sensation. Galileo claimed to have seen mountains on the Moon, a multitude of tiny stars in the Milky Way, and to have seen four small bodies orbiting Jupiter.

Like Bruno, Galileo supported Copernicus' heliocentric view of the universe. According to the Catholic Church, this was clearly in contradiction with Scripture, and in 1616, Galileo was given some kind of secret, but official, warning that he was not to defend Copernicanism. Just what was said on this occasion was to become a subject for dispute when Galileo was accused of departing from this undertaking in his Dialogue concerning the two greatest world systems, published in Florence in 1632.[6]

Galileo, who was not in the best of health, was summoned to Rome, found to be vehemently suspected of heresy, and eventually condemned to house arrest, for life, at his villa at Arcetri (North of Florence). He was also forbidden to publish. By the standards of the time, he had gotten off rather lightly. Galileo is said to have uttered, *sotto voce*, the words '*Eppur Si Muove*' ('And yet, it moves'), after leaving his meeting in Rome.

It was not Galileo's theory about the movement of the Earth, and the planets around the sun that got him into trouble with the church. As long as theories merely *described*, in mathematical form, a part of reality, scientists were in safe territory. It was different when they tried to *explain* reality. Revelation, as expressed in the Bible, provided the ultimate explanation of reality. Therefore, any attempt to give alternative explanations of reality were considered heresy, and hence, had to be punished. This gap, between the acceptance of the descriptive function of scientific theory and the acceptance of its explanatory power, persists until today. The continued existence of this gap is one of the reasons why widely accepted scientific theories did not lead to a paradigm shift, a change in our view of the world, a *change in our mindset*.

Orbiting Jupiter, October 29, 2002

What compels us to explore Jupiter? Jupiter holds clues to help us understand how the Sun and planets formed over four-and-a-half billion years ago. One of Jupiter's moons has active volcanoes, and others have strange icy terrain. How does this compare with Earth?

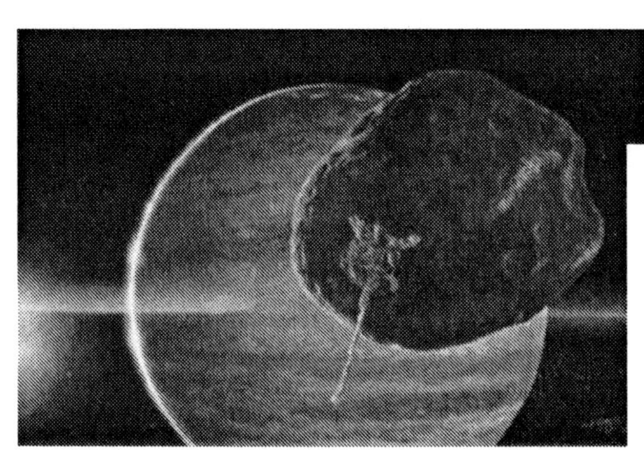

Illustration 4: Artist's rendition of NASA's Galileo spacecraft (foreground) passing Jupiter's small inner moon Amalthea. Image Credit: Michael Carroll, courtesy NASA[7].

As Galileo approached Jupiter, it skimmed past Amalthea on Nov. 5, 2002. Amalthea is one of four small moons closer to Jupiter than the four large moons -- Io, Europa, Ganymede and Callisto -- that Galileo has inspected during more than 30 encounters since late 1995. Navigators have set a course for the orbiter to pass about 160 kilometers above Amalthea's cratered surface. After this flyby, Galileo will be on course to hit Jupiter in September of 2003. Its propellant supply, needed for pointing the antenna toward Earth and controlling the flight path, is nearly depleted. While still controllable, the orbiter is being steered into Jupiter to avoid any risk of the spacecraft hitting Europa in years to come. That precaution stems from Galileo's own discoveries of evidence for a hidden ocean under Europa's surface, heightening interest in Europa as a possible habitat for life.

The eyesight of the real Galileo was failing by the time of his house arrest. He needed the help from his pupils to write up his studies on motion and the strength of materials. The book, Discourses on two new sciences, was smuggled out of Italy and published in Leiden (in the Netherlands) in 1638. Galileo died in 1642. One of his discoveries was that bodies do not fall with velocities proportional to their weights. This was another blow to the mindset of his time, but it would put future scientists on the path of new amazing discoveries.

Four centuries later, Einstein would explain Galilei's conclusion, when he discovered that the mass of a body is equivalent to gravity, and that gravity was equivalent to inertia. Inertia, the resistance of a body to change its velocity, might have a deeper meaning yet: In 1992, 450 years after Galilei's death, Pope John Paul II declared that "Galileo Galilei was unjustly condemned by the Roman Catholic Church for promoting Copernican cosmology," after a special Vatican Commission finished its investigation of the matter. As might be expected, Mr. Galilei was not available for comment. Inertia is a property of the human mind as well, at least judging by the 450-year journey of the Catholic Church to synchronize its mindset with a 'changed' reality.

Four centuries ago, Galileo risked loosing his life by convincing the world that the planet Earth was moving. Today, another Galileo is in pursuit of the search for life on far away moons and planets. A dream Galileo could not have dreamt, when he exposed Copernicus' dream as reality, and reinforced its foundation by new observations.

Illustration 5: The cathedral of Pisa and its bell-tower, the famous leaning tower in Pisa, Italy. In popular lore, Galileo Galilei refuted Aristotle's laws of motion by dropping unequal weights from the Leaning Tower of Pisa, to show that bodies with different weights fall with the same speed.[8]

The expression on Galileo's face might be one of disbelief if he could have watched the launch of his name-twin in 1989, or if he could have commented at the Pope's declaration in 1992. However, he might have been most amazed by the outcome of a National Science Foundation survey adults' knowledge of science. In late 2001, 16 percent of men and 32 percent of women surveyed in the United States, failed to correctly answer the question: "Does the Earth go around the Sun, or does the Sun go around the Earth?"[9]

"Who cares what goes around what, JJ? My life doesn't change a bit, one way or the other."

"Typical Alvas. He is *so* resistant to change. Once he has an opinion, I need the longest time to change it. Panta Rei is 'Greek' to him, if I may say so. It is easier for me to have him repeat the alphabet backwards, than to change his mind. It is about your mindset, Alvas. Since you cannot see reality as it is, you keep running around in the treadmill of your own mind, trying to fit in what I give you. Then, miraculously, either you twist it around to fit your liking, or you simply ignore it."

"Hah! Sure, I can look outside. You are my window on the world, remember? I see enough to know that it was the likes of you that burned poor Bruno, and doomed Galileo to stay home until he died, not me. Still, what's the importance of what moves around what?"

People had formed a mindset over thousands, no, millions of years. With a few exceptions, like Heraclites, the first philosophers largely confirmed the view of a permanent universe in which the human was the master. When science started to show evidence to the contrary, the world that had settled on the idea of permanency refused to change its mind. Therefore, it is not about the importance of the Sun, it is about the attack on the supremacy of humans. Even more important, it is about an attack on the supreme explanation of reality, as given in the Holy Scriptures.

"Can I add something Mr. J?"

Go ahead, Novare.

"How should people, back then, and today, change their minds when it concerns phenomena they cannot perceive directly? People, back then, even if they would look through a telescope, would not be able to figure that the earth moves. Alvas is right in that regard. Even today, I have to assemble a new worldview from scientific concepts, the evidence for which even scientists cannot observe directly. Therefore, I have to believe them, or study hard to understand them. Shouldn't science be more responsible for communicating new insights? Especially when these insights lead to a change in our view of the world?"

Good point Novare.

"To me, it doesn't make a difference. I've to take care of the inside too, remember? I've enough to do as it is. That's all I'm saying JJ." "Alvas just made my point, Mr. J."

Illustration 6: Launching of ARIANE-5 in Kourou, French Guiana. Image courtesy European Space Agency. Insert: Isaac Newton.

Kourou, French Guiana, December 10, 1999

"The European Space Agency (ESA) launches ARIANE-5 from its facility in Kourou, French Guiana, It is a picture perfect event, broadcasted live around the world via satellite television and web casts on the Internet. Twenty-nine minutes af-

ter takeoff, it releases XXM NEWTON, the most powerful X-ray telescope ever placed in orbit. Its Mission? To help solve many cosmic mysteries, ranging from enigmatic black holes to the formation of galaxies. Many celestial objects generate X-rays in extremely violent processes. But Earth's atmosphere blocks out these X-rays, messengers of what occurred in the distant past when stars were born or died, and clues to our future."[10]

By the calendar in use at the time, Isaac Newton was born on Christmas Day 1642, more then 350 years before the European Satellite would carry his name as the XXM-NEWTON into a highly eccentric orbit, traveling out to nearly one third of the distance to the Moon, to watch the events on distant stars. Without Newton's scientific work, his mathematics, his discoveries in optics, and his laws of gravity, XXM-Newton might not exist today.

Newton would devote only the first half of his life studying the heavenly bodies, long enough to produce his ground-braking theories. During the second half, Newton worked as a highly paid government official in London with little further interest in mathematical research. He died a rich man, a feat that not many scientists would repeat after him.

He was appointed to the Lucasian chair of mathematics at Cambridge University in 1669. He was 27 years old. An outbreak of the plague had kept him at his home in Lincolnshire for the previous 4 years. His work during this period would be the start of his revolutionary advances in mathematics, physics, and astronomy (today, Stephen Hawking, renowned scientist and author of the best-selling book *The History of Time,* holds the same Lucasian chair at Cambridge University). In 1672, he published his first scientific paper on light and color, in which he attempted to prove, by experiment alone, that light consists of the motion of small particles rather than waves; one of the principles at core of XXM-NEWTON's technology. To explain his observations, he had to use both, a wave theory for light, and a theory of light as particles.[11]

"Did Newton have a premonition of the wave-particle duality, Mr. J?"

You can put it that way, Novare. It would take almost 350 years before quantum theory would solve, at least largely, the mysteries of this phenomenon.

"Do we really need to talk about this, JJ? To me, that quantum stuff is the mother pretzel of all mental pretzels. Waves or particles, who cares? The effect on what I do is the same anyway."

Since it lies at the foundation of all behavior, and might play a role in the mysteries of life and human consciousness, we need to point out its existence. But don't worry, Alvas. We'll say just enough about it, to point out its unbelievable strangeness. And how it effects our thinking, once we know about its existence.

"If Alvas would only understand that he might *be* such a mental pretzel."

We are not sure about that Novare.

Newton's intuition that light is the propagation of tiny particles, or corpuscles, as he called them, did not gain broad acceptance. The established view was that of a universe filled with a mysterious substance called 'ether'. Light moved like waves through this ether, just like sound waves. Just as sound needs the air to propagate, ether was supposed to be the substance in which light created the disturbances that we call light-waves. Many of us might still – intuitively - experience the world that way today. Doesn't Alvas have a point when he says that it is the effect on our lives that count? After all, whether you get a tan from light-waves or from light-particles, who cares? It would take a few centuries, before scientists would see light in a different 'light'. For most of us, this view did not change much, although our TV-sets, our lasers for eye surgery or our computers would not have existed without changing our understanding of light.

Las Palmas, Spain, June 2000

The Spanish Observatorio del Roque de Los Muchachos is nestled in the mountains of Las Palmas, overlooking the Mediterranean Sea. The observatory hosts a series of telescopes that form the Isaac Newton Group of Telescopes (ING). It is owned and operated jointly by the Particle Physics and Astronomy Research Council (PPARC) of the United Kingdom, the Nederlandse Organisatie voor Wetenschappelijk Onderzoek (NWO) of the Netherlands and the Instituto de Astrofísica de Canarias (IAC) of Spain.

Illustration 7: Spanish Observatorio del Roque de Los Muchachos. Picture credit: Nik Szymanek and Ian King, 1999. Courtesy: Isaac Newton Group of Telescopes.[12]

Newton's greatest achievement was his work in physics and celestial mechanics, which culminated in the theory of universal gravitation. According to the well-known story, it was when seeing an apple fall in his orchard at some time during 1665 or 1666 that Newton conceived that the same force governed the motion of the moon and the apple. Once again, imagination came before mathematical deduction, when Newton imagined that the earth's gravity could influence the moon. By 1666, Newton had conceived the early versions of his three laws of motion. From his law of centrifugal force and Kepler's third law of planetary motion, Newton deduced his famous inverse-square law that described the orbits of celestial bodies. After some persuasion, Newton finally wrote a full treatment of his new physics and its application to astronomy. In 1687, Newton published this work as Philosophiae naturalis principia mathematica or Principia, as it is commonly known. It is recognized by many as the most important scientific work ever written.[13]

The image in illustration 8, taken by one of the telescopes of the Isaac Newton Group, shows the gravitational pull of the main galaxy 'M51' on a smaller one, disturbing and extracting material like a hurricane. However, Newton did not know that such movement of large-scale masses would change the shape of the universe itself.

Illustration 8: M51 galaxy, known as The Whirlpool Galaxy, is a bright spiral galaxy fairly close to us (25 million light years). To the north of M51, at the top of the picture, is a companion galaxy, which is being disrupted by the gravitational tidal forces of the main galaxy. Picture taken in: June 2000. Courtesy: Isaac Newton Group of Telescopes, Las Palmas, Spain.[14]

In Newton's worldview, time and space remained to be the absolute and unchangeable dimensions that had existed since the Divine Creation some 6 to 10 thousand years ago, and they would remain unchanged forever. In this sense, Newton brought back peace of mind. Although the earth was no longer the center of the universe, human superiority remained intact. People could continue to live on the earth like the actors on a stage. After all, it is about the play, not the stage. The stage provides the props and the ambience that the players control and modify at will. This stage doesn't change itself, and is certainly not an intricate part of a larger whole.

The image of an earth that had a fixed position in the cosmos was replaced with the image of neatly arranged celestial bodies that dance to the baton of Newton's precise mathematical rules; rules that were assumed to be valid for all times, past and future. The skies continued to be part of the ambience for that big stage of life: the planet earth. Space continued to have a 'two-and-a-half' dimensional nature. Even today, the stars, and the movement of planets and galaxies, are just part of a different world 'out there' that serves as a romantic backdrop for lovers; at least those that are still able, or capable, to see the beauty of a starry night. However that might be, Newton restored the peace of mind to science, and to religion of his time. Once again, the universe was just the static backdrop to the play of politics, social interaction, and the chore of daily living. Peace of mind was back, once again. This would last for the next two centuries. After that, the mindset of permanency would be shattered. Now, it would be forever.

Living Proof

Geneva, Switzerland, June 18, 2002

"The last five days have witnessed the unprecedented announcement of 25 new planet discoveries. These discoveries are split almost evenly between European and American astronomers. Didier Queloz and his colleagues at the Observatoire de Genève, Switzerland, have found a dozen of the new planets. Their discoveries include the most tantalizing one yet: a planet that closely resembles Jupiter in our own Solar System. The find brings astronomers another step closer to detecting an Earth-like world."[15]

The European space agency established a scientific program, called the Cosmic Vision 2020 science programme. One of the missions within this program is the launch of a flotilla of eight spacecraft that will fly in formation and combine their observations to detect the light from Earth-like planets around other stars. By analyzing that light, astronomers will be able to deduce the chemical compositions of distant planets' atmospheres and search for the telltale chemicals related to life. The name of that program is DARWIN.[16]

The North American Space Agency (NASA) is planning an ambitious mission, called the Terrestrial Planet Finder. This mission "may put multiple telescopes on separate spacecraft to fly in formation. Working together, these telescopes would then be able to take "family portraits" of entire solar systems and would analyze spectra to look for the chemical fingerprint life would leave on the reflected light from the planet."[17]

The search for life in the universe is on. And so is the search for the origin of life on our own planet. DARWIN[18] is the name of just one of those missions. Darwin was the man who changed our mind about that very origin of our own lives.

Devonport, England, December 27, 1831

"After having been twice driven back by heavy southwestern gales, Her Majesty's ship Beagle, a ten-gun brig, under the command of Captain Fitz Roy, R. N., sailed from Devonport on the 27th of December, 1831. The object of the expedition was to complete the survey of Patagonia and Tierra del Fuego, commenced under Captain King in 1826 to 1830, -- to survey the shores of Chile, Peru, and of some islands in the Pacific -- and to carry a chain of chronometrical measurements round the World. On the 6th of January we reached Teneriffe, but were prevented landing, by fears of our bringing the cholera: the next morning we saw the sun rise behind the rugged outline of the Grand Canary island, and suddenly illuminate the Peak of Teneriffe, whilst the lower parts were veiled in fleecy clouds. This was the first of many delightful days never to be forgotten. On the 16th of January, 1832, we anchored at Porto Praya, in St. Jago, the chief island of the Cape de Verde archipelago."

Charles Darwin[19]

This is Darwin's own description of the start of his historic voyage; a voyage that would change our thinking about nature. Darwin was born on February 12, 1809 in Shrewsbury, England. His father, Robert Darwin, was a physician, the son of Erasmus Darwin, a poet, philosopher, and naturalist. Charles's mother, Susannah Wedgwood Darwin, died when he was eight years old[20].

"I wonder what Darwin was thinking when he saw the Canary Islands."

Why Novare?

"Don't you see JJ? Years later, one of those Islands -Las Palmas- would be the home of the Isaac Newton group of telescopes. That made Novare think about the difference between Darwin and Newton's way of thinking."

True. I wondered how Darwin could sense the evolution of biological life forms over time, while Newton never thought about a universe that evolved over time. Just curious.

"I can shed some light on that JJ."

Really Alvas?

"In order to change my image of reality, I need new input. An idea, an intuition, an observation, a hunch, is necessary for me to recreate your mindset. Sometimes I can help with a gut feel, but the main thrust has to come from Novare. However, she always has to deal with what I already know, that's why it is difficult sometimes."

"Alvas is right, at least in that he resists my ideas so often. I am the trigger for a new idea. Something has to catch my attention, whether it comes from the outside, or whether it comes from Alvas. In Darwin's case it worked, while in Newton's, it didn't."

"I don't agree. Both had an idea first. Just remember Newton's apple. Newton was looking to explain movement, the change of a body in space, while Darwin looked for change that happens over time."

Good thinking guys. Scientists, as well as artists, or anybody else have to come up with an idea first. Only then can they search to create renditions for their ideas, be it in the form of scientific theories, painting, poems, or musical symphonies.

"Always knew it all starts and ends with me JJ. It's hard to be a genius."

The event took place on November 24, 1859 that shocked the world out of the reverie of the historic permanence of life. It was the day that Darwin published On the Origin of Species, in which he proposed a rock solid theory for the biological evolution of all living organisms, offering an alternative for the common view of Genesis chapters 1 and 2 that assumes the simultaneous creation of all living species, as we know them today. The view of creation had been an implicit part of everyone's mindset, scientists and common folks alike. His theory created a fire of controversy that keeps burning fiercely until this very day. Over the centuries though, some had already imagined the evolving nature of everything in the universe. Heraclites' Panta Rei was not the only vision of an ever-changing world. And it was not the first one either.

Greek philosopher Anaximander (611-545 BC) was a student of Thales, the "father of philosophy'. Anaximander believed that the universe began as a chaotic mass, out of which everything emerged by the separation of opposites. Anaximander imagined that life originated in the sea, and that a decrease in the sea level forced animals to adapt to dry land, and some of them would manage to develop respiratory systems.[21] His voice too, was quieted by the views of Plato and Aristotle, who established the permanent and absolute nature of the universe, that prevailed throughout the millennia. Ernst Mayr, Professor Emeritus at Harvard University, expresses the importance of Darwin's theory as follows:

"This event represents perhaps the greatest intellectual revolution experienced by mankind. It challenged not only the belief in the constancy (and re-

cency) of the world, but also the cause of the remarkable adaptation of organisms and, most shockingly, the uniqueness of man in the living world. But Darwin did far more than postulate evolution (and present overwhelming evidence for its occurrence); he also proposed an explanation for evolution that did not rely on any supernatural powers or forces. He explained evolution naturally, that is, by using phenomena and processes that everybody could daily observe in nature ... No wonder the Origin caused such turmoil. It almost single-handedly effected the secularization of science."[22]

The turmoil about Darwin's theory continues, and keeps billions of people from acknowledging the forces of evolutionary change, and from adjusting their mindset to the reality of change. At the time, society was involved in the making of a revolution that kept them from worrying too much about evolution: the industrial revolution was starting. The invention of the steam engine would produce the railroads, and change the perception of space by connecting cities and oceans, and the motorized ships, connecting the shorelines of far-away countries. Thousands would find new security in the factories 'of the future', while creating a separation between family life and 'professional life' that had never existed in the agricultural era of the last twenty thousand years. Life started to change quickly, people grew in number, and they grew closer together.

Illustration 9: the *DARWIN project* and similar programs are designed to look for extra terrestrial life in solar systems that might resemble our own. This illustration shows a newly discovered Jupiter-like planet and its hypothetical moon circling the star 55 Cancri, in an artistic rendition by artist Lynette Cook. The large object is the Jupiter- like planet, orbiting around its own Sun (the bright spot). The small object in the front is what might be one of the planet's moons. Courtesy European Space Agency [16]

At the time of Darwin's publication of 'the Species', our planet had 1.2 billion inhabitants, and was well on its way to double that number in the next one hundred years. Today, we send space probes into space to find life elsewhere in the universe, like the DARWIN project, named after the man who showed us that life had a beginning, as depicted in illustration 9.

Darwin showed that, in biological evolution, time meant change. However, time and space themselves remained constant and absolute. Darwin reached out in time, studying living species on the workbench that was the planet. His tool was the microscope that extended his senses to the small variations inside plants and animals. A concept had formed in his mind. Now his mind had to make the connection between concept and reality, thus embarking on the treacherous road that we call scientific discovery, keeping the proper distance between conception and perception, both being products of the mind. Newton reached out into space, using his intuition about the force that attracted the apple to the earth as his guide. His tool was the telescope that extended his senses into the universe of faraway stars, planets and moons. Now his mind had to connect the dots in the picture of his imagination and relate them to the way the celestial bodies interact with each other. The concept of an attracting force in his imagination turned out to be the reality of the gravitational force in the universe.

Both Newton and Darwin reached out into a world beyond the known, proving that the unknown is not the same as the unknowable. Both used an initial concept that had formed in their mind through intuition and imagination. This concept served as their bootstrap to bring order in their observations of reality, and evolved along with their grasp of that reality. Newton's mind didn't grasp the ultimate significance of gravity, although he noticed some discrepancies between his theory and his observations. It would take two more generations and another brilliant mind to figure out that gravity was not a force as Newton had proposed, but that it was much more than that, and more mysterious. In 1896, the mind that would figure out this mystery had to produce an essay to gain entry to a prestigious university...

Changing Times

Zürich, Switzerland, 1900 AD

...That mind belonged to Albert Einstein; the University was the Eidgenössische Technische Hochschule (ETH) in Zürich, Switzerland. The young Einstein, who failed an admission exam before, had to write an essay on 'his plans for the future' to seek entry into the school. Here is what he wrote:

"If I were to have the good fortune to pass my examinations, I would go to Zürich. I would stay there for four years in order to study mathematics and physics. I imagine myself becoming a teacher in those branches of the natural sciences, choosing the theoretical part of them. Here are the reasons which lead me to this plan. Above all, it is my disposition for abstract and mathematical thought, and my lack of imagination and practical ability."[23]

Einstein graduated from ETH in 1900; he was 21 years old. He tried to get a job as a research assistant at the ETH, but he had not impressed enough to be successful. He obtained the Swiss nationality, but managed to avoid Swiss military service on the grounds that he had flat feet and varicose veins. In 1901, he found a job as a mathematics teacher at the Technical High school at Winterthur, Switzerland, and later at a private school in Schaffhausen. He wrote, "I have given up the ambition to get to a university." Marcel Grossman's father recommended him to the director of the patent office in Bern, where he started to work as 'technical expert third class', and was promoted in 1906 to 'technical expert second class'. He continued to work at the patent office until 1909. Marcel Grossman had been Einstein's classmate at ETH and would help him later with the mathematical formulation of the theory of general relativity.

Einstein's time at the patent office proved amazingly productive. In 1905, a thesis that he dedicated to Marcel Grossman earned him a doctorate from ETH. In the same year, he published three papers, including his theory of special relativity.

Pacific Ocean, Summer 2002

You got lucky. Your frequent flyer status earned you an upgrade to first-class. You feel good, comfortably settled in seat 2A, flying about 12 kilometers above the Pacific Ocean somewhere near the equator. Your seat reclined slightly, window shades drawn and enjoying the movie, and a glass of wine. No cell phones, no interruptions; you love it. The ride is smooth and your glass of wine doesn't move. It doesn't move at all. Or does it? You wonder.

The plane is moving with a speed of 1000 kilometers per hour, as you can read from the screen in front of you. When you lift the window shade slightly, you wonder about the plane's speed, because the ocean is moving below and the airplane is not moving at all.

To solve this dilemma, we talk about the speed of the airplane *as measured in relation to the surface of the earth.* But wait! The earth itself is rotating around its own axis, once every 24 hours. This means that seat 2A is moving with a speed of about 1600 kilometers per hour, even when standing still at the gate. Therefore, your wineglass is moving with a total speed of about

2,600 kilometers per hour (you are flying in the direction of the rotation of the earth) in *relation* to the earth's axis of rotation!

But wait! That's not all either. The Earth is rushing around the sun with an average speed of 29 kilometers per second or roughly 105,000 kilometers per hour. And we are not there yet. Our solar system is a suburb in the Milky Way galaxy, which is spiraling around its own axis with a speed of 250 kilometers per second or 900,000 kilometers per hour ...

Let's assume that you are traveling at a time when the directions in which all of these bodies move are lined up. Your glass of wine has a speed of a daunting million kilometers per hour -give or take a few thousand kilometers- and you are not spilling a single drop!

"What's the big deal, JJ?"

It is quite a big deal Alvas. Speed can only be measured in relation to another object. You have no way to prove, or even notice that you are moving, unless you refer to the motion of another object.

"What if there is no other object?"

You got it. If we would continue to relate the speed of the wine-glass to ever-larger configurations in the universe, we would ultimately end up with the universe itself. That brings us to one of the important foundations of Einstein's theory. For Newton, the universe was a box, albeit a big one. It was the ultimate *frame of reference.* All phenomena, like the movement of bodies, were events that could be related to that frame of reference. With Einstein's theory, that went out the window. Every body in the universe constitutes its *own* frame of reference.

"Like the airplane?"

Just like the airplane. Within the airplane, I can safely measure things. If the flight attendant walks up to me, I can safely say that she 'walks slowly', say with a speed of 3 kilometers per hour. However, for somebody on earth, her speed would be 1003 kilometers per hour.

"How big does an object have to be to form its own frame of reference?"

"Alvas, can't you *think*? Theoretically, every tiny particle is an independent frame of reference. For practical purposes, it is safe to say that any coherent object is an independent frame of reference: an airplane, a train, a human being, the planet earth or a star. As long as you remember that there is no *preferred* or *absolute* frame of reference. Every single one has its own measurements. Any observer that is not part of your reference frame will obtain measurements that are very different from your own. Somebody standing still (if that would be possible at all) at the center of our galaxy will measure the speed of our flight attendant to be more than a million kilometers per hour. Not the right speed to serve coffee."

"I still don't see how it affects me Novare. When I'm traveling in an airplane, I feel exactly like standing still, except on takeoff. On the runway, even when I close my eyes, the stomach tells me, I'm moving. To take a sip from a wineglass while taking off doesn't work too well. So acceleration is different?"

"Exactly right. When you accelerate, you feel as if something is pushing you back. That is called the *inertia* of a body, you see; inertia is your resistance to acceleration. While traveling at a constant speed, you simply have no way to tell whether you are moving or not. When you look out of the airplane's window, you don't know whether it is the earth, or the airplane that is moving."

"So, what causes this inertia?"

Good question, Alvas. The inertia of a body is equivalent to its mass; inertia *is* the mass of a body. Another way to say this is that mass is the inertia of energy to accelerate. The bigger its mass, the bigger a body's resistance to changing its velocity. That's why heavier airplanes need more powerful engines to accelerate them to the right speed for take-off.

"Is this acceleration like in one of those fast elevators, when you feel heavier when it starts going up, or lighter when it starts going down? Gives me the same feeling in my stomach."

Exactly.

"I thought that had to do with gravity."

Now we are getting to the core of the mystery of gravity, Alvas.

"I like this Mr. J. May I explain the big mystery to Alvas?"

Sure Novare, go ahead.

"Einstein imagined such an elevator, out in empty space, away from the earth and the influence of gravity. Then he assumed that Alvas' feeling of 'being heavier' would exist in the same way, when the elevator is accelerated. As we know today, that is exactly what happens. There is no difference between the gravity that keeps our feet on the earth, and gives us a feeling of 'weight', and the force we feel when we accelerate. The conclusion was simple, yet mysterious: gravity and acceleration is the same thing!"

"Wow! I can follow that, but it still doesn't tell me what gravity *is*."

The answer to that is simple: *nobody knows*. Scientists are torturing their brains to find out, but we don't know the answer today. Two words existed that each had an independent meaning: the word *gravity* and the word *acceleration*. Each described a very different phenomenon. As Einstein proved, both are but manifestations of the same thing.

"Let's move on to the real mystery!"

Novare is very excited.

"Remember the energy that works so hard to accelerate the airplane?"

"JJ, when does she finally realize that without me she wouldn't remember anything?"

"Anyway, that energy *adds* to the mass of the airplane as it accelerates. Since mass is inertia, the energy has to increase continuously to further accelerate the object. That process has a limit though."

"I remember. The speed of light."

We are really getting ahead of ourselves guys. Let's return to the airplane and 'look' at the speed of light.

Back to our seat 2A on our intercontinental flight. We determined that the glass of wine was moving at a speed of about a million kilometers per hour. We obtained this speed by simply adding all the different speeds. As we learned in physics class, this is the scientifically correct way to do it, as long as these speeds are all measured in the same referent frame. Day has turned into night during our pacific flight. We now change the glass of wine for the strongest flashlight that money can buy. Then, we proceed to the cockpit and shine our light through the front window of the airplane, and into the pitch-dark night. Novare's observer, the one standing still at the center of our galaxy can observe the plane approaching at a speed that we measure at 1 million kilometers per hour. Just to illustrate what happens, let us assume for a moment that the speed of light would be just 2 million kilometers per hour.

At what speed will our light travel to the observer at the center of our galaxy, the one mentioned by Novare? It will be 1+2=3 million kilometers per hour, right? Wrong. It will be exactly 2 million kilometers per hour. It doesn't matter where the observer is, it doesn't matter how fast or in what direction the observer or the observed are moving. The speed of light remains the same, no matter who is looking, from where you are looking, and no matter what the speed is of the object emanating the light. As we know, the speed of light is 300,000 kilometer per second. However, that doesn't change the fundamental truth of Einstein's postulate that the speed of light is constant everywhere in the universe, and independent of the speed of the light-source.

"JJ, I understand how the speed limit on the highways affect people, but I still have trouble seeing how the speed limit of light influences *me*."

"May I explain it to Alvas?"

Go ahead Novare.

"Imagine that the speed limit of light would be lower; even much lower than the one we assumed in the airplane example above. Imagine it would be just 1,000 kilometers per hour. Now Alvas, what would happen if Mr. J were on the phone with a friend in Europe, some 8,000 kilometers away?"

"What does that have to do with light?"

"Nothing can travel faster than light, Alvas. Not phone calls, not remote controlled airplanes, and no information going over the Internet, even if you use the fastest equipment possible."

"Well, in that case it would take 8 hours for the phone to even ring on the other side! I got it."

"There you go. With the enormous speeds that light travels with, we don't realize its impact. But when scientists want to control a robot on Mars, it becomes a different story, doesn't it?"

"I guess, it will."

Nothing can travel faster than light, including gravity.[24] This meant that gravity could not be what Newton's laws predicted, namely that a change in the mass somewhere in space would be immediately noticeable anywhere else in space. In his famous Principia Isaac Newton stated, "Absolute space, in its own nature, without relation to anything external, remains always similar and immovable."

Einstein showed that such a universe did not exist. The universe is a dance. The dancers are the galaxies, stars, planets, solar systems, comets, and cosmic dust and radiation. The choreography of the dance is an evolving one, dictated by a mysterious interactive attraction called gravity, its dance patterns only limited by the speed of light. The configuration of the dancers, in time and in space, is totally dependent on the viewpoint of an observer.

Not unlike your own experience in a ballroom. If you are 'waltzing away' with your partner, the other dancers seem to be moving around you. When you observe the dancing people from a high vantage point, you see the swirling pattern of the moving crowd. As you change your vantage point and watch the dancing crowd, time and space seem to change. In the universe, time and space are interwoven with each other into perspectives that are personal: they are only valid for a specific observer at a specific position in space, at a specific point in time.

The Einstein Simulator

It is the fall of the year 2010. The class of 12 is ready for their fieldtrip. A new operating system had been installed to cope with the complexity of the application to simulate the Milky Way and the surrounding cluster of galaxies. The language used to develop the software is QUBE[25] (from 'QuBit-Extension language). QUBE can handle the quantum capabilities of the new generation of computer hardware. Virtual reality gear is no big deal these days. The regular clothing of the students already contains all the basic functionality such as touch, acceleration and position feedback, temperature and control and sound transmissions. Those with contact lenses or glasses just had to switch them to virtual reality mode, while the others received special glasses for the trip. Some have signed up to beta-test the optical neuron connectors that bypass the retina altogether. The early devices show severe problems, because of the software. The algorithms to render the signals in the correct format are still problematic. Attempts to bypass the brain's own image rendering mechanism have failed thus far. Today's project was to learn about the effects of the speed limit of light.

As an instructor, I am at the controls of our spaceship. We have to go to place way beyond our solar system to avoid the influence of gravity of the sun and planets. We are heading to a place about 30 million light-years outside of our Milky Way, because that gives us the most astonishing view of our galaxy. At that position, it appears exactly on its edge, as you can see from the picture (of a galaxy similar to ours) in Illustration 10.

Illustration 10: The picture shows a unique image of a galaxy 'on edge'. It is the galaxy called 'NGC 891', a spiral galaxy. If we could view the Milky Way from a distance of about 30 million light years, it would look something like this image. From this unusual vantage point, we can see in NGC 891 the surprising narrowness of the obscuring dust lane, a dark slightly irregular band across the galaxy. Also similar to the Milky Way is the prominent central 'bulge' corresponding to the rich star clouds in Sagittarius. The image was taken using the Isaac Newton Telescope at the Las Palmas observatory in 1992. Picture courtesy European Space Agency.[26]

After we settle into our position, I bring the spacecraft to a standstill, at least in relation to the surrounding stars. To minimize the time for the two experiments, I instruct the software to set the speed of light to 2 million kilometers per minute. The students do not know this, since it is exactly what they have to figure out. I give them the following instructions:

"Please watch the dashboard in the upper left corner of your field of vision. It is now exactly 11.55 o'clock am. Five minutes from now, at 12.00 noon, two explosions will occur. One is located exactly 6 million kilometers behind us, and the other one at exactly 6 million kilometers out in front of the spacecraft. When the explosions occur, the software will generate the sound which you'll hear right away. Cameras mounted on the front and the rear of the spacecraft register the light of the explosions. These cameras will project the explosions next to each other on your dashboard. I want you to log the times you notice the explosions and calculate the speed of light."

"Simple JJ. Even I can do that in my sleep. They will log a time of exactly 12.03! Since the students know that the light has traveled 6 million kilometers, it's easy to figure that the light of the explosion has traveled with a speed of 2 million kilometers per minute."

Well done Alvas. Indeed, all students figure this out.

The second experiment is designed to show the students what happens when the spaceship is on the move. Here is what I tell them:

"We will now make a right hand circle and go back to a point where we have the positions of the explosions in front of us. Then, we will accelerate our spacecraft to a speed of 1 million kilometers per hour and keep that speed constant. When we reach the same position as the one before, the software will reset the clocks in your dashboards again to exactly 12.00 noon. Again, two explosions will take place at the same positions as before, one 6 million kilometers behind, the other at 6 million kilometers in front of the spacecraft. As before, I want you to log the times at which you see the explosions."

"You'll need Novare for this one, JJ."

OK, but without any formulas[27] please, Novare.

"Sure Mr. J. The students will see the front explosion at exactly 12.02. At that time, the spacecraft has traveled 2 million kilometers. The light from the front explosion will have traveled toward the spacecraft, at 2 million kilometers per minute, exactly 4 million kilometers. That is where the light from the explosion will 'meet' the spacecraft, therefore the students can see it at that time. The light from the rearward explosion has also traveled 4 million kilometers toward the spacecraft, but is still 4 million miles behind it. The students will log the second explosion at 12.06, 4 minutes after the first one! At that time, the spacecraft has traveled 6*1 million kilometers, and the light from the rear explosion will have traveled 6*2 million kilometers."

"I'll have to think about it, JJ. I'm not sure I get it."

Take your time Alvas.

Some of the students find this simple to comprehend, while others find it 'weird'. As we wrap up our field trip and travel back at warp speed, I take the time to explain the significance of what we just experienced. The plain fact that the speed of light has a limit, 300,000 kilometers per hour, means that we have to be very careful with our observations. Whether two events take place at the same time totally depends on our position, our speed, and the position and speed of the events we observe. Every event is defined by its position in space and its position in time (where and when does the event take place). A student being at a certain place at a certain time is an event. The appearance of the light of an explosion on the student's dashboard is another event as observed by the student at that time. The student's observation establishes a relationship between both events, the student's position in space-time, and the position of the observed event in space-time.

As we unplug our virtual reality gear from the simulator, I wonder. Will students think about the experiment when they enjoy the next sunset? Will they realize that it is just the light that they see, while the sun already sunk behind the horizon before they got to the beach?

The simulator we used is the *Einstein simulator*[28]. The (imaginary) project, initiated by the European (ESA) and the North American Space agencies, later joined by Japan, China, Russia and India, provided an educational and research tool for cosmology and astrophysics. It started out as the 'digital sky' project to map the skies in digital format. Before, there had been no way for young people to get a sense of the relative nature of the universe. As the awareness in the world had grown because of the simulation capabilities of digital technology, the value of these tools became obvious. We humans are too small to experience the true nature of the very big, and too big to experience the world of the very small. Digital Technology removes that barrier. It was the hope of the funding parties to advance the awareness of young people, and to expand their mindset. To demonstrate, and to investigate the intimate connections between all living and non-living systems in the universe, and the relativity of phenomena in it, had been impossible without the possibilities of the digital world. With it, we now can adjust our own perception to the scale of the time-space continuum under consideration, from the very small to the very big, and everything in between.

At the Airport, August 29, 2002

On the large scale of the universe, the effects of relativity are dramatic. During the billions of years of evolution, there was no need to be aware of this, because these effects cannot be noticed on the scale of a human life, or the lives of animals and plants. To look at the distortions of space and time, we'll return one more time to our airplane. However, this time, your friend takes seat 2A. You brought your 8-year-old son to watch the airplane take off. The airplane lines up at the start of the runway, three kilometers away from the outdoor observation deck. Engines start roaring, and just seconds later the big airplane, thunders by in front of you, lifting its nose, and taking off.

As the plane climbs into the sky, it soon becomes smaller and smaller. The skies are clear and you can follow it all the way up. The incredible speed that we noticed as it passed on the runway seems to be slowing, as it gets further away. Now, all the way up, against the backdrop of the blue sky, it seems to be moving slower and slower, as the airplane itself seems to be shrinking even further.

Illustration 11: "Are people getting smaller too?" Departing airplane by James[29]

After a minute or so, the only remaining thing is -hardly visible against the sky- a tiny spot that seems to be standing still, seemingly just hanging there. Your son, who had been watching in silent awe, asks you a question. Waking from your focus on the airplane, you ask him what he said. "The plane became so small. Did Uncle Max become small too?"

Imagine that in the time we looked at the departing airplane, its speed would have increased to something close to the speed of light. What was just a matter of optical illusion earlier, is now about to actually happen. The airplane will become smaller and smaller, and Uncle Max will become smaller too! Time will slow down further and further. Right at the speed of light, the airplane and everything in it will collapse into itself, and a tiny speck of matter with an intense density is all that will remain, forever. Time will come to a halt. This scenario describes a physical reality; it is not just a fancy metaphor, like the departing airplane. The airplane transforms into a spacecraft, and continues its journey. Max is timing the journey. The spacecraft keeps its speed, to avoid collapse under its own mass, just below the speed of light. When the calendar on board the spacecraft shows the year 2003, the spacecraft reaches its destination and heads back for earth. The return flight takes the same amount of time, and Max's calendar shows today's date:

For your friend, it is August 29, 2004.

You are at the airport to welcome him back to earth. Since you know about relativity, you are prepared for what is coming. It had made you wonder about life and death this morning, but hey, you were alive. You had asked your son to join you for the reunion, and so he did.

Max arrives. He has aged only 2 years! He gives you a hug, and makes a comment on your graying hair. Then he looks at your 28-year-old sun with amazement, and finally, turns to your 4-year-old grandson. All of you have aged 20 years! Your own calendar shows your own truth:

Today's date on the planet earth is August 29, 2022.

While traveling at speeds so close to the speed of light, Max's clock slowed down, his metabolism slowed down as his body shrunk, while your own life went on, a day at a time, as did his. Upon his return, as the plane slowed down, he regained – relative to you –his former proportions of space and time. This is the reality of relativity.

By 1919 Einstein had established his celebrity status, something he never sought, and would never succumb to. When British experiments in astrophysics confirmed Einstein's predictions, the London *Times* ran the following headline on November 7, 1919:

"Revolution in science. New theory of the Universe. Newtonian ideas overthrown!"

"Why do I need to know about relativity at all, JJ?"

Because relativity is one of the most important properties of evolution, as we will see in future chapters, Alvas. We cannot assign an absolute value to the things we see, or the things we experience. For cosmic phenomena, we have the theoretical proof by Einstein, and the scientific evidence to support it. Because of that, the universe offers the only scientifically sound opportunity to explain the essence of relativity. You don't need to fully grasp its theoretical foundation, as long as you got a 'feeling' for its meaning.

"What I see in the universe, is dependent on my own point of view; that's how much I understand."

Correct.

"I also realize that matter and energy are two sides of the same coin; a coin that is always on the move, never standing still. There is no such thing as standing still. What I notice are things that happen: events, experiences and so on. I can only assign a value to such an event, or measure any of its qualities, in relation to another event."

That's great Alvas.

"The feeling I get is there is nothing that I can latch on to. The firmer I try to grasp something, the faster it evades me. The whole thing feels like a handful of sand: the firmer my grip on the sand, the faster it runs away, JJ. Now you know why I always go with the flow."

Strange Events

Brussels, Belgium, 1927

Illustration 12: This photograph of famous scientists was taken at the international Solvay Conference in 1927. Among those present are many whose names are still known today. Front row, left to right: I. Langmuir, M. Planck, M. Curie, H. A. Lorentz, A. Einstein, P. Langevin, C. E. Guye, C. T. R. Wilson, O. W. Richardson. Second row, left to right: P. Debye, M. Knudsen, W. L. Bragg, H. A. Kramers, P. A. M. Dirac, A. H. Compton, L. V. de Broglie, M. Born, N. Bohr. Standing, left to right: A. Piccard, E. Henriot, P. Ehrenfest, E. Herzen, T. De Donder, E. Schroedinger, E. Verschaffelt, W. Pauli, W. Heisenberg, R. H. Fowler, L. Brillouin. Illustration used with the kind permission of the Solvay Institutes.[30]

Twenty-nine scientists attended the *Solvay Conference on electrons and photons* in Brussels, Belgium 1927. Amongst the participants, depicted in illustration 12, were some of the scientist that changed our understanding of the world, including Max Planck, Albert Einstein, Louis de Broglie, Paul Dirac, Niels Bohr, Werner Heisenberg, and Edwin Schrödinger. They were in Brussels to discuss the discovery of the strangest phenomena in the mechanics of the universe: the behavior of the quantum.

All of the attendants at the Solvay conference had a common scientific concern: the world of the very small. Only recently, around 1900, atoms were found not to be the smallest particles of matter, as had been thought since the ancient Greeks. Rather, they contained a nucleus with electrons orbiting around them. Attempts to look at the atom as a mini-solar-system and to use the Newtonian laws to describe the movement of its components failed, and a new theory was needed. Apparently, the micro-world of energy and particles did not work the same way as the macro-world of solar systems and galaxies.

The research of some of the men at the conference had led to amazing discoveries and to controversy between them. As Richard Feynman would say half a century later:

"Working out another system to replace Newton's laws took a long time because phenomena at the atomic level were quite strange. One had to lose

one's common sense in order to perceive what was happening at the atomic level."[31]

"Here we go JJ. Quantum behavior. I'll never understand it, no matter *how* you explain it."

I don't understand it either, Alvas, nobody does. However, Novare and I have a feeling of its strangeness, and of its existence. It is the most fundamental mechanism in every physical entity that exists, every 'body' that lives, and in every thinking human being. The only thing we need to achieve is to give readers who are not familiar with quantum behavior, a sense of its reality, and a sense of the strangeness of this reality.

"I'll warn you, when you run into a mental pretzel."

All right.

Max Planck, born in 1858 in Kiel, Germany, postulated in 1900 that energy behaves in quanta. Planck shocked the scientific world because he suggested that energy did not just 'flow' as a variable wave, propagating through 'ether', but existed only in discrete quantities. Such a quantum (quantum is the Latin word for amount) of energy can never have a value that lies in between these discrete values, just like a shot of your favorite whisky in your drink. You ask for "a single" or a "double" shot, but not for a 1 ½ shot. You might ask for a "one-and-a-half" shot, but the bartender would probably charge you for two anyway. In the world of energy, there is no such thing as a 1½ shot. Energy only exists as Quanta. This is the reality that forms the basis of the strangest theory of the 20th century: quantum theory. Whenever we wonder about the understandability of the behavior of such quanta, let us remember the words of Richard Feynman:

"No, you're not going to be able to understand it... You see, my physics students don't understand it either. That is because I don't understand it. Nobody does."[32]

Albert Einstein used Planck's quantum hypothesis to describe the electromagnetic radiation of light, and postulated the equivalence of energy and matter in 1905, in his famous formula $E=MC^2$. A particle was nothing more, or less, than a quantum of energy. The photon as a particle, or a quantum of light, saw the 'light of day'.

Niels Bohr, born in 1885 in Copenhagen, Denmark, used the quantum ideas due to Planck and Einstein's quantum ideas in his work published in 1913. Although he had suggested a Newtonian view of atoms, he proposed that an atom could exist only in a discrete set of stable energy states. This provided further evidence for the validity of Planck's theories. The quantum behavior of energy became accepted fact.

Louis de Broglie was born in 1892 in Dieppe, France. In 1924, he finished a doctoral thesis in which he postulated that electrons behave like waves, not unlike the waves caused by plucking the string of a guitar. Einstein was asked to give his opinion on this extremely controversial thesis. Einstein, having discovered the duality of light, was sympathetic to the idea. If photons could behave like waves or particles, why couldn't electrons behave like waves? He replied 'It might sound crazy, but it really is sound'. De Broglie received his degree. His theory was later confirmed by experiment. Now, the theory of quantum behavior, the equivalence of energy and matter, and the duality of waves and particles, were extended to all elementary particles. The remaining question was: how did these particles-waves behave?

Werner Heisenberg, born in 1901 in Würzburg, Germany, discovered in 1927 that the position and momentum (velocity and direction) of a particle couldn't be determined with equal certainty at the same time. This phenomenon, which baffled and frustrated many scientific minds, made Heisenberg famous. Its formal description became known as Heisenberg's *Uncertainty Principle*. Heisenberg formulated a quantum theory, based on matrix mechanics.

Paul Dirac was born in 1902 in Bristol, Gloucestershire, England. In his 1926 doctoral thesis Quantum mechanics, he was the first one to propose a mathematically consistent general theory of quantum mechanics, inspired by reading the proofs of a paper by Heisenberg, in the summer of 1925.

Erwin Schrödinger, born in 1887 in Erdberg, near Vienna, Austria, had read some of Einstein's comments on De Broglie's wave mechanics, and decided to investigate further. To do this, he took flight to a villa in the Swiss Alps in 1925, leaving his wife behind and inviting a former Viennese girlfriend. The result of this 'quiet' period would change the view of the microworld of physics. Schrödinger published his revolutionary work relating to wave mechanics and the general theory of relativity in a series of six papers in 1926.

The equations as proposed by Schrödinger were to become the most important formulation of quantum theory, as far as the behavior of electrons in atoms and molecules is concerned. Planck described it as epoch-making work and Einstein wrote: ... the idea of your work springs from true genius... [33].

Today, several approximations are used to calculate the behavior of particles within an atom, and atoms within a molecule, blurring the boundaries between theoretical chemistry and physics. Only the most powerful supercomputers in existence today are capable of calculating the behavior of even a dozen of atoms. A very interesting aspect of Schrödinger's equation is that it is not derived from other, more simple mathematical constructs. A joint production of intuition, imagination, and logic was proven correct through experimental evidence.

The Solvay conference was an event that can be seen as a symbol for a phenomenon that emerged in the 20th century. It was a 'meeting of minds', made possible by the increased mobility of people through new means of transportation and the advent of communication technologies. Most of the attendants had already met, had a teacher-student relationship, or had been working together at the same institutions. However, the improved means of communication would speed up the interaction between these minds, which in turn produced an incredible acceleration in the generation of new ideas. Later, in 1961, Heisenberg wrote about the 1927 Solvay conference:

> *"To those of us who participated in the development of atomic theory, the five years following the Solvay Conference in Brussels in 1927 looked so wonderful that we often spoke of them as the golden age of atomic physics. The great obstacles that had occupied all our efforts in the preceding years had been cleared out of the way, the gate to an entirely new field, the quantum mechanics of the atomic shells stood wide open, and fresh fruits seemed ready for the picking."*[34]

Einstein was a troubled man when he attended the Solvay conference. He did not like the statistical nature that was now the accepted foundation of quantum mechanics. He expressed this with his famous remark: "God does not play dice!" As the story goes, Bohr replied "Einstein, stop telling God what to do." The discrepancy between the mechanics of the very small and the mechanics of the very big kept Einstein's mind occupied until the end of his life. After him, the unification of these two worlds by one single theory remains at the top of the scientific agenda of physicists. Until today, the mystery remains.

This brief description of the main personalities that uncovered the secrets of atoms and the behavior of elementary particles is but one of many threads that link them (many others have been totally omitted). In reality, it was a merging and recombination of ideas, thoughts, and theories, starting with the philosophers in ancient Greece, and proceeding through the centuries, accelerating in the mid-nineteen hundreds, and culminating around the middle of the 20th century. After that, a new technology would add a new dimension to the communication between minds. Digital technology would provide a platform that gave new meaning to the expression 'meetings of minds'.

At the Casino

Quantum mechanics describes the behavior of elementary particles as quanta of energy. Schrödinger's theory describes this behavior as a statistical process. We never know, at any given point in time, the exact position and the velocity of a particle, as Heisenberg found out. Particles behave like a wave spreading out in all directions. This is the same as saying that particles take every route possible when they travel through the universe. This is exactly what quantum theory does say.

The statistical nature of the micro-world is not some fancy metaphor, or a metaphysical construct. It is our physical reality. Stephen Hawking compares the statistical nature of the micro-world to a casino operation:

> *"It is the same [as the casino operation] with the universe. When the universe is big, as it is today, there are a very large number of rolls of the dice, and the results average out to something one can predict. That is why classical laws [of physics] work for large systems. But when the universe is very small, as it was near in time to the big bang, there are only a small number of rolls of the dice, and the uncertainty principle is very important."*[35]

In the small world of photons and electrons, all of the paths these particles can take, should be thought of as a wave of probabilities. However, when somebody makes a 'conscious' observation of the wave, things change. At the very moment of observation, an event shows itself. Something 'happens', like the manifestation of a particle. At the moment of observation, the wave (called the quantum wave) seems to *collapse* (that is how it's called in quantum 'language'), and exposes the true 'nature' of the event.

"Mental pretzel JJ! I still don't follow."

Novare comes to the rescue.

"Here is an analogy, Alvas. It is simple and incomplete, but maybe it helps. Imagine plucking a string on a guitar. Then, after plucking it, leave it alone and watch it. What do you see?"

"Actually, nothing. Except that I see something that looks like a fuzzy band around the place where the string was. It is the string swinging and vibrating...*like a wave*!"

"See? A quantum wave is the totality of probabilities of a point on your guitar string being at certain positions and having certain velocities at a specific moment in time."

"Thanks Novare. Still a pretzel, but at least I can sense its strangeness."

"To me, this quantum wave sounds like getting a new idea, Mr. J. When I am trying to come up with something new, I will look to Alvas to find stuff based on the few clues I get from you. Then, seemingly out of the blue, or a whirlpool of possibilities, I get this new insight!"

You mean it feels as if a quantum wave of all sorts of possibilities is brewing. Then, when you look, you collapse the wave of possibilities, and the idea is born?

"You allow her to compare me to a quantum wave JJ? As if I don't know what I'm doing? I admit that I have my ways, but they have more to do with the way in which I organize my stuff, see. In many cases, it's simply a matter of waiting for Novare to make up her mind. I'll have the answers, as soon as I know what her question is."

"QED, Mr. J?"

Like everything else, we too, are built from the matter and energy that makes up the universe; we feed ourselves with it, we are born from it, and we die from it. Bacteria, grandfather clocks, hamsters, lawn chairs and grandchildren, all built from the same substance, with the same properties, some of which are known, while some remain mysterious. Some scientists believe that quantum behavior can play a role in the emergence of life, and even in the emergence of consciousness.

Quantum mechanics is 'counter-intuitive', because of its statistical nature and the 'impossible' interactions of tiny particles. These particles defy the natural flow of time, while 'waves of probable movements' seem to make up their 'mind' before they reach their destination. On the other hand, a host of non-fiction publications makes 'intuitive' use of it, no matter what the subject matter is. When somebody wants to express the appearance of some effect 'out of nothing', reference is made to the 'quantum dynamics' of the underlying process. When a truly innovative product appears in the market, we say that the new product is 'a quantum leap' ahead of its competition, referring to the energy jump of an electron orbiting the nucleus of an atom. It seems to me, that the 'weirdness' of quantum mechanics resembles our lack of understanding of intuition, gut-feel, imagination, and the creative 'moment'.

While the discoveries of quantum theory and the theory of relativity made the headlines of the 1920's, the view of the universe remained to be one of a static and permanent nature. In fact, Einstein had added a factor to his equations to *make* the mathematical description fit a static universe.

After all the scientific discoveries by Copernicus, Galileo, Newton, Einstein, Planck, Schrödinger, Heisenberg, Dirac, De Broglie and many others, the universe was still a permanent dance in a dancehall that never changed. Darwin's theory had not sunk in. The link from the evolution of life to the possibility of an evolution of the universe had not been made.

That view was about to change very shortly. It would expose the beginning of existence itself; and it would kick off the slow process of re-assessing the reality of the world we live in.

"JJ?"

"You want to know, *why* you need to know about quantum behavior, right?"

"Right, Novare. I do understand that very strange things are going on in the micro-world of electrons and photons. But why care about that small stuff, when the world is full of things that I can *see*?"

"That is exactly the point Alvas. The things we can perceive shape our mindset. Our senses don't perceive evolution, so we don't care about it. We can't see distant galaxies, so we ignore them. We are too big to observe atoms with our eyes, so we take them for granted. Every atom in the universe, and every atom in every cell of your body, depends on quantum behavior for their existence. Without it, nothing would exist for very long."

"So what?"

"What if quantum phenomena play a role in the large scale behavior of living beings, or even in consciousness? After all the small and the big are parts of one space-time, continuum and we are still looking for their connection. Relativity concerns the scientific frontier of the big, and quantum theories are the frontier of the small. I thought you wanted to know."

"I'm not sure that I care."

"I should have known: it's not in your nature to care about the outside world."

"Novare, I am the outside world. I am the whole history of the outside world, as far as you are concerned. I'm all you have. When do you understand that? Of course I care, because I need to accurately reflect that outside world."

"Then you might understand that digital technology will change your world. With that technology, we can model the micro-world of atoms, electrons and photons. We are then able to untangle its remaining mysteries. Already, we can design new structures using single atoms. We will be able to build new products - any product you can imagine - from the ground up. Atoms are the design elements for the products of the future. We will be able to build tiny machines, that can enter your bloodstream and repair your brain, and bigger machines that can build the small ones. Computer Aided Design software is the drawing board, and nanotechnology is the production mechanism. Similarly, digital technology is the key to biotechnology. Because of digital technology, Alvas, the human scale has changed.

We can now observe, and manipulate anything. We can even change the speed of light, as in the Einstein simulator, if it suits the purpose of understanding. The small world is no longer small. It will provide the Lego-blocks for the technologies of the future."

"So I have to prepare to see things, experience things, and store things into memory that I never dreamt of before?"

"Indeed, Alvas. And by the way, I like you anyway."

Chapter 3: New Beginnings

"Evolution in the most general terms is a natural process of irreversible change, which generates novelty, variety, and increase in organization: and all reality can be regarded in one aspect of evolution. Biological evolution is only one sector or phase in this total process. There is also the inorganic or cosmic sector and the psychosocial or human sector. The phases succeed each other in time, the later being based on and evolving out of the earlier. The inorganic phase is pre-biological, the human is post-biological."

Sir Julian Huxley[1]

The Beginning of the Past

Kennedy Space Center, March 1, 2002

"Following a flawless final countdown, Shuttle Columbia lifted off at 6:22:02.080 a.m. EST today on the STS-109 mission to service the Hubble Space Telescope." This was NASA's official 'status report' issued at 6.30 in the morning of March 1, 2002, after the launch from launch pad 39A at the Kennedy Space Center.[2]

Illustration 13: Hubble Space Telescope during repairs and upgrades in March 2002. Courtesy NASA.[3]

During its 12-day mission, the crew installed new equipment that increased the optical capacity of the Hubble space telescope ten-fold, extending its capacity to look into the past of the universe by billions of years. In April of 2002, Hubble sent its first images of far away galaxies to Earth.

The most advanced telescope in space bears the name of the man who delivered a final blow to the view of a static and permanent universe. Edwin Hubble was born in 1889 in Marshfield, Missouri, USA. Young Edwin Hubble was fascinated by science and mysterious new worlds from an early age, having spent his childhood reading the works of Jules Verne. After studying law at Oxford, he returned to the United States and decided to devote his life to astronomy. In the early 1920's, he tried to answer the question: what exactly are galaxies? Scientists wondered about some of the fuzzy clouds of light (called *nebulae*) that are visible in the night sky. Are they part of the Milky Way or are they separate galaxies, much farther away in the universe? One of these 'clouds', the nebulae Andromeda gave Hubble the first clue.

In 1924, he measured its distance to be a hundred thousand times bigger than the distance to the nearest stars. He measured distances to several other galaxies using their brightness as a guide. His measurements showed a shift in the spectrum of the emitted light that was proportional to the distance of the observed galaxies. This Doppler shift of their light[4] could only mean one thing, which he announced to the world in 1929. Galaxies are moving away from us with a speed proportional to their distance. The explanation is simple, but revolutionary: the Universe is expanding. If the universe is expanding into the future, it must have been evolving in the past. The conclusion is inevitable: *our universe had a beginning.*

In the 1920s, Hubble used the best telescope available at the time: the 100-inch telescope on Mount Wilson in southern California. The telescope that carries his name today is used to reach out ever further into the past. Scientists use the Hubble telescope to explore the events at the beginning of the universe, and to look for signs of life in distant places, distant in space, as well as distant in time, as they try to shed light on the beginning of life here on earth. Hubble's name remains to be the symbol for our ongoing quest to answer the questions: where does human life come from, and, are we alone in the universe?

Cambridge, United Kingdom, February 4, 1922

On this day, John Haldane presented his paper *DAEDALUS or science and the future,* in which he presented his views on the future of science considering the latest discoveries such as Einstein's theory of relativity, quantum mechanics, and the theory of evolution. Talking about the subject of new biological discoveries, he said:

"Darwin's results are beginning to be appreciated, with alarming effects on certain types of religion, those of Weismann and Mendel will be digested in

the course of the present century, and are going to affect political and philosophical theories almost equally profoundly. I need hardly say that these latter results deal with the question of reproduction and heredity. We may expect, moreover, as time goes on, that a series of shocks of the type of Darwinism will be given to established opinions on all sorts of subjects. One cannot suggest in detail what these shocks will be, but since the opinions on which they will impinge are deep-seated and irrational, they will come upon us and our descendants with the same air of presumption and indecency with which the view that we are descended from monkeys came to our grandfathers. But owing to man's fortunate capacity for thinking in watertight (or rather idea-tight) compartments, they will probably not have immediate and disruptive effects upon society any more than Darwinism had."[5]

In 1931, Haldane published his book *The Causes of Evolution,* in which he proposed his theory on the evolution of life, based on Darwin's theory and the work of Mendel. His was the first of two serious theories (the other one by Russian scientist Oparin) about the beginning of life. Darwin himself had perceptively speculated on the beginning of life when he wrote in 1859:

"all the conditions for the first production of a living organism ... [could be met] ... in some warm little pond with all sorts of ammonia and phosphoric salts, light, heat, electricity, etc. present."[6]

In the late 1700's Frenchman Antoine Laurent Lavoisier, the founder of modern chemistry, theorized that 'La vie est une fonction chimique' (life is a function of chemistry). We do not know whether this was a prophetic intuition, or Lavoisier's bold assumption.

Today, several theories exist that can be considered to have serious potential for explaining the emergence of living organisms from inanimate matter. However, as Mayr points out:

"In spite of all the theoretical advances that have been made toward solving the problem of the origin of life, the cold fact remains that no one has so far succeeded in creating life in a laboratory ... many more years of experimentation will likely pass before a laboratory succeeds in producing life."[7]

Some scientist assume that the basic living compounds were imported to the earth, contained in the interstellar dust that reached the early planet; some, like famous scientist Francis Crick (one of the discoverers of the structure of DNA), even assumed that a distant civilization could have sent the basic germs of life to the earth using their space-ships.

Most scientists today agree that it is a matter of *when*, not *if*, we unveil the secret of creating life out of inorganic matter. They assume that it is a matter of the right basic ingredients, subjected to the right set of energetic circumstances. Whatever the difference might be, most scientists agree: *life had a beginning.*

Cambridge, USA, 1892

The product of science is the recreation of the world, by, and in, the human mind. No matter how hard we try to describe the reality of the world outside of the human mind, the resulting rendition is a mental picture. To arrive at a description, or a view, of reality that reaches a state of independence of the idiosyncrasies of individual interpretations, or individual states of mind, is the quest of science. Objectivity is the label we use to describe this independence. Subjectivity on the other hand is the label we use to qualify the idiosyncrasies of individual minds. Psychology is the (new) science that studies these mechanisms and tries to identify the universal laws that explain them. Saying that psychology is a 'new' science does not do justice to the efforts through the ages to understand the human mind, especially to philosophy, the greenhouse of psychology. At the end of the 19th century however, psychology was an emerging science, at least according to William James, a 'maverick' who never took a class in psychology. Yet, he was the first 'professor of psychology' in the United States when he started teaching courses on the subject in 1875 at Harvard, and published his masterwork on the subject, Principles of Psychology, in 1890. To sum up his view on the state of affairs in psychology as a science in the late 19th century, he wrote in 1892:

> *"It is indeed strange to hear people talk triumphantly of 'the New Psychology', and write 'Histories of Psychology', when into the real elements and forces which the word covers not the first glimpse of clear insight exists. A string of raw facts; a little gossip and wrangle about opinions; a little classification and generalization on the mere descriptive level; a strong prejudice that we have states of mind, and that our brain conditions them: but not a single law in the sense in which physics shows us laws, not a single proposition from which any consequence can causally be deduced. This is no science; it is only the hope of a science."*[8]

"Why talk about psychology Mr. J? This book is not about *psychology*, is it?"

It is not Novare. But I want to draw the attention to that which makes us human. And isn't the human mind the place where that humanity finds its expression?

"Am I not the living proof of that?"

Yes you are.

"What about me JJ? I am the one that gathered all the experience; I had billions of years to do so. Novare is just the icing on the cake. Actually, she makes life a lot more difficult, if you ask me."

"Only since I came on the scene, Alvas, new things are happening. You think that Solvay conference in Brussels could have happened without me? You really believe that Newton, Darwin, Einstein, Planck and Hubble could have developed the insights into the nature of nature as they have, without me? The great apes, as smart as they are, did not invent $E=MC^2$, the most beautiful butterflies did not write poems, and the might and the grace of whales did not result in gravity-defying flight. Lasting creative thought is human, and I made that happen."

"You couldn't have done it without me, either. I was the one who gave you the yarn for intuition and imagination in the first place; you just had to weave it together."

"I will grant you that, Alvas."

That settles that.

Whatever it is that produces creative human thought, most of us agree that it is something new that distinguishes human beings from the rest of nature, although Psamtik I, king of Egypt concluded differently, during the second half of the 7th century BC. He conducted the first recorded experiment in psychology:

"The Egyptians had long believed that they were the most ancient race on earth, and Psamtik, driven by intellectual curiosity, wanted to prove that flattering belief. Like a good psychologist, he began with a hypothesis: if children had no opportunity to learn a language from older people around them, they would spontaneously speak the primal, inborn language of humankind – the natural language of its most ancient people –, which, he expected to show, was Egyptian.

To test his hypothesis, Psamtik commandeered two infants of a lower-class mother and turned them over to a herdsman to bring them up in a remote area. They were to be kept in a sequestered cottage, properly fed and cared for, but never hear anyone speak so much as a word. The Greek historian Herodotus, who tracked the story down and learned what he calls 'the real facts' from priests of Hephaestus in Memphis, says that Psamtik's goal 'was to know, after the indistinct babblings were over, what word they would first articulate.'

The experiment, he tells us, worked. One day, when the children were two years old, they ran up to the herdsman as he opened the door of their cottage and cried out 'Becos!' He sent word to Psamtik, who at once ordered the children brought to him. When he too heard them say it, Psamtik made in-

quiries and learned that Becos was the Phrygian word for bread. He concluded that, disappointingly, the Phrygians were an older race than the Egyptians."[9]

As we know today, human beings have no innate language; at least not the languages we characterize as human. Whatever it is, and however it works, the capacity for the creation and the expression of thought and emotion is what distinguishes us as being human. We call it 'soul', 'consciousness', 'self-awareness', 'sentience' and give it a host of other labels. Today, apart from psychology, scientists from virtually all branches of science chime in with new hypotheses or opinions on this mysterious phenomenon that makes humans human.

Pope John Paul II changed the encyclical Humani generis of Pope Pius XII, which contains the position of the Catholic Church on the issue of evolution. In 1996, Pope John Paul II stated that:

"Today, almost half a century after the publication of the Encyclical, new knowledge has led to the recognition of the theory of evolution as more than a hypothesis. It is indeed remarkable that this theory has been progressively accepted by researchers, following a series of discoveries in various fields of knowledge. The convergence, neither sought nor fabricated, of the results of work that was conducted independently is in itself a significant argument in favor of this theory."[10]

Just like Galileo, Darwin, or Teilhard for that matter, were not available for comment either. However, they might have been delighted by this progressive stance from the Catholic Church. However, as Francis Fukuyama explains:

"But the Pope went on to say that while the church can accept the view that man is descended from nonhuman animals, there is an 'ontological leap' that occurs somewhere in this evolutionary process. The human soul is something directly created by God: consequently, 'theories of evolution which, in accordance with the philosophies inspiring them, consider the mind as emerging from the forces of living nature, or as a mere epiphenomenon of this matter, are incompatible with the truth about man'."

Hubble showed the way to a possible beginning of the universe and Darwin led the way to a beginning of life. Finally, all evolutionary evidence, as well as our own observations, and even the Catholic Church lead us to assume the beginning of human thought. Whether this happened in the form of an 'emerging consciousness, or in the form of a 'soul' that was bestowed on us from the outside, it must have started to happen at some point in the past: human *thought had a beginning.*

The Beginning of the Future

During the first half of the century, scientists did little to make their work accessible to the broad public. Virtually nothing was done to make the case for the implications of new scientific insights for society. According to C.P. Snow, in his book Two Cultures, the self-proclaimed literary intellectuals of the first half of the 20th century excluded scientists such as Hubble, Einstein, Heisenberg, Niels Bohr and others mentioned in the previous chapters from the definition of "intellectuals." In his book, C.P. Snow calls the literary intellectuals "natural Luddites," proclaims that scientists "have the future in their bones," and that "the traditional culture responds by wishing the future did not exist"[11]. In the second half of the 20th century with the appearance of digital technology, especially the Internet, this would slowly start to change. To look at the cradle of this new technology, we will return to 1944.

Bletchley Park, UK, 1944

We are back in Europe at the end of World War II. Despite all of the physical force deployed in the war, the planes and the ships, the bombs and the tanks, the politicians and the soldiers, the final stages of the war might have been a long time away, if it weren't for a new invention. A new device, quietly infiltrating the war, would change the course of history. It was called a 'computing machine', a machine that could compute numbers faster and more accurate than a human being. In the 1930's and 1940's a 'computer' was the job title for a person doing calculations.

In 1944, the first programmable, electronic, digital computer in the world was deployed to help decipher German military codes. Its name was *Colossus* and it was used at the "Government Code and Cipher School" (in short: "GC&CS") at Bletchley Park in England. Its existence and the secrets of its purpose and inner workings were not officially revealed until 1983. According to F.H. Hinsley, official historian of the GC&CS, the war in Europe was shortened by at least two years because of the intelligence operations at Bletchley Park.[12]

The Beginning of Digital

Alan Mathison Turing[13] was an important witness to the success of the Colossus Computer in 1944. He was now 32 years old and was expanding his electronic experience by building, with one assistant, an advanced electronic speech scrambler. In fact, he was to receive a large portion of the credit for shortening the war. He had joined the intelligence agency at Bletchley Park in 1939 and took on to decipher the German Naval communications code, part of the German military code called "Enigma." Based on the work of a Polish mathematician, against all hope, Turing cracked the code. Together with Gordon Welchman, he developed the "bombes," dedicated electromechanical machines to speed up the deciphering process. Thanks to the bombes, by early 1942 GC&CS was decoding about 39,000 intercepted messages each month, rising subsequently to over 84,000 messages a month - approximately two every minute. Turing's work on the version of Enigma used by the German navy was vital to the battle for supremacy in the North Atlantic.[14]

Turing's impact on the war might, or might not have been a deciding one. However, his influence on information technology is a defining one; and extends beyond the technology itself into fields like cognitive science, neuroscience and the philosophy of consciousness.

In 1936, he had written the paper that would give him the lasting reputation as one of the founders of computer science and the designer of the blueprint for digital computing: *On Computable Numbers, with an application to the Entscheidungsproblem*. In this paper Turing introduced an abstract machine that could follow a precisely defined set of rules fed to the machine, that operate on a number fed into the machine as well. The paper was not published until 1937, after Newman who had been one of his teachers argued the case before the London Mathematical Association. Such a machine would become known as a "Turing Machine"; a machine was capable of storing, interpreting and performing any set of instructions (what we now call the program or the software) on numbers (or symbols) fed to it: the modern computer, modeled after Turing's understanding of the human mind.

Newman had not been the only one influencing Turing's mind. Turing has been quoted to say that the book *Natural wonders that every child should know* had a seminal influence on his life. His fascination for science would make his headmaster exclaim in a note:

> *"If he is to stay at Public School, he must aim at becoming educated. If he is to be solely a Scientific Specialist, he is wasting his time at a Public School."*[15]

Turing's life can be seen as one long quest for understanding the human mind and for designing and building a universal machine that would be capable of equaling the intelligence of the human mind. At the age of twenty, he still expressed his belief that there was a human spirit that was separate from the mechanism of the body. Later he seemed to have changed his mind: things like creativity and intuition might be solved by a machine, given the right amount of complexity in the machine's instructions, which would give it learning capabilities. As Andrew Hodges explains:

> *"The postwar Turing claims that Turing Machines can mimic the effect of any activity of the mind, not only a mind engaged upon a 'definite' method."*[16]

Computing before Turing.

The world had been 'pregnant' with the idea of computing machines since another mathematician, the eccentric and visionary Charles Babbage, conceived the idea of a general-purpose computing machine in 1834. This machine would use the punch card, already successfully deployed by the Frenchman Joseph-Marie Jacquard in 1804, as its input and controlling device. Jacquard used the punch card in his famous looms to weave intricate patterns into textiles. Different patterns merely required the design of different punch cards. Another French inventor, Jacques de Vaucanson, invented the concept in the century before. The holes in the cards provided a means to invoke an action or not: the basic on/off notion of modern digital computing (The development of the binary system goes back to Leibniz in 1679). Jacquard's looms angered weavers, the Luddites of his time, for fear of their jobs. His looms were burnt, but in 1806, his invention was declared public property and he received a royalty from each loom produced from then on.[17] For almost a century only special purpose machines were built and used, until electrical technology could be exploited. In the early 1880's one evening, over 'after dinner' tea, Dr. John Shaw Billings said to Herman Hollerith, son of a German immigrant: "There ought to be a machine for doing the purely mechanical work of tabulating population and similar statistics." They were talking about the processing of census data for the US Census Bureau. Hollerith, who at the time was working for this government agency, remembered later that Dr. Billings mentioned "something on the principle of the Jacquard's loom." One day he saw a ticket inspector punch holes in his train ticket, the idea clicked, and he would become known as the inventor of punch card processing. Hollerith's system was first tested on tabulating mortality statistics in Baltimore, and New Jersey, in 1887, and again in New York City. This punched card system was in use by the time of the 1890 US census but it was not the only system to be considered for use with the census. It won convincingly in competition with two other systems considered for the 1890 census showing that it could handle data more quickly. The counting was completed by 12 December 1890 having taken about three months to process instead of the expected time of two years if counting had been done by hand. The total population of the United States in 1890 was found to be 62,622,250. Speed was not the only benefit of using Hollerith's system. It was possible to gather more data, and data such as the number of children born in a family, the number of children still alive in a family, and the number of people who spoke English was part of the 1890 census.[18] The appearance of new technology stimulated the generation of new ideas for the use of that technology. Technology changed the human mindset. In turn, the new mindset was a source of inspiration for new technologies: that seems to be the way of Homo sapiens; in the past, the present, and in the future.

The first one to combine the potential of electro-mechanical technology and a *binary* system of arithmetic into a working computer was Conrad Zuse, in Germany. The "Z3," completed during the war in 1941, would later be recognized as the world's first programmable electro-mechanical calculator. It used old movie film as the punch tape as input medium. In the US, Howard Aiken had begun building an electro-mechanical machine in 1937 at Harvard. The "Mark I," weighing in at five tons, was also finished a year earlier, in 1943. This was when Conrad Zuse tried to convince the German authorities to fund the next generation -all electronic- machine. It was rejected with a quite interesting motivation: "The war would be won, before Zuse could finish his project!" Alas, that is how German arrogance lost and British awareness won the race to build the world's first digital electronic computer. The Computer as we know it today had been born, its gestation accelerated by a World at War, and in turn shortening the war by its deployment to crack the knowledge embedded in German communication codes. Cause and effect were entangled in one act of renewal: the era of global destruction of the physical world was ending; the era of the creation of a global digital world had begun.

In 1950 Turing published "Computing machinery and intelligence," the paper that not only would make him famous, but that would be the subject of scientific debate until today, and well into the future. In this paper he describes his famous test that is designed to determine the distinction (if any) between human and machine intelligence. Here is the core question that this test, called the Turing Test, should answer:

> *"Instead of considering the question 'can machines think?' Turing explains, 'I shall replace the question by another, which is closely related to it and is expressed in relatively unambiguous words':*
>
> *The new form of the problem can be described in terms of a game, which we call the 'imitation game'. It is played with three people, a man (A), a woman (B), and an interrogator that stays in a room apart from the other two. The object of the game is for the interrogator to determine which of the other two is the man and which is the woman…*
>
> *…We now ask the question, 'What will happen when a machine takes the part of A in this game?' Will the interrogator decide wrongly as often when the game is played like this as he does when the game is played between a man and a woman?"*[19]

The 'Turing test' determines whether computers equal the intelligence of human beings. Turing predicted that computing machines would pass his test within the next sixty years. In June 2002, fifty-two years after Turing's publication of the test, people are still making bets on its outcome. In Wired Magazine 34, June 2002, another famous scientist renewed this bet, and predicts that machine intelligence will surpass that of humans by 2029. The man who makes this prediction is Ray Kurzweil, inventor of speech synthesis and speech recognition, and author of the visionary and controversial book The Age of Spiritual Machines, published in 1999.[20] In this book, Kurzweil predicts, like other Artificial Intelligent advocates, the supremacy of artificial intelligence over human intelligence. He further predicts that this will happen well within the 21st century.

Will we hand the baton of the conscious human mind to the digital machines we have just created? Are we just an intermediate, short-lived runner in the evolution of the universe? It took the universe 11 billion years to produce life, and 4 billion more years to produce the human capacity for knowledge. Up until today, we spend thirty to a hundred thousand years of thinking as human beings. Are we done thinking? Or are we just thinking of another excuse *not* to think?

"You do not seem to agree with the predictions of Ray Kurzweil Mr. J?"

I'll have more to say about that Novare. However, we need to look at evolution itself first, before we venture into any predictions ourselves, don't we?

"You bet JJ. I'd love to listen to the history of my own development, because even I lost track. And Novare is so young, she simply has no clue."

OK then. But let us first look at the man who was the first, at least to my knowledge, to put all of evolution in perspective.

"Teilhard de Chardin? I know that you admired his vision all of your life."

Precisely. I read *The Phenomenon of Man* as a young man, but I had a problem with the last quarter of the book.

"The part about the thinking layer and its relation to his religious beliefs?"

Indeed Novare. It left me with the impressions that Teilhard was describing the beginning of the end of something. I could never accept that feeling.

Beijing, China, 1944

In April of 1944, an American visitor passing through Beijing, China, on his way to Washington offered Teilhard de Chardin[21] to take a manuscript entitled The Phenomenon of Man to the Vatican. Teilhard, a French Jesuit priest and renowned paleontologist, co-founder of the Peking Institute of Geobiology in 1940, was looking for a way to get the result of his life's work to Rome to seek approval for publication. In August of the same year, Teilhard learned that ecclesiastical permission had been withheld.

The Phenomenon of Man, or 'Man', as Teilhard referred to it, describes the human being in its relationship to humankind, humankind in relation to biological life, and life in relationship to the universe. Teilhard describes the past and the future of humanity as the course of one evolutionary river, starting at the well of atomic particles, gathering strength through the evolving universe from atoms to galaxies, gaining versatility during the evolution of life from molecule to cell to living organisms, and culminating in the diversity and creativity of human thought. The human phenomenon manifests the center and the expanse of the universe, and all the qualities in it as they emerge over the course of the evolution of 'pre-life', 'life', and (human) thought.

Teilhard sees the future of humankind as the emergence of a 'higher form of life', forged from the interactions between, and the union of, individual human thoughts. If we see the contributions of individual human thought as a multitude of rivers that evolution created, the future might be seen as a vast ocean of thought that represents the future of a higher form of life. Teilhard coined a new word for this future universe of human thought, that 'ocean of thought' in our metaphor of evolutionary rivers. He calls it the 'Noosphere'. He imagines this Noosphere as a new spherical layer around the planet Earth, around the existing layers of the lithosphere, the biosphere, and the atmosphere.

The growth of each individual 'contribution' to this Noosphere is the result of an immense increase of structural and functional complexity of the individual that – over the course of evolutionary times – leads to an ever increasing 'within' of individual entities, living creatures, and human beings. This increased 'within' leads to ever-higher levels of consciousness and reflection. Teilhard speaks of this as the 'cosmic law of complexity-consciousness', now sometimes referred to as 'Teilhardian law'.

In 1941, Teilhard read a newly published work by British scientist Sir Julian Huxley (Teilhard had already finished most of his manuscript for 'Man' at that time), and he wrote to a friend:

> *"I am continuing to work towards a better presentation, clearer and more succinct, of my ideas on the place of man in the universe. Julian Huxley has*

just brought out a book...in a way so parallel to my own ideas (though without integrating God as the term of the series) that I feel greatly cheered."

In contrast to Huxley, Teilhard's vision of the ultimate future is one in which science and religion mutually transform each other in an emerging 'super-consciousness' or 'super-soul', bringing forth an ever closer proximity to, and unity with, God. This ultimate state of a new form of consciousness, the emergence of a divine state of being, is called the "Omega-point."

Sir Julian Huxley, considered as one of the most prominent biologists of his time, maintained, like Teilhard, that humankind must attempt to achieve a unity of knowledge. Both believed in the unity of nature, saw evolution as the direction of greater complexity, and acknowledged the innovative nature of evolution. However, in contrast to Teilhard, Huxley saw humankind as the instrument of future evolution, the shaper of human destiny, and opposed all 'theological, magical, fatalistic, or hedonistic views of human destiny." According to Huxley, only science can provide a potentially universal type of knowledge.[22]

Teilhard expressed his passionate faith in a universal energy, underlying the evolutionary process, coming from 'within' ever more complex organisms, inevitably leading to the appearance and growth of human beings, a process he called hominization. Many publications in the last decade explore, and use Teilhard's vision in their explorations of the future. Some see the Internet as a materialization of Teilhard's Noosphere, and see the World Wide Web as an emerging 'global brain'.

Not only did Teilhard acknowledge the Darwinian theory of the evolution of life, he describes the evolution of the tiniest particle to the human being, which he calls cosmogenesis, and beyond as the fundamental process of expanding manifestations of energy that ultimately leads to his point Omega. It is not surprising, that he never received approval from the Vatican to publish his book. Publication had to wait until 1955, almost a year after his death.

Teilhard de Chardin and Sir Julian Huxley are probably the most prominent scientists of the twentieth century to establish the awareness of the wholeness of nature, including human thinking, and the awareness of reality as a process. For the first time, since Darwin, Einstein, Planck, and many others changed our static view of a permanent universe, they looked at nature as the evolving, ever changing products of evolutionary history. A history in which the evolutionary events can be ordered in a sequence of increasing complexity of individual entities, organisms, or human thought-constructs, as well as increasing complexity of the evolving structures of interacting individual entities, living creatures, or thought-constructs.

In Teilhard's vision, the human thought constructs will converge, and unify, toward a connected Noosphere of thought, or 'intelligence'. Many contemporary thinkers see this connected system of thought as the emergence of a 'global brain'. Just like biological evolution culminated in the appearance of the human brain, the evolution of thought will culminate in the emergence of this mental super-brain. Julian Huxley wrote the following about Teilhard's (and his own) vision of the evolution of human thought, in 1958:

> *"I had independently expressed something of the same sort, by saying that in modern scientific man, evolution was at last becoming conscious of itself, a phrase which I found delighted Père Teilhard. His formulation, however, is more profound and more seminal: it implies that we should consider interthinking humanity as a new type of organism, whose destiny it is to realize new possibilities for evolving life on this planet. Accordingly, we should endeavor to equip it with the mechanisms necessary for the proper fulfillment of its task – the psychosocial equivalents of sense-organs, effecter organs, and a coordinating central nervous system with dominant brain; and our aim should be the gradual personalization of the human unit of evolution – its conversion, on the new level of co-operative interthinking, into the equivalent of a person."*[23]

With the visions of Teilhard de Chardin and Sir Julian Huxley, evolutionary thinking was added to the arsenal of tools for the interpretation of humankind's place and meaning in history. More importantly, the process view of nature allowed revealing some of the properties of the future. Many current beliefs, practices or attitudes that chain us to the comfort of the status quo, present amazing barriers for a common acceptance and application of evolutionary process thinking. Religions cling to the reality of the (prophetic) *words* that no longer match the reality of the world; and the political-economical alliance is too busy maintaining the past and extending its one-dimensional, monetary interpretation of progress.

Evolutionary Beginning

The middle of the 20th century marked the emergence of a solid scientific foundation for the evolutionary perspective on the past, and on the future, and it marks the beginning of the technology that would provide humankind with 'the nervous system' for the communication of thought as Teilhard and Huxley envisioned. However, digital technology would provide much more than they asked for. Its 'bits and bytes' would prove to be the amazing new particles of a new universe of reality. Not only would it allow us to 'reverse-engineer' the world as we know it and re-create it in digital format; the technology would allow us to create new realities, and thus new possible futures.

"Mr. J?"

Yes, Novare.

"I do realize your problem now, when you read *The Phenomenon of Man*, many years ago."

You do?

"It had to do with the friction between Teilhard's scientific mind and his religious beliefs. He tries to bring the two together, but finds himself in the vagueness of a mystical definition of his *Point Omega*, as the ultimate state of human affairs. As it seems to me, he must have been fighting an enormous inner struggle."

Maybe he did, Novare. Your description of 'my problem' comes close, but there was something else.

"I know. I just couldn't put my finger on it back then. Now I can. You had a problem with the convergence of thought. At least, you didn't like the feel of that."

True, Novare. The connection of human thought is accelerating through the advent of digital technology. I also believe that human knowledge will reach new depths, and new heights; this is possible with digital technology, and other technologies that can only advance with digital tools. In my opinion, the future will show a new explosion of diversity, rather than an implosive conversion or confluence. For the record, Novare, Teilhard saw a union of thought, rather than a convergence. Together with this divergence at the new frontiers of thought, we will see coalescence at 'the bottom'. What I mean with this, is the strengthening and broadening of the body of knowledge that is common to all human beings. I'm referring to scientific knowledge, as well as ethical understanding. Convergence will occur in as far as it concerns the strengthening of a common base for a global humanity, but beyond that, we will see a universal expansion of human thought and all of its manifestations.

"That's quite a bold assertion, Mr. J."

Yes, it is Novare. Call it an assertion, a speculation, a contention, or a hypothesis. I will do the best I can to clarify it.

"Will you explain what you mean by 'strengthening of the common basis'?"

Later, yes, I will.

Teilhard's *Phenomenon of Man* radiates the spirit of the beginning of an endgame, albeit an elaborate and mysterious one. However, I prefer to see the middle of the last century as the opening moves of a new game: the game of creative human thought. The most important tool in this game, in my opinion, is the emerging digital reality. All other instruments, such as biotechnology or nanotechnology, can only advance on the workbench of that same digital reality, and as they advance, in turn, change digital technology itself.

By the end of World War II, the reality of the world had changed substantially. Inventors, engineers, designers were quick to translate the new scientific insights into marketable products. The military mindset of governments and the business mindset of entrepreneurs were the driving forces behind the creation of innovative products serving the creation of new realities, and serving the destruction of existing ones.

By the end of World War II, evolutionary thinking had started. Huxley and Teilhard were the pioneers of a new way of thinking that matched the discoveries of a century of imagination and scientific exploration. These pioneers had clearly 'changed their mind' about how the world works and how it evolves and changes. However, the world at large still maintained the same mindset as the generations before. The permanency of the world, the static nature of everything in the universe and on the planet earth continued to be the foundation of people's mindset. Everybody noticed dramatic changes all around, but these changes were like the changing props and the changing décor on an otherwise permanent stage. Reality was changing, and people took notice. The mindset to deal with this changing reality followed the same patterns that it had in the past. New organizations, like the United Nations, created the excitement of new hope. New expectations took hold in the minds of the world.

Wouldn't the enormity of the new insights in the history of the universe, and the history of biological life, result in a different perspective on human thinking? A new mindset would take hold, as Huxley and Teilhard pointed out, resulting in new ways of dealing with the reality of different societies, different cultures, different nations, and new ways to deal with the future of the planet, and the future of human interaction. Hadn't two World Wars taught us that killing human bodies does not eliminate human mindsets?

Surely, during the remainder of the century, not only the reality of the outside world would change, but also the human mindset would catch up. Surely, by the year 2000, we would live with a peaceful mind that would match the peaceful reality in which we would live. Or wouldn't we?

Changing Realities

"Economically, politically and technologically, the world has never seemed more free – or more unjust.

Human Development Report 2002, United Nations Development Programme.[24]

"According to received wisdom in the international community, we are told that during the last 20 years, the period of globalization, there has been a large increase in world inequality. I find the opposite to be true. Not only has inequality not increased, it has actually fallen, and by the end of 2000 was at its lowest level in 50 years. Moreover, by the end of this decade, the level of inequality is likely to be equal to that prevailing 100 years ago."

Surjit Bhalla, 2002[25]

Two recent publications by respectable organizations, and two different worldviews. The United Nations agreed on a set of 'Millennium Goals' at its Millennium Summit in 2000. At the March 2002 UN Conference on Financing for Development in Monterrey, Mexico, world leaders and policymakers assessed the progress towards the development and poverty eradication goals set at the UN Millennium Summit, and pledged an unprecedented global effort to achieve those goals by 2015. One of the important goals was to reduce poverty to fewer than 15 percent of the world population by 2015.

In his book *Imagine There's No Country: Poverty, Inequality, and Growth in the Era of Globalization*, Author Surjit Bhalla reaches some conclusions about poverty that differ sharply from the findings reported in the Human Development Report 2002, published by the United Nations Development Programme. According to Bhalla, global poverty decreased from 44 percent of the global population in 1980 to 13 percent in 2000, therefore achieving the UN's goal set for 2015. The UN reports a poverty figure of 23 percent in 1999. Both figures are the top of enormous icebergs of the statistical machinery, reflecting the difference in the view of the world. However, both publications agree that *world poverty is shrinking*, and that is encouraging.

Rainbows of Change

Therefore, between the last World War and the first Global War, between the Terror of War and the War on Terror, my generation did do well, didn't it?

"We achieved many great things in that period, Mr. J. Just think back. You want me to mention some of them?"

Go right ahead Novare

"In 1953, Francis Crick and James Watson discovered the structure of the DNA molecule. Today, we have revealed the human genome, and can start to find cures for many diseases on the genetic level. In 1955, Erwin Muller[26] looked at an actual atom for the first time. Today, we can manipulate single atoms using nanotechnology, opening up the possibility to create the tiniest of machines, and machines that can build other machines. This technology will totally change our business paradigm, when you can build artifacts of your own design in your backyard. You just need the right software on your PC to do it."

"Excuse me, JJ. I have a totally different perspective."

You do, Alvas?

"Yes I do. Sure enough, we maintained the warm feeling of peace under the chilly umbrella of a cold war. Indeed, we taught many how to fish but we kept the fish. We gave our colonies back their autonomy but not their independence. We invented drugs to cure disease but introduced the disease of drugs. We can kill bacteria in our gardens, but we don't care about malaria. We increased our mobility a thousand-fold and burn up the planet to do so. We invented robots to make our lives easier and declared many fellow humans their equals. We pray to the Lord for mercy and kill in His name. We forgot the old when we were young, and depend on the young when we are old..."

"Stop it Alvas. You are just being ironic. You ignore the enormous progress we have made. After all, we eradicated polio and smallpox. Life expectancy almost doubled in those 60 years. Soon, we will extend our lives by 1 year every 10 years. New technology gives people back their hearing with cochlear implants with neural connections. Soon, we can give visually impaired people back their vision with retinal implants. In that period, we learned how to transplant hearts and a variety of other organs. In the future, we will be able to use genetically engineered animal hearts, or even manufacture organs for use in humans..."

"Novare, it is you who is ignorant. You forget how people feel about things. In the last 50 years, we went from milking Bella to cloning Dolly, from confession to psychotherapy, and from penance to penalty. We went from the joy of little things to the maintenance of abundance. We went from hand-written birthday wishes to preprinted kisses, from family dinner celebrations to drive-through feeding frenzies, and from a shortage of cash to an abundance of credit. Granted, we had little money for living and now live for making lots of it. We went from manually controlled hugs to remote controlled wars, and from job security to security jobs. I ask you whether you prefer the trails of smoking horse manure in the street, or the frozen trails of airplane manure in the skies. To sum it up, we went from poetry to Prozac. Your turn Novare, if you still feel like it."

"I sure do, and I do not like your melancholic mood at all. With the advent of digital technology, we made globalization possible. We have erased barriers, and opened the doors to increase wealth around the world. The important thing is that we did teach 'people to fish', you just put it in the wrong light. In doing so, we increased the potential of many countries in the world, just look at China and India. As both publications pointed out, poverty did decrease. The global spread of the free market idea works. Also, you forget the enormous advances in space technology. The first human being on the moon in 1969 was an example of what we can do if we put our minds to it. Because of space technology, we are now exploring the universe for the search of life, and the exploration of possibilities for terra-formation out in space. We are exploring possible future places for the expansion of life Alvas. And what about all the new materials and products that we derived from space-research?"

"You still don't get it Novare. Agreed, we have all that new stuff. Does it make us better? Or happier? You say that the choice we have for filling our lives with gadgets is so much greater, but what about the freedom to choose our way of living? What did we gain? We went from mating in the automobile to 'auto-mating' in mobile labs. How many moved from loveliness to loneliness and from secure shelters we call home to secured homes, we call shelters? We went from the conveyor-belt of the factory worker to the belted conveyors we call knowledge workers, and from the security of the homeland to the department of homeland security. I think we went from a sense of destination to a feeling of destiny, instead of the other way around."

"No Alvas, you are simply wrong. Don't you understand? Our technological progress is nothing less than amazing, and everybody benefits from it. People, at least an increasing number of them, have more choices, and more freedom to make those choices, than ever before. Right Mr. J?"

Right, and wrong. Right, because so many more options are available. Wrong, because how many can truly choose from the available options? Also, Alvas has a point. There seems to be something missing in our collective mindset. Our way of thinking seems to be stuck in the 18th century, when we still thought that the world would never change in any fundamental way.

That mentality served us well in the previous centuries, while a connected world with 6 billion people evolved; a world that eliminated the barriers of space and time through digital technology. A world that renders its thoughts, its scientific knowledge, and its emotions in digital format, accessible to everyone, at any time. Our individual and collective mindsets, the way in which we manage the world, hardly changed. We still use the old paradigm of a static universe that Aristotle and Ptolemy established. That static world is divided into fixed geographic territories in which people live for themselves, serving their own interests only. The 'rest of the world' is periphery, providing resources in the forms of materials, energy, labor, and consumers. That is a major cause of the frictions that Alvas refers to: the friction between a new reality, and outdated individual and collective mindsets. To change that will require changes at the most fundamental level of the institutions of social management. As Manuel Castells, exploring the vast expanse and penetration of the networked society on the organization of humanity, sums up the great challenges:

> *"until we rebuild, both from the bottom up and from the top down, our institutions of governance and democracy, we will not be able to stand up to the fundamental challenges we are facing. And if democratic, political institutions cannot do it, no one else will or can. So, either we enact political change (whatever that means, in its various forms) or you and I will have to take care of reconfiguring the networks of our world around the projects of our lives." … "If you do not care about the networks, the networks will take care of you, any way. For as long as you want to live in society, at this time and in this place, you will have to deal with the networked society. Because we live in the Internet galaxy."*[27]

In 1999, Gallup International's Millennium Survey asked more than 50,000 people in 60 countries if their country was governed by the will of the people. According to the UN report, less than 1/3 of the respondents said yes. Only 10 percent said that their government responded to the people's will. A look at some of the findings in the UN report reveals many promising developments that are the beginnings of the bridges that can span the chasms of inequality; and at the same time expose the depth of the chasms that have developed over the last few generations.

38 peacekeeping operations have been set up since 1990 – compared with just 16 between 1946 and 1989, while 200,000 people were killed by genocide in Bosnia in 1992-1995, and 500,000 were killed in Rwanda.

220,000 people died in inter-nation conflicts in the 90's, a sharp decline from the 80's, were nearly three times as many were killed; at the same time 3.6 million people died in wars *within* nations in the 90's, half of them children.

57 countries, with half of the world's people, have halved hunger, or are on track to meet the UN's goal, set for 2015, but at the current rate of improvement, it would take 130 years to rid the whole world of hunger.

140 of the world's nearly 200 countries now hold multiparty elections, more than at any time in history, while only 82 countries, with 57% of the world's people, are fully democratic.

During the 1990's extreme poverty was halved in East Asia and the Pacific and fell by 7% in South Asia, while extreme poverty rose from 242 million to 300 million in sub-Saharan Africa.

"Maybe the elimination of poverty should be more than a goal Mr. J. Maybe it should evolve into a constraint."

A constraint Novare?

"Even I can understand that, JJ. Poverty should simply be impossible. The basic needs of everybody on the planet should be taken care of, before any surplus is distributed. In the same way that evolution has constantly raised the bar, giving all human beings the same bodily functions, and the same basic mental abilities, *we* should raise the bar, and expand that common basis by ensuring the fulfillment of basic needs for a human existence."

"Good thinking Alvas. Sometimes you surprise me. We should consider basic things like food, education and healthcare as a fundamental part of the evolution of the human being. That is what I mean by constraint. Whatever we do, whatever our intentions are, whatever the economic or political system deployed: basic human needs should be taken care of first. It would be a natural continuation of evolution by means of human mentality."

"We would have to change our intentions first."

"Exactly Alvas"

"To do that, we need to change our mindset."

"And we would have to change it globally."

"We would have to start measuring progress by well-being, not only by the increase of monetary gain."

Precisely.

"Novare, you're naïve. That will never happen."

"What is the alternative?"

The Axle of Change

Within a time span of less than a century, we discovered that the universe, life, and human knowledge had an actual beginning. We discovered that the universe started from a single event that contained everything that is knowable today. The cosmos evolved like the stem, the kingdoms of fungi, mosses, plants and animals like the branches, living creatures as the leaves, and, finally, human thought as the blossoming of flowers. We use words like 'consciousness', 'self-awareness', or 'intelligence' in our attempts to define what makes us human. We see ourselves on the top of a mountain, superior beings that control all of the rest in nature. That mindset hardly changed, even with the astonishing discoveries about the evolution of a world that sprang from a single event. If the past was the evolution of a physical world and of living creatures, why do we fail to see that the future is the evolution of humankind, whether you define that as consciousness or otherwise.

We claim to be conscious beings, but what is it that we are conscious *of*? Are we indeed aware of the wholeness of nature? The wholeness of our own being? The wholeness of ourselves as a totality of human thought and emotion? Many of those who believe in our creation by a divine being see humanity as the crown of nature, residing right up there, on top of the mountain, closest to God. Many, who have accepted evolution, see the mountain as the slopes of evolution, where humans happen to be the best climbers, and thus ending up on top. Others notice the reality of evolution and view technology as the continuation of the evolution of life. Robots will now take the baton away from us and occupy the mountaintop. Ray Kurzweil explains how this will happen in the coming decades:

> *"One approach to designing intelligent computers will be to copy the human brain, so these machines will seem very human. And through nanotechnology, which is the ability to create physical objects atom by atom, they will have humanlike – albeit greatly enhanced – bodies as well. Having human origins, they will claim to be human, and to have human feelings. And being immensely intelligent, they'll be very convincing when they tell us these things."*[28]

Are we done as human beings? Is our human mind getting close to that singularity where we are no longer capable of dealing with a reality that is smarter than we are? The answer I suggest in this book is no. Not if we wake up to our own potential. Does that mean we can continue on the road we are on? Again, the answer is no. Digital technology provides us with opportunities, and challenges the world has never seen before. We have to make a 180 degree turn in our mindsets, from the reactive nature of Darwinian thinking, to the proactive creativity our imaginative minds are capable of. We have to invent a new language to interact differently, and define a new set of values that guide us on the road ahead. The awareness of an ever changing, expanding universe is less than 75 years old. Most of us did not catch up with the significance of this discovery, and do not see that we are at the very beginning of the evolution of the human mind. Back in 1962, Julian Huxley said:

> *"Our feet still drag in the biological mud, even when we lift our heads into the conscious air. But unlike those remote ancestors of ours, we can truly see some of the promised land beyond. We can do so with the aid of our new instrument of vision – our rational, knowledge based imagination ... All that is required – but that is plenty – is for us to cease being intellectual and moral ostriches, and take our heads out of the sand of willful blindness."*[29]

Willful blindness or blissful ignorance? The result remains the same: our mindset still seems to be one of accepting destiny, instead of defining destination. It is one of defense against predators, driven by the fear and mistrust that enhanced the chance of biological survival. The ironic thing, of course, is the self-perpetuating nature of the situation: as long as everybody has that mindset, everybody will need it for mere survival.

Two forces drive change. One force is the way reality evolves, and the other is the way our minds deal with that reality: our mindsets. It is like the rear driving axle of a vehicle that is driven by electric motors on each side (like those new electric vehicles). On one end, the axle is driven by the power of the changing reality, on the other end by the motor that is driven by the human mindset. Reality is everything that surrounds us in the physical world: lampshades and high-speed trains, flowerpots and the flowers in them, computers and garden hoses, lions in the zoo and deer in the wild, cell phones and Lego-blocks, the sunset and the morning smog, and grandfather clocks and space shuttles are all part of this reality. Reality contains the laws and 'best practices' that rule societies and businesses; it contains the movies, the works of art, the videogames, and the lawyers that do our bidding. It is the totality of the world outside ourselves. It is the world we live in, as it has been changing ever since we can think.

The change of this reality began to accelerate around the middle of the 19th century with the start of the industrial revolution. The discoveries of the 20th century provided the knowledge for the creation of even more profound changes. In fact, so many new insights became available by the middle of the century; we replaced the whole motor on that side of the axle with a new, more powerful one, and further accelerating change.

The driving force on the other end of the axle is the human *mindse*t. Our mindset is the way we view the world. This view might be in total agreement with reality; it might be in total disagreement, or anything in between. This mindset might be an individual's view on his environment, e.g. the company he works for, or it might be a collective mindset, e.g. a nation's view of the world. This view determines how the individual, or the community, interacts with its surrounding reality. If we are in disagreement with that reality, the motor stalls; if we agree with reality, a perfect balance exists between the two motors. Now, the vehicle of change is easy to control and can be steered in any direction. If not, we have a problem.

Both motors need to perform at equal speed for the axle to propel the wheels on each side with equal force. If the motors perform differently, one of two things will happen. If the vehicle has no firm steering mechanism, the vehicle will go into a circular motion, and our changing landscape will turn into a repetitive pattern. If the vehicle has a forceful steering mechanism and a firm, powerful driver, it will continue to proceed in a straight line. However, the difference in power transmitted to both ends of the axle will put enormous stress on the axle itself, on the vehicle, and on the driver. The vehicle will go straight ahead for a while, until finally, sooner or later, the axle, the vehicle, the driver, or all of them, will break down.

"Isn't that stress eliminated by the differential gears in the axle of the vehicle, JJ?"

"That is the core of the problem, Alvas. That differential gear is the human mind. The axle of change runs right through it. You and I have to pick up that stress and deal with it. Since you have such a hard time changing, you are the cause of the friction!"

"That's unfair, Novare. First, how can I deal with things I don't know about? It is *your* responsibility to keep me informed. Second, I confess that sometimes I can't agree with the things you propose. Against my principles, you see. You might be well advised to follow my instincts sometimes."

"My guess is, it concerns both of us, then."

I guess it does, Novare. The important observation is that the axle of change runs right through our mind. Our mind is the differential gear that has to process the stress.

"Is that why so many people need alcohol or drugs? To lubricate the gears?"

Maybe so, Alvas. Maybe so.

In its December 2002 issue, TIME magazine announced its 'Person of the Year choice' for 2002. Three women shared the honor: Enron's Sherron Watkins, WorldCom's Cynthia Cooper, and the FBI's Coleen Rowley. All of them were labeled 'whistleblowers' for exposing wrongdoing within their respective organizations. Financial wrongdoing in the case of Enron and WorldCom, non-action in the case of the FBI, the US law enforcement agency. In a commentary, American ethicist Rushworth M. Kidder assumes 'a thirst in society for moral courage', when he comments on the underlying reasons for Time magazine's choice. It demonstrates the phenomenon of friction between people's mindset and the surrounding reality; the friction in what I call 'the axle of change':

> *"But there's something else, I think, driving this thirst for moral courage. Call it the Age of Disjunction. These days there's an unusual disconnect between words and action, theory and practice, assertion and demonstration. Increasingly, it seems, there's an inertia that keeps goodness in a state of suspended animation while badness rolls on of its own momentum. It's an age fixed on show and surface, a two-dimensional televisual culture that militates against depth and penetration. Result: an almost hypnotic inability to bring things to conclusion."*[30]

My assertion is that we are creating a growing imbalance in the axle of change, because we fail to recognize the evolutionary reality of the living and the thinking inhabitants of the planet, and the changing reality of the planet itself. While our reality keeps changing at a breathtaking pace, our mindset remains the same as it has been for centuries.

The reality is a totally connected world. People in Los Angeles, Beijing, the Saharan plains, London, Afghanistan, and Sidney are neighbors. The reality of McLuhan's 'global village' is here[31]. Yet, our mindsets still strengthen the boundaries between religions, nations, and cultures, like pet-dogs that mark an imaginary territory with a mindset of life in the wild, long gone by.

The reality is that the evolution of nature equipped all of us with the same arsenal of physical and mental capabilities, and with the same physical needs, and mental hopes and fears. Evolution brought all of us to the threshold of modern society. Yet, our mindsets did not cross the threshold into a world of global human thinking. We should be well beyond satisfying the physical needs of everybody, and be on our way with the evolutionary expansion of global human mentality. Instead, we are fortifying the walled cities of our isolated mindsets and consider the world outside – at best - as a resource for the beautification of our own illusionary thrones. Globalization obtained the singular meaning of expanding markets, not of expanding community.

We still act as if we are in the middle of the era of biological evolution, where physical forces are the hallmark of the strongest and the proud possession of the 'fittest'. We continued to apply the biological rules of the 'struggle for life', we did not wake up to the power of our mind as Sir Julian Huxley expressed so eloquently 40 years ago, and Teilhard de Chardin expressed by pointing in the direction of a future of mindful creation. No wonder that most people – literally – still struggle for life.

The Expanding Glass

The axle of change is an axis that runs right through the human mind. Our mind is the synchronizing gear between the way we think and the way we (are forced to) operate. In the same way, collective minds –of nations or cultures- are the differential gears that synchronize the realities of 'the rest of the world' with their own ways of thinking, their own collective mindsets. We can deal with a lot of stress because our minds are the most flexible synchronizing devices ever invented. When the gear fails to cope with the friction, the axle breaks. We fix it with war, murder, divorce, lay-offs, drugs, and anti-depressants. Less education, hunger, and poverty are the casualties of the conflicts we create in the name of progress of the prevailing mindset.

"That sounds very somber, JJ. You took my remarks on the last 50 years too serious. You should know better. I am an optimist, always have been, always will be. It's my nature."

"You call yourself optimistic? You just take everything lightly, Alvas. These are serious things. It is as Bill Joy said in 2000, 'We are being propelled into this new century with no plan, no control, no brakes. Have we already gone too far down the path to alter course? I don't believe so, but we aren't trying yet, and the last chance to assert control - the fail-safe point - is rapidly approaching.'[32] There are too many reasons for pessimism, Alvas."

"Is the glass really half empty, JJ? Both of you certainly sound that way. What do you really think?"

What I really think? That we live in the most fascinating time imaginable. That we have the knowledge to *create* our future. That we have the instruments to transform destiny into destination. I believe that we should recognize that it is not a matter of the glass being half-full or half empty, but the glass itself getting bigger all the time. It is the glass of opportunity, the glass of choices. It is the glass of evolutionary expansion. Expansion is what evolution does. It is what evolution is. I believe that we still have to wake up to that, and I strongly believe that we will. Of course, the pessimists, when they wake up, will see an even emptier glass, while the optimists see new opportunities to fill it. Lastly, I believe that digital technology is the instrument that will wake us up, and that will make new futures attainable. Bill Joy's concern is a serious one, because it presents challenges that we can only deal with if we change our mindset. This involves changing priorities and mechanisms for social management. And that is hard to do.

"I like the expanding glass metaphor, JJ. Right up my alley."

It is precisely what evolution does, Alvas. The world is an expanding glass of opportunity, and evolution does the expansion.

"Like the growing tree with the blossoming flowers that you mentioned above?"

Yes, indeed.

"I was wondering about that tree."

How come?

"You didn't mention *the roots* of the tree."

There is a good reason for that, Alvas.

"Which is?"

We don't know, what the roots are, and we don't know the soil they are grounded in. For many, the trunk of the tree is rooted in religious faith, for others in – yet unknown - scientific phenomena.

"What is your view, Mr. J?"

To me, God is in the future, not in the past. At some point, we will discover that time is a human tool to bring order to our lives, and that 'sequence' has no meaning. At that point of discovery, the point that Teilhard de Chardin and Frank Tipler call 'Omega', time will be exposed as an illusion. Beginning and end, roots or blossoms of trees are but part of the texture of being.

"Is that heaven, JJ?"

Maybe so, Alvas.

"What gets us there, JJ?"

Evolution is the progression of the emerging reality of the universe. To me, as such, it is the continuing revelation of an unfolding truth. Evolution will get us there.

"We should be aware of evolution then?"

Yes. Evolving is what the world *does*. Evolution will continue, with or without us. If we want to be part of its future, we have to pay attention to it, because our mindsets determine the direction of our own future.

"Part II of the book explains evolution, then?"

As best as we can with our current knowledge, Alvas.

In Search of the Roots

We made two 'nested' whirlwind tours in the previous chapters. One, from World War II in the 1940's to the War on terror in the 2000's, and another one from Ptolemy, Copernicus, Galilei, Newton and Darwin to the great scientific minds of the first half of the 20th century. The discoveries of the parade of scientists shattered the prevailing view of the world. No such thing as a static reality, with any permanent features whatsoever, exists.

However, the resulting paradigm shift rang through in the minds of a few (scientific) insiders only. The vast majority of the world-population still lives with the illusionary mindset of an assumed permanency. This vast majority includes leaders of all sorts, including business leaders, political and religious leaders.

Technology, on the other hand, did not stand still. Technological innovation, based on the new scientific insights, be it molecular biology or quantum behavior, accelerated, and a plethora of new products, methods and services saturated the planet, especially the western world.

A lot of us who wonder 'what in the world is going on' do not see that it is not the world that has estranged itself from us. It is our own mindsets that didn't keep pace with the world as it really is: a reality with an evolving universe, evolving life, and evolving human intelligence. We have estranged ourselves from the reality of evolution, while producing new technologies based on the very mechanisms of evolution. Our minds hurt because they are in the middle of two different worldviews: the 'permanence' view of the past, and the evolutionary view that matches reality.

That is why I invite you to join me and watch the performance of evolution, called Songs from the Universe in three acts, and discover what actually happened. Because, whatever happened in that past, is likely to proceed into the future.

PART II: SONGS FROM THE UNIVERSE

"The most beautiful experience we can have is the mysterious. It is the fundamental emotion, which stands at the cradle of true art and true science. Whoever does not know it and can no longer wonder, no longer marvel, is as good as dead, and his eyes are dimmed. It was the experience of mystery – even if mixed with fear- that engendered religion. A knowledge of the existence of something we can not penetrate, our perceptions of the profoundest reason and the most radiant beauty, which only in their most primitive forms are accessible to our minds –it is this knowledge and this emotion that constitute true religiosity."

Albert Einstein[1]

"Up to now, most scientists have been too occupied with the development of new theories that describe what the universe is to ask the question why. On the other hand, the people whose business it is to ask why, the philosophers, have not been able to keep up with the advance of scientific theories."

Stephen Hawking[2]

"We must admit that if the neo-humanisms of the twentieth century dehumanize us under their uninspired skies, yet on the other hand the still-living forms of theism-starting with the Christian – tend to under-humanize us in the rarified atmosphere of too lofty skies. These religions are still systematically closed to the wide horizons and great winds of Cosmogenesis, and can no longer truly be said to feel with the Earth whose internal frictions they can still lubricate like soothing oil, but whose driving energies they cannot animate as they should."

Teilhard de Chardin[3]

We are about to embark on the journey of evolution, because we are experiencing something so fundamental, something with such profound and far-reaching consequences for the future, that only the perspective of 15 billion years of evolutionary expansion can begin to shed light on the potential of the human mind as the co-creator of that future. Humanity is about to submerge itself in a new domain of existence, a new domain of being. While we are playing in the surf of technological gadgets that science produces, we are missing the depth of an approaching tide that has already started to engulf the planet, and the universe.

We want to travel the road of evolution with a sense of awe and wonder about the mysteries that only deepen as we unravel more of nature's secrets. At the same time, we want to acknowledge the mystery of our own knowledge about the history of the universe. We are entering an era, in which the human mind is at the helm of its own future. Our 'thought leaders', in every institution of social influence, from politics to religion, will have to change their mindsets, and their institutions, or become obsolete. These are the messages that emanate from the universe, as it performs its ever-expanding symphony.

Chapter 4: *Act One, "To be or not to be"*

"I am satisfied with the mystery of the eternity of life and with the awareness and a glimpse of the marvelous structure of the existing world, together with the devoted striving to comprehend a portion, be it ever so tiny, of the Reason that manifests in nature."

Albert Einstein.[1]

"...cosmology is also the grandest environmental science, and its ... aim is to understand how a big bang described by a simple recipe evolved, over 15 billion years [Martin Rees uses 13 billion years as the age of the universe. I will use Hawking' number of 15 billion years], into our complex cosmic habitat: the filamentary layout of galaxies through space, the galaxies themselves, the stars, planets, and the prerequisites for life's emergence. No mystery in cosmology presents a more daunting task than the task of fully elucidating how atoms assembled – here on earth and perhaps in other worlds – into living beings intricate enough to ponder their origins."

Martin Rees[2]

Prelude

What makes the music of your favorite composer unique? Mozart's music? U2's? Bach's? The Beatles'? We don't know what that is, but we feel it. Through intense and repeated listening, we might understand. Then we may recognize some of the properties that make our body and soul resonate to its vibrations. Thus, by listening, we may be able to build a conceptual framework that encapsulates its unmistakable qualities.

The Script

This is what we plan to do in the coming chapters. We want to listen to the symphonies of evolution, the evolution of the universe, the evolution of biological life, and the evolution of human thought. At each stage, nature composed and performed its own melodies, themes, and variations, while the emerging players choreograph their own ballet to the rhythms and sounds of their own symphonies. As we absorb the beauty of nature's performance over the ages, we wonder about the fundamental message that it conveys. What makes it unique? What defines its continuity? Which themes does it use consistently? What is the *story* it tells us? Finally, what does that story mean for our future?

By watching nature's performance, we want to grasp the mysteries of its astonishing variety to look at the *possible*. We want to expose the boundaries of its variations, as they define the contours of the *feasible*. Finally, we want to discover nature's ways to evolve in space and in time, so that we can identify the *valuable*. Our mind is the gate between the past and the future. If our mind can see the continuum that is the past, it might shine new light across the threshold of the present into the future. A future that is worthy of anticipation and active pursuit; a future that transforms destiny into destination, and apocalypse into rebirth.

"So, you assume the story of evolution has a script, Mr. J?"

"Don't you see Novare? The story of evolution has no script. It just is what it is. Evolution *is* the script."

"Alvas, you are so short-sighted! What Mr. J is after, is the story *within* the story, if there is such a thing. Right Mr. J?"

Alvas is right Novare. Science looks at evolution as the unfolding of reality, and we ought to pay more attention to the available knowledge. We have to do a better job at educating people around the world about these insights, because it changes the way we think.

"To balance the axle of change?"

Indeed. We have to absorb that knowledge into our mindset. In the coming chapters we will watch the evolvement of the universe, and we will make it part of our perspective on the future.

Novare is right about the 'story within the story', because there is a story within evolution as it unfolded, and as it still unfolds every day. We will adhere to scientific fact when we explore this unfolding. Sometimes, to find the story within, we'll have to reach beyond scientific fact, because science doesn't have the answers yet. We will be dealing with the flow of events over time, and the unfolding of time itself. We have no choice but to enter a domain where current scientific paradigms might be subject to change themselves. Not unlike the enormous transition from the permanency of a Newtonian mindset, to the dynamic world of relativity, quantum-behavior and the Darwinian evolution of biological life.

Songs from the Universe is the name of the performance we are about to watch. The language for the narrative comes, by necessity, from a variety of different worlds. Usually I can rely on the language of science -the world outside ourselves- to describe what's going on. We need to reach out to various scientific renditions to grasp the underlying message by combining them like the pieces of a puzzle, while attempting not to bend the pieces. Sometimes, knowing that science is incomplete, some pieces have to be assumed, at least in their outline. Just imagine looking at a building under construction. Once the foundation is in, and the first walls are going up, you might be able to discern its ultimate shape, even its purpose. From various components and properties, you might conclude that it is going to be a warehouse, a hospital, a residential home, a shopping mall, or a church. To describe the missing pieces, to render the 'whole story' the 'language of everyday life', as Einstein calls it, will do just fine. It is the language of our perceptions: the language of our mind that connects the perceived reality to our frame of mind. We have to make sure that the concepts our mind uses to integrate our perceptions into this frame of mind, are in harmony with accepted scientific knowledge, because the connections themselves are made intuitively; they escape logic. As Einstein puts it:

> *"The connection of the elementary concepts of everyday thinking with complexes of sense experiences can only be comprehended intuitively and it is inadaptable to scientifically logical fixation. The totality of these connections – none of which is expressible in conceptual terms – is the only thing, which differentiates the great building, which is science from a logical but empty scheme of concepts. By means of these connections, the purely conceptual propositions of science become general statements about complexes of sense experiences."*[3]

Our objective is to identify those properties of evolution that apply equally to the evolution of the cosmos, the evolution of biological life, and the evolution of human knowledge. As we set out to do this, we have to remind ourselves that we have no coherent language to do this. Teilhard had to invent new words to describe some of the processes or properties of evolution. Existing words that describe the realms of cosmic, biological evolution and the evolution of thought –Huxley calls it psychosocial evolution – are loaded with meaning that isolates the different realms. Words like 'intelligence', 'consciousness', 'awareness', 'intelligence', 'the senses' or 'sensation', carry the burden of a static world view in which the human being is the sole possessor of such properties. Accusations of being 'animistic' or 'anthropomorphic' lurk when using such words. Only poetic license allows one to say such things like 'the car was screaming', 'the ocean whispers', or 'the dolphin knows what it is doing'. When we talk about evolution, things get even worse. We have to avoid words that imply 'intention', 'design' or 'purpose', as we will. Fortunately, science itself provided a new tool to avoid any danger of being accused of proposing a 'plan'.

That tool is '*the anthropic principle*'. This principle provides a way to assign a classifying label to some of the astonishing recipes found in many stages of the evolution of nature. Stephen Hawking explains the anthropic principle as "the idea that we see the universe the way it is because if it were any different, we wouldn't be here to see it."[4] I think of it as a reverse way of predicting the world. We cannot predict the future from any moment in time (past or present), but at least we can predict the past.

"Talking about mental pretzels. Why is the anthropic principle important, JJ?"

The universe showed some amazingly fine-tuned mechanisms and properties over the course of its history. Without this fine-tuning, you, Novare, or myself, would not be here to talk about it. The anthropic principle gives us a scientifically accepted way to talk about such 'recipes'. We will watch several of them as they unfold.

"Smells like 'design' to me, JJ."

Some religious folks try indeed to derive proof from it that the world is following a Divine design. On the other hand, some scientists use it to support their vision of the existence of multiple universes, each of which might follow different 'recipes'. I want to show that there are universal properties of evolution, and that these principles are at work today, even when we ignore them. Some of the manifestations of the anthropic principles are important to our discussion, since they do support the notion of a direction of evolution. To be sure, it is a direction that only becomes evident in hindsight.

"You are not suggesting any design, Mr. J?"

I am suggesting that evolution has a direction, Novare, and that it is the direction of a multi-facetted expansion. The universe produces expanding boundaries of our *destiny*. I suggest that, within those boundaries, thinking human beings are the co-creators of their own *destinations*.

Watching the Show

Let's sit back and relax. We are about to watch the biggest show ever performed. The script is written by nature, improvising as the show goes on.

The name of the show is *Songs from the Universe*. It is the unfolding performance of the evolution of the universe spanning 15 billion years. To watch it, we'll have to use our imagination, as a virtual reality simulator like 'Einstein's simulator' is not available yet. When such a simulator becomes available and accessible over the Internet, everybody in the world will be able to watch the show of the universe, in color, and surround sound. For now, words will have to do the job.

In act one, we'll follow the dance of galaxies and stars and planets, as they emerge from primordial clouds of dust in a fresh new universe that we call ours. In act two, we will watch the emergence of life and its evolvement into a harmonious tapestry of species covering the earth. Finally, in act three, we will travel the unfolding road of the independence of human knowledge.

Three stages of evolutionary production, three songs from the universe, three acts of the grand performance that is the history of our genesis. As it unfolds, the performance itself will be our guide in the search for underlying and overarching themes. Then, we might be able to give better answers to the questions we have. Where, in the grand scheme of evolution, are we today? Are we at the end of act three? Are we approaching a singularity of the human mind? Is act three just an interlude between biological life and a new world of artificial intelligence? Is 'Robo Sapiens' coming: the intelligent robot? Are we creating Hans Moravec's '*mind-children*'[5] that have no choice but to abandon their parents as the biological ancestors of their man-made bodies of artificially forged atoms and molecules? Or is something else going on, something that is entirely different from visions of programmed atomic beings? Is it possible that we are just crossing the threshold to the creative power of the human mind? Is it possible that human consciousness as we proudly call it is just waking up to observe the unity of the universe, to observe the power of evolution, and to observe the depths of its own 'intelligence' mechanisms? Where in the world are we, and where in the universe are we going?

We are about to find out.

"Very ambitious, on the verge of presumptuous, and speculative, Mr. J. Frankly, I'm not sure I like that."

Novare wears her frown. The look in her eyes is dark. Not a good sign.

"Speculative, Novare? Aren't you supposed to be the one using *imagination*? Usually, when you wear that frown, you are in your scientific mood, and you would call it *hypothetical*. She is just in a bad mood, JJ. As far as I'm concerned, you're all right. One warning though: be careful what you ask for, because I can find any evidence, and make any association you are looking for, being the genius that I am."

"I am glad you admit to forgery, Alvas! If I don't give you precise instructions, you give me garbage, not evidence. I am just concerned that you allow me to stay objective, Mr. J. Alvas cannot be trusted on his own."

"Forgery? If you don't know what you want, I have to use my own intuition, don't I?"

"Alvas, it is time you understand something. I am the driver, you are the car. If I would not tell you precisely where to go, you would hit the next tree."

"Hear, hear! Who is *doing* the driving? Who keeps the tires on the road? Who switches gears? Who deploys the airbag? You just decide the road I'm on, or even the road I'm *off*. The rest you leave to me."

Slow down, guys. I will heed your warnings, and promise to be careful; as I'm sure, you'll watch me. Now, it's time for some teamwork.

Episode 1, "The Birth of Existence"

The Beginning,
11 billion years before the delivery of life,
15 billion years before the awakening of knowledge.
15 billion years before today.

At the Edge of Nowhere

It requires a special state of mind to watch the opening-scene of the story of evolution. It is the state of mind we deploy every day, and one that is impossible to fake. But we'll try anyway. We have to imagine nothingness, because that is how it was right before the start of space and time. No seasons, no sunshine, no good days, nor bad ones. No day or night. No time to worry, no time to be happy. There was no *time* at all. Just like in a dreamless sleep.

Who can remember last night, *after falling asleep*? First setting the alarm, messing with the sheets and straightening the pillow. The last signs of existence. Then *falling*. Then *sleep*. Nothingness. No time and no space...

Then the alarm goes off. On the way to the bathroom. Climbing the stairs to conscious living. The first cup of coffee brings back smell and vision and sound, then the whole world. What do we remember from those hours of uninterrupted sleep? When we crossed that threshold from waking to sleep, we entered ... what? Whatever *what* is, that is how it was when the story of evolution began...

The stage is empty...
There is no stage. *There* does not exist.
Then it happens. *Then* happens. *There* happens.
Nowhere becomes Now-Here.

In a trillionth of a trillionth of a trillionth of a second, a gigantic ball of fire appears out of nowhere. Nowhere becomes here, and once upon a time acquires meaning. Time and space make their appearance as the canvas on which a new universe paints itself. This unimaginable inferno is a billion trillion times hotter than the center of our sun.

"It is no ordinary fire Mr. J. It contains all the material and all of the logic ever needed to create the stars, the planet earth, human beings, and coffeepots. It is a trillion trillion trillion trillion times denser than the densest rock on this earth and seething and swirling like the eruptions from the archetype volcano. Watch the appearance of tiny particles, right in the fire. They appear, and then disappear again. Until, again, something astonishing happens. As a new particle appears, so does another particle that is very similar, yet slightly different! We call those slightly different particles 'anti-particles' today. Emerging particles collide with their anti-particles and annihilate each other."

"Don't be so negative, Novare. Something very productive is happening. The newly formed particles find companions. As they meet and mate, they cease to exist as independent entities, while producing new energy in the process. New and different particles and anti-particles appear, meet, and produce the radiant energy that forever warms the background of the universe and keeps it from death. Each potential kind of particle has a chance to exist, a once-in-a-universe-time opportunity, during this first fraction of a second. Never again will this opportunity come back in this universe, because the conditions will never be the same ever. It is now or never."

We are witnessing the genesis of the place we call 'home' for everything we do; the source of everything we know. The first act has started, the first song is emanating from the universe. The performance of evolution is on its way:

Evolution crossed the First Threshold, and gave Birth to Existence.

Expansion

This inferno of chaos erupted 15 billion years ago. It happened in a tiny sliver of time that unfolded *itself* while creating the space it needed. Anaximander's chaotic mass, out of which everything emerged by the separation of opposites, has emerged. There was no way, back then, to verify or clarify Anaximander's musings. No wonder nobody listened.

The *Expansion* that Edwin Hubble would discover in 1929 had begun. As recent as 1970, Roger Penrose and Stephen Hawking provided proof that the universe, time, and space, had an actual beginning. The debates over the scientific, as well as the spiritual meaning of this beginning never stopped since, and are likely to continue for a long time to come.

Today we know that our galaxy is one of at least a 200 billion galaxies, each one existing of some 200 billion stars. Friedman in the years before Hubble's discovery had predicted the existence of a multitude of galaxies, in theory. Over the past few decades, measurements of the Hubble expansion have led to estimated ages for the universe of between 7 billion and 20 billion years, with the most recent measurements giving a range of 10 billion to 15 billion years.

In 1965, Arno Penzias and Robert Wilson of Bell Labs discovered the existence of a radiation that was omnipresent in the universe. It was a mysterious phenomenon. The only valid explanation was that it was the "afterglow" of the beginning of our universe.[6] Scientist George Gamow predicted the existence of this 'Background Radiation' in a 1948 paper, written together with Ralph Alpher, but could not prove it. This background radiation was the final evidence for a hot-cloud-of-fire beginning of our universe. In 1970, Stephen Hawking and Roger Penrose proved, using Einstein's theory of relativity, that there was such a thing as the beginning of time. Between them, they shared a $ 300 prize for this paradigm changing work![6] In 1991, the Cosmic Microwave Background Explorer (COBE) satellite confirmed the background radiation; it has the exact spectrum as predicted by a Big Bang origin of the universe.

A variety of theories describes the way the universe crossed the threshold into existence and started its journey. The most common is the inflation theory, according to which the universe expanded to its current size in the first 10^{-36} seconds, like the inflation of a balloon. This inflation had to be driven by the same expulsive forces that cause the continuous expansion of the universe today. For the inflation theory to hold true, however, these repulsive forces must have been 10^{120} times higher than today. Some scientists like Roger Penrose doubt the validity of this theory: "most theorists regard inflation as a beautiful idea, which they will cling to until something better comes along."[7]

Did our universe arise out of nothing? This seems hard to swallow, because it assumes the existence of something called 'nothing', which is the flipside of 'something', and therefore, well, something. Was it the creation of a Divine Being? That does not solve the problem either, because that too, assumes a previous existence, and merely shifts the question of a beginning to the Divine Being itself. As Stephen Hawking puts it:

> *"In Newtonian theory, where time existed independently of anything else, one could ask: What did God do before He created the universe? As Saint Augustine said, one should not joke about this, as did a man who said 'He was preparing Hell for those who pry too deep.' It is a serious question that people have pondered down the ages. According to Saint Augustine, before God made heaven and earth, He did not make anything at all. In fact, this is very close to modern ideas."*[8]

Other hypotheses exist that have our universe spring up like a new branch on a tree of many universes, like a new rosebud on the stem of an existing presence. According to such theories, inflation could occur in certain regions of an existing universe, and under certain local conditions spin off a new world in a dimension not accessible to us. The same thing might happen deep down in black holes. According to these theories, our universe would be one of many universes, or a 'multiverse'. Several other theories, like 'string theory', or 'brane theory', exist, and attract serious scientific consideration. Their 'pretzel nature' protects them from being presented here, hence only their mention for completeness.

A glance at various theories suggests that we might not have reached the end of conceptual shifts that the likes of Copernicus, Darwin, Einstein and Planck brought us. Are our concepts of time and space still faulty, or incomplete? Are they just properties of something else altogether, a domain where beginning or end are meaningless words? Should we expect a conceptual shift in science itself, a transformation of the way science looks at the world? There might be more to physics than computation alone, as Roger Penrose suggests:

> *"Yet without such an opening into a new physics, we shall be stuck within the strait-jacket of an entirely computational physics. Within that strait-jacket, there can be no scientific role for intentionality and subjective experience. By breaking loose from it, we have at least the potentiality of such a role."*[9]

As new theories might change our understanding in the future, we'll consider our universe, as we know it today, as the universe that crossed a threshold at the beginning of its existence. This threshold was the first one that nature crossed in the history of our world. Crossing this threshold marks the boundary of a new domain of existence. At the moment of crossing, the universe contained the information for everything that would follow. Nothing was added, or taken away later: our universe is a closed system. We will call this new domain of existence the *cosmic domain*, and the threshold that gives passage to it, the *cosmic threshold.*

Within the first instance of its existence, the information in this domain manifests itself as energy that produces the first particles of matter that seem to condense out of its heat. Here, at the quantum level of particles and energy, we find one of the frontiers of our current knowledge. All the known laws of physics, and all the mathematical equations describing the behavior of the universe, stop working at this threshold.

From the very beginning, right out of the gate, our universe *reaches out* into nothingness, and carves out its own existence in time and in space. It establishes the core of what reaching out means: going beyond the known and the knowable, creating new existence. This marks the beginning of *expansion* that continues to this day. *Expanding is what the universe does.* It defines the direction of cosmic evolution, the direction of history, hence, the direction of time as it unfolds. If expansion is the arrow of time, then reaching out is its arrowhead.

Chaos

The newborn universe manifests itself as a gigantic amount of *chaos* that fills the totality of its own existence. With the appearance of chaos, the second law of thermodynamics, made its powerful entry; it is the most fundamental law of physics. The law postulates that 'The entropy in a system never decreases, and has the tendency to increase'.

"What *is* entropy, JJ?"

Let's first look at two generalized definitions[10,11] , by scientists who stood at the cradle of measuring entropy. Rudolph Clausius introduced the thermodynamic notion of entropy in 1854:

"Entropy measures the energy dispersion in a system, divided by temperature."

The second description is based on Boltzmann's definition:

"Entropy measures the energy distribution by molecules on quantized levels in a system: the number of ways in which energetic molecules in a system are distributed among the energy levels available to them."

"That doesn't help me at all, JJ."

"OK Mr. J. Let me try to explain. Both definitions use the word *energy* in. Entropy is about the distribution of energy. As we saw in the initial Big Bang cloud of fire, the universe started as an infusion of energy. As we know from school, energy is the capacity of a system to perform work. Every particle, atom and every molecule is a 'vessel' carrying energy."

Thanks Novare, energy is indeed the 'substance' of entropy. Something in the universe, something that we do not understand, forces energy to spread out, to diffuse, to become less concentrated. Entropy is a measure for this diffusion of energy. For all practical purposes, it is also a measure for the spreading out, the diffusion, of the carriers of that energy: atoms, molecules and other physical structures.

"I can understand that, JJ. What about chaos?"

In everyday life, we use words like 'chaos', 'disorder', or 'randomness' when we, in fact, mean entropy. I use the word chaos, because it implies both, disorder and randomness. It is important, however, to keep in mind that entropy is about energy. Entropy is not a force that causes diffusion, but only a measure of the diffusion.

"Do I correctly assume then, JJ that the law of entropy states that everything has the tendency to fall apart, so to speak?"

You can put it that way, Alvas.

"Is that why a bicycle left in the rain starts rusting, or why the heat of a furnace spreads all through the room?"

Exactly, Alvas. When you put ice in your drink, the ice will melt, distributing the heat and the molecules of water evenly throughout your drink, thus increasing entropy. If you boil a pot of water, the water will escape in the form of vapor and distribute evenly throughout the room, again, increasing entropy.

"Can I think about entropy as increasing confusion, Mr. J?

Can you explain, Novare?

"The larger the entropy of something I'm considering, the bigger my confusion when I try to find some order in the diffusing substance. Not unlike playing pool. When I start off, the balls are neatly ordered in the shape of a triangle. After the opening punch the balls are all over the place; it is hard to detect any pattern in them. The larger the difficulty to discover a pattern, the bigger the entropy, and the bigger my confusion."

If you want to form a mental picture of the effects of entropy, that is a good way of doing it.

"I think I get it, JJ. But why is it so important?"

Since entropy in the universe never decreases, and tends to increase, it is the only physical phenomenon in the universe that exhibits a direction. Therefore, it gives *direction* to history itself.

"Wow. Now I understand why you keep hammering on this entropy-thing."

Just remember, Alvas. When I use 'chaos', I mean entropy.

"I'll try, JJ."

The law of entropy resides at the core of the 'logic' of the universe. Relentlessly and continuously, the entropy in the universe increases. As such, the universe acts like a gigantic recycling machine that continuously breaks up ordered systems, and redistributes its 'debris' throughout its own expanse of space and time. This constant renewal of materials generates opportunities for the formation of new ordered entities. The law of entropy ensures that these opportunities are distributed evenly throughout the available space, just look at the vapor of boiling water spreading throughout the kitchen.

The perpetual creation of entropy is nature's way of providing new opportunities for the formation of novel systems by creating new order out of the apparent chaos. This is what I mean, when I define *chaos as opportunity.*

We just have to think of the opportunities hidden in the infernal chaos of the first second after the Big Bang. As it turned out, it contained the potential for the creation for everything that followed. No scientific theory is necessary to deliver proof that this initial recipe existed in the seemingly chaotic cloud of fire that we call the Big Bang: honeybees, dolphins, spiders and the laughs, dances, and wars of human beings are living proof. I don't imply that this recipe contained a design for the future. However, I do imply that this recipe contained the algorithm for the process of evolution. Random experimentation with every possible combination of emerging parameters might be an important component of such an evolutionary algorithm. The question *why this initial recipe* was the prevailing one remains unanswered by science, philosophy or religion. Was it pure chance? Roger Penrose answers this question as follows:

"What is the probability that, purely by chance, the Universe had an initial singularity [the state of the universe before the first 10^{-36} seconds, that cannot be described by current scientific laws] looking even remotely as it does? The probability is less than one part in 10^{1200}... In order to produce a universe resembling the one we live in, the Creator would have to aim for an absurdly tiny volume of the phase space [the total volume of possible recipes] of possible universes – 1 in 10^{1230}."[12]

"Is this an example of the anthropic principle, JJ?"

Yes, Alvas.

"Our very existence is dependent on this initial recipe?"

In the form we exist today, yes, indeed.

"Amazing, JJ. Thank God, it was that way. I don't know how else to say it."

Relativity

Existence was born without boundaries and without rulers, grids or guides. The universe doesn't have the convenience of pre-arranged reference points that we can use to position anything in it. No signposts, no X-, Y-, or Z- axis, no light towers, no template. The universe is like an ocean without shores, and without anything, which indicates time. No discernible rhythm, no clock, no conductor swinging a baton, no metronome that is ticking away time.

"Therefore, we have no way to tell where or when that first particle appeared, JJ?"

"You can only tell by relating that appearance to yourself, being the observer, Alvas. Or you can relate it to the appearance of another particle."

"But you need something to take measurements, in space and in time."

"Right! That's why we invented the 'meter', and the 'second'. They are just arbitrary yardsticks we developed to bring some order in the chaos surrounding us."

Which one of the particles that we notice in the cloud of fire is moving? Which one is standing still? What is time? How do we observe *anything*, other than by the *differential* between the *events* that we watch? Once we have observed two different events, how do we measure the differential? The differential between two events is their relationship. Four coordinates are necessary to describe such a relationship in the universe. Three coordinates describe their relation in space, and one describes their relation in time. Our human memory gives us a sense of sequence, and we called it time (1 temporal dimension). Our body gives us a sense of volume (3 spatial dimensions), and we called it space. Our human senses tell us how to position events in time and in space; it is what our senses do. Hence, a sense of time and space is a vital part of the interface between ourselves and the world around us. We are built to live in time and space. Other intelligent life-forms, if they exist in the universe, might, or might not have different ways of positioning events in the universe, or have a different notion altogether, of what we call the universe.

We may now be better able to imagine that the positioning of events in four dimensions can only be done relative to another event, and relative to ourselves, the observers of the events. We have to remember the limit of the speed of light to see that events do not take place when we 'see' them. There is no stage on which we can position the actors in the show. The players are the events that take place as energy transforms into particles and particles transform into energy. Together, the players form their own stage. The universe has no lifelines attached to some fixed boundary, no handrails to latch on to. The universe has no East or West, no longitude or latitude, and no city-blocks. It is ruled by the relationships between events taking place in it. In hindsight, relativity might appear too obvious for us to have missed until Einstein came along. Relativity is inherent to the Universe from the very beginning. It is a fundamental property of the universe, right from the beginning of existence.

Harmony, Part 1

What would happen if the initial force of the Big Bang had been stronger? What if it had been weaker? Had it been stronger, all matter and energy would have been blown far out, and spread thinner and thinner. The density would be too low for the formation of local structures. Gravity would be too weak at this stage to pull together particles of matter, and to form atoms or stars. A vast 'nothingness' would have been the result.

Had it been weaker, all matter would collapse inward, back on to itself. The result would be one clump of matter with an enormous mass: the universe would be just one big black hole.

Harmony is born. The harmony between the expulsive and the attractive forces within the initial cloud of fire reaches an exact balance within the first second of existence. As we learn from Martin Rees,[13]

> *"Back at one second (remember we have good grounds for extrapolating back that far), it's [the universe's] kinetic and gravitational energies must have differed by less than one part in a thousand million million (10^{15})."*

Existence was born with the balance of a finely tuned machine. This harmony between the forces that drive everything in the universe apart and the forces that hold it together established itself, right from the very beginning.

"Another example of the anthropic principle, I take it?"

Yes Alvas, and there are more to come. However, they should be self-evident. I should mention, that some scientists advocate a strong version of this principle.

"Strong version, Mr. J?"

Frank Tipler[14], physicist, and author of *The Physics of Immortality: Modern Cosmology, God and the Resurrection of the Dead,* claims that life and intelligent life are destined to pervade and dominate the entire universe. An important presupposition of the thesis in his book is the anthropic principle in its strongest form, claiming that the appearance of life and intelligent life in our universe are *necessary*. He calls the point at which this happens (billions of years in the future), the *Omega-Point*.

"Just like Teilhard de Chardin?"

Indeed, Novare. Teilhard might have been happy with the scientific insights that Tipler provides.

Episode 2, "Energy That Matters"

3 minutes after the beginning,
11 billion years before the delivery of life,
15 billion years before the awakening of knowledge.
15 billion years before today.

Matter: the Dance of Energy

The episode we are about to watch will take all of three minutes. In contrast to the first episode, the observed phenomena here are well within the realm of current scientific theory.

"The inferno cools down to about 10 billion degrees, JJ, equivalent to the temperature of an exploding H-bomb. Millions of millions of the tiny particles and anti-particles emerge, collide, and produce enormous amounts of light. It's better than fireworks."

"Alvas, these tiny particles are called *quarks*. They meet their fate when they encounter a companion in the form of an *anti-quark*. They produce that quantity of light that Albert Einstein would identify as a *photon*."

I am happy that you remember our visit with Einstein, Novare. Now I want you to watch carefully, because something astonishing is about to happen; something that will determine the fate of the future.

"Something has to happen, JJ. If the number of quarks is the same as the number of anti-quarks, the result will be a continued show of light that would fade away when all the quarks are consumed. Then the show will be over, and no stars, no planets, no life, and no human intelligence will ever happen. I would never get the chance to tease Novare."

"Now I can see, Mr. J. Some of the quarks survive!"

One in a billion quarks does not find a mate in this dance at the primordial party. They are here to stay. Everything in the universe, from Christmas ornaments to human hearts, will be built from these leftover 'Lego-blocks', while the main dish turns into the energy to keep the cosmos warm, forever.

"These left-over quarks are going to work right away, JJ. As soon as they survive, they start to organize themselves into larger bodies."

What you are watching is the emergence of the first *protons and neutrons,* Alvas. They are the building blocks of every atom we know.

"What 'glues' these particles together, JJ? Why doesn't this force glue *all of them* together?"

As we wonder about this, the universe cools down to a billion degrees, and now we can see how some of the protons and neutrons bond together too. That same mysterious force seems to work on them. The newly formed objects are the first nuclei of helium, lithium and deuterium atoms. The process of the creation of nuclei –nucleosynthesis- has started.

Strong nuclear force is the name of the mysterious 'glue' that bonds particle together. The fusion of particles produces large amounts of energy, as we would find out billions of years later when we unleashed the enormous power of nuclear fusion (and its opposite: nuclear fission), in nuclear reactors to generate energy, and in nuclear weapons to generate destruction. We still do not know exactly what this strong nuclear force *is,* but we do know how to make use of it.

After watching three minutes of the biggest show of light the universe has ever produced, the swirling clouds of photons gather into a tornado that sweeps away the tiny protons, neutrons, and nuclei, as soon as they form, spreading them throughout the expanding universe, thus preventing them to assemble into entities that are more complex.

The universe is calming down, cooling down further, and spreading its clouds of energy and freshly generated particles throughout its own, ever-expanding existence. The next major transformation is three hundred thousand years away…

Quarks. American physicist Murray Gell-Mann (born in 1929) named the newly discovered smallest particle a "quark," referring to the phrase "three quarks for

Muster Mark" from James Joyce's Finnegan's Wake (he thought there were only three different quarks, bur more have been discovered since).

Protons. Particles, built from three quarks. They have a positive charge, and hence the ability to attract electrons. An electron is another elementary particle with a negative charge. The nucleus of a hydrogen atom consists of a single proton; it is the lightest and most abundant element in the universe. Hydrogen is the basic fuel for the nuclear reactions that are at the source of all energy production in stars, including our sun. The nuclei of other atoms consist of protons and neutrons.

Neutrons. Particles that are also built from three quarks (different combination of slightly different quarks). Neutrons have no electromagnetic charge; they bond with protons to form the nuclei of atoms.

Harmony, Part 2

Our future was hanging on the thin thread of surviving quarks. The surplus of matter over antimatter had the exact ratio necessary to produce everything in the physical world. This harmony is characteristic for the mechanics of evolution throughout its history. The recipe turned out to be exactly right. Had the strong nuclear force, holding together the elementary particles in the nuclei of atoms, been just a few percentage points stronger or weaker, we would never have been here to invent the anthropic principle.

"Does antimatter still exist, JJ?"

Yes, Alvas, scientists agree it does. Some even ponder the possibility of using matter and anti-matter as the fuel for future spacecraft.

"Makes sense, after I watched the power of the quarks' energy-producing mating ceremony."

"What about these electrons, JJ. Do we have any use for their quantum behavior today?"

"We would not be able to live without them, Alvas, especially today. Electrons run through the wires in our homes and provide the electricity to our lights and appliances. Electrons bombard our TV screens to light up the phosphor and produce the images. There would be no lasers to perform microsurgery without them. Electrons carry the bits in your computer; they even make geckos climb walls."

"Electrons and Geckos?"

"Recently it was discovered that their ability to climb walls or walk across ceilings has nothing to do with suction cups or glue. You know how they do it? Their feet are covered with billions (indeed, billions) of hair tips. Momentary shifting of electrons within these tiny 'spatula' creates adhesive forces (called van 'der Waals' forces).[15] Might be an idea for the next generation of 'Spiderman'."

"Something for you, JJ?"

Order

The quarks appear in the midst of chaos, in a maelstrom of turbulent energies. Then, they bond together in different combinations and organize themselves into larger entities called protons and neutrons. In turn, these form entities that are more complex yet, synthesizing themselves into the bodies that would become the nuclei of atoms. Against the odds of the law of increasing entropy, particles form relationships ruled by nuclear forces, and survived as new, individual entities.

Order was born. The prevailing theory assumes that local variations occur in the entropic diffusion of particles. Within some of the local regions of higher density, nuclear forces prevail and new aggregate entities emerge. The battle between *order* and *chaos,* between *organization and entropy,* has begun. This battle assures the maintenance of a dynamic balance between the two. It would turn out to be the most fundamental mechanism that drives evolution.

Diversity

Together with order, even at this earliest stage of evolution, we watch the emergence of *diversity*. Why are *different* particles emerging? For the emergence of just *some* form of universe, there is no reason at all. The crucial point is that it *did* happen this way.

Diversity is born. Within the first 3 minutes of evolutionary expansion we can see that not 'more of the same', but diversity is a defining property of the universe we live in. Without the thrust towards increasing diversity, no plants, vacuum cleaners, hummingbirds, human hearts, or different cultures would ever appear. The universe goes to work on order and diversity right away.

Synergy

Particles with different properties establish relationships to produce novel combinations, thus producing new entities with uniquely new qualities that transcended those of its constituting parts.

Synergy is born. Protons and neutrons would continue to combine and, later, go on to form the nuclei of all 92 naturally occurring chemical elements in the periodic table.[16] Each one of them has properties that make up its unique behavior that goes beyond the sum of the behavior of its parts. Order, diversity, and synergy are essential ingredients of what we call complexity. They are determining factors for the increased functionality that is usually associated with increased complexity. The recipe for the evolving universe embraced the value of synergy, which it creates by the formation of relationships between a variety of different components.

Awareness

"I'm still wondering about that glue, JJ."

The strong nuclear force?

"Yes. Neutrons and protons are combinations of different kind of quarks, right?"

Certainly, Alvas.

"How do the right quarks get together? How do they know what the right ones are to bond with?"

"They don't *know*, Alvas. It is a matter of their properties, such as electric charge or spin."[17]

"Novare, there has to be something that makes them choose?"

"I can't answer that one, Mr. J. You have to help."

Frankly, I don't know how to answer that one either. There doesn't seem to be the language to explain it. But I'll try.

Certain properties define the 'compatibility' of certain particles, or of more complex entities like atoms, or even human beings. Depending on these properties, the particles or objects form a relationship or they do not. Before such relationships happen, or do not happen, there is a moment of 'recognition'.

"Are you implying any form of consciousness here?"

Definitely not, Novare. I'm only trying to give a name to that moment where an object acknowledges, or detects, the existence of another object. After that, one specific action will be carried out. At the moment of detection, the objects form a relationship, or they do not enter into a relationship: it depends on the properties of the objects.

"You are on thin ice here, JJ."

I know.

Thin ice, or not, some form of algorithmic procedure is at work, imbedded in the nature of particles, or maybe better, in the nature of the underlying energies. Some arche-recipe that emerges from the swirling energies of the Big Bang. For the lack of a generic term that can describe this feature throughout the history of evolution – the capacity to sense the presence of an object – I will call it the most primitive form of awareness. This unconscious awareness has an algorithmic nature, like two numbers brought together at the right time and the right place in a computer, and whose fate is determined by a mathematical equation that is going to establish a relationship between them. Without a mechanism for the detection of a presence (of something else), no new entities will ever emerge.

Awareness, defined as 'the detection of a presence', is the first and necessary step for any action to follow, whether the encounter of two or more entities is of a random nature, or otherwise.

Episode 3, "Weaving Existence"

300,000 years – 3 billion years after the beginning,
11 - 8 billion years before the delivery of life,
15 – 12 billion years before the awakening of knowledge.
15 – 12 billion years before today.

The Dark Age

The universe has cooled down. The fire is gone. Particles have condensed out of the energy like the drops of rain caused by a sudden temperature drop on a sunny day.

Are you ready, Novare?

"Not quite yet, Mr. J. The universe is completely dark, you see. Pitch-black, and filled throughout with invisible infrared radiation. I need to switch to infrared mode to see anything at all. What we will see, takes place over a period of three billions years, so I'll set the simulator to advance time at twenty million years a second ... Here we go."

Thanks, Novare. A spectacular transformation is taking place, right there in the dark. The nuclei that originated over the course of millions of years now start to attract the right amount of energy that form a permanent wave around them. These waves are like the 'bodyguards' of the nuclei. We know them as the electrons that orbit the nucleus and protect them like a shield that we call 'quantum wave'. The first atoms appear. The number of protons in their nuclei and the number of electrons tell us what they are. The positively charged protons attract an equivalent number of negatively charged electrons.

"Hydrogen, helium and lithium are the first atoms to complete their formation, Mr. J."

"Why are they the first ones to form, Novare?"

"Because they are the 'lightest': they require the least amount of components. Trillions of the new atoms gather into local clouds, gaining mass that grows exponentially as their number increases. Atoms are millions of times larger than the components within them, and we can see them now with out virtual zooming feature. The accumulated mass of these clouds starts to influence the cloud's own density, as well as the behavior of other clouds."

"How come?"

"That is *gravity*, Alvas." The mass of those clouds of matter is becoming big enough for gravity to have a visible impact on itself, and on other concentrations of matter."

"When does this 'dark age end?"

"You need to pay attention, Alvas. It is about to end. We just passed the one billion year time-mark, when you could see the first clumps of matter break loose from the cloud of dust and energy. These clumps form the beginning of emerging galaxies. We call them proto-galaxies. In local regions within these galaxies, new, even denser concentrations of matter start to synthesize heavier nuclei within their cores under the shear pressure of their own gravity. The pressure increases the heat and helium and hydrogen atoms speed up, until the concentration of matter frees itself from the proto-galaxy, gains autonomy from the background tapestry, and finally starts to glow like a ...."

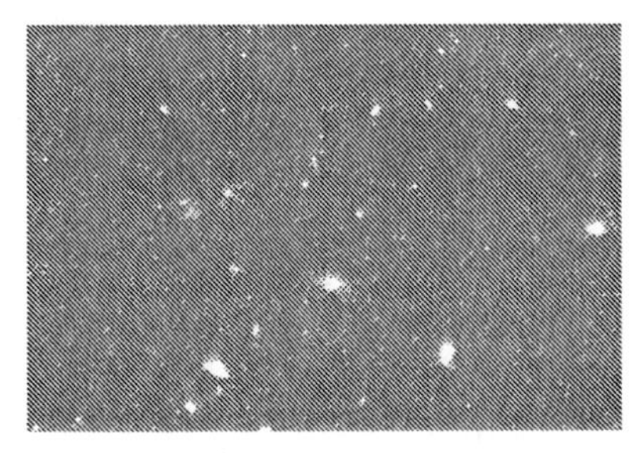

Illustration 14: Image taken by Hubble telescope on January 9, 2003. Researchers believe they are seeing the conclusion of 'dark age' where young proto-galaxies started to shine, about 1 billion years after the Big Bang.[18] Courtesy & Credit: NASA, H.-J. Yan, R. Windhorst and S. Cohen (Arizona State University).

"...Star! I can see them now, Novare, millions of them, in a matter of just millions of years, shining like beacons in the dark ocean of the universe."

"Can you see how they move around each other, acknowledging each other's existence?"

"I can. It's like a heavenly space-time ballet. It's awesome!"

"You better move on, Mr. J, Alvas is having a romantic attack."

What Alvas saw as stars where the first proto-galaxies, shining like stars against the blackness of the universe. As recent as the beginning of 2003, we received the first glimpses of that universe, as shown in illustration 14. The universe is two billion years old (I will continue to use a total age of 15 billion years, following Stephen Hawking number used in *The Universe in a nutshell.* If the increase of this number over the last decades is any indication, the estimates might extend even beyond that). Within the next billion years, galaxies will take their final shape, forming out of the clusters of prototype galaxies.

All the fundamental forces of nature (that we know of today) have now manifested themselves: the nuclear forces that rule the formation of nuclei, the electromagnetic forces that rule the interaction between electrically or magnetically charged particles, and gravity that rules the interaction between bodies of mass. Whether it is the particle's 'spin'[18], electric charge, a combination of both, or properties still unknown that determine the compatibility for a match, the particle has the innate capacity to determine the right course of action. The properties of energy and matter are the parameters in a *natural algorithm* of the universe.

Individuality

When particles form a novel entity, a unique phenomenon manifests itself that keeps our minds occupied since we can think. An *inside world* emerges, and with it, the distinction between the 'inside' and the 'outside' of an individual object, or, much later in the evolution, a subject. From now on, every entity has *internal* and *external* relationships. The internal relationships manifest themselves by the interactions between their constituting components, and result in the coherence of the entity, as well as its vulnerability. Externally, the new entity maintains new kinds of relationships with the outside world.

Individuality is born. As entities form that behave as one unit, we can distinguish them as individual entities. These individual entities have distinct boundaries with the outside world. At this boundary, we find the *interface* of the entity, through which it interacts with the outside world. Every individual entity has an internal order that requires energy for its maintenance. The mechanism ruled by the law of entropy will exert a continuous pressure that causes the entity to decay, be it an atom that decays while radiating radioactive gamma rays, or a star that burns its own fuel and continuously decays into heavier and heavier elements, until all fuel is used, and a lump of dead matter remains. Unless an entity can draw upon the chaos in its environment to supply itself with the energy to restore its order, the individual entity will decay and finally 'die'.

Environment

Individual entities, and the interactions between them, form a larger organization that we call an environment. This environment influences the behavior of individual entities, and vice versa, individual entities influence the behavior of the system that we have called environment. The universe consists of a hierarchy of such environments. We can look at an atom as an individual entity and consider the star, or even the galaxy, as the environment with which it interacts, or we can consider a star to be an individual entity and the galaxy its environment. At the highest level of organization, we find the universe itself. We may isolate certain portions of the universe and consider them as environments for certain practical purposes. However, this can be quite dangerous if we forget to consider the artificial boundaries of the isolated environment with the rest of the universe, as the deterioration of the natural state of our planet shows.

Episode 4, "Matter that Matters"

3 - 10 ½ billion years after the beginning,
8 - ½ billion years before the delivery of life,
12 - 4 ½ billion years before the awakening of knowledge.
12 - 4 ½ billion years before today.

Building a Home

"This is the time when matter really begins to matter, Alvas."

"How come, Novare?"

"Emerging stars are now - after 3 billion years of evolution - producing more of the chemical elements found in the periodic table. A variety of components, and more complex compounds evolve under the influence of the enormous pressures, and the high temperatures of nuclear reactions within stars."

"How do these compounds get into the universe, JJ?"

Each proto-galaxy transforms itself into a vast spiraling ensemble of stars. They first emerge and shine, then slowly build up internal pressure over millions of years as they burn hydrogen, then helium, while forging new and heavier atoms deep inside themselves to use as fuel. Some of the stars finally collapse under their own mass and die. However, within some stars, explosions of gigantic dimensions occur. These will eject the newly formed atoms into the universe, on their way to form planets and asteroids. These gigantic explosions of stars are called *supernovas.* They are the fountains of large amounts of building materials for future asteroids and planets.

"Will the universe continue to expand?"

We do not know that either Alvas, although many think it will. Others believe that at some point in the future, the universe will 'reverse direction' and begin to contract. At that time life and living intelligence will start to control the universe. Later yet, life becomes impossible and intelligence will have to create other means of survival. At the end, the material universe will converge, while intelligence diverges, as it transforms the material universe. As the universe approaches the endpoint, time slows down and the last moments will last an eternity. That is Frank Tipler's point Omega.

"Wow, I'll call that a double pretzel."

It seems to be scientifically sound. Several renowned scientists agree... You seem to be irritated, Novare?

"You bet I am. Both of you are missing the show. The universe is now ten billion years old. Billions of years went by while you were talking, and billions of galaxies, each with billions of stars have emerged. Now it is time to watch, because we are going *home.*"

"Home, Novare?"

"Yes, home. One of the new galaxies is of particular interest to us. In one of the outer regions of this galaxy, a cloud has gathered that is seeded with the materials of supernovas. Part of the cloud collapses onto itself and ignites a new furnace of nuclear fusion: a new, mid-sized star is born. That galaxy is our *Milky Way,* and that star is our *Sun.*"

Thanks Novare. I want to leave the description of the final portion of the journey to the authors of *A walk through time:*[19]

"As this cloud collapses and evokes its fusion reactions and becomes our Sun, a remnant of the cloud continues spinning around the new star. This remnant breaks into ten bands of matter that cool and accrete into Mercury,

Venus, Mars, the asteroids, Jupiter, Saturn, Uranus, Neptune, and Pluto. And Earth. The supernova that reached its end has now become a new beginning – a planetary system composed of all the elements and their dazzling possibilities. Rising up into existence, the Sun and earth together are poised to give birth to a new kind of beauty, one that grows out of all the universe has given birth to thus far."

Two thirds of cosmic evolution has passed and the emergence of life is just one billion years away. Five billion years to go before we see the appearance of a human that can build an 'Einstein simulator' that can play back its reverse engineered history. From the first quark to the planet earth, emerging on a horizon of billions of stars.

"How can anybody not be convinced of the evolution of the universe, Mr. J? And still maintain such a static view of the world?"

It takes time to look at the evidence, Novare. After looking at the evidence, you have to absorb that evolutionary thinking into your mindset.

"Simple JJ. Do you know the three most important things for a successful retail business?"

That would be location, location and... location.

"Similarly, JJ, for any society, the three most important things are *education, education*, and... *education*. I know from my own experience. And that's what's missing."

That is an important factor, Alvas.

"See, JJ. If leaders, political, religious, or otherwise, stick their heads in the sand, they think they communicate directly to the world. They have to create their own illusions of the changes above ground. No wonder things don't change."

"Alvas, you are inconsiderate and simplistic. But you might have a point."

Harmony, Part 3

We have seen the law of entropy at work, as trillions of particles and trillions of trillions of energy quanta spread through the universe. During that same time, evolution created trillions of new organizations of new entities, from atoms to clusters of galaxies, while maintaining a dynamic balance between chaos and entropy. None of the two prevailed over the other. Chaos stayed just enough ahead of order to prevent a gravitational collapse of a universe that would coalesce into one ordered body. The universe kept creating and spreading new opportunities by expansion, and by recycling ordered system. Constantly, the universe followed the opportunity by reaching out with its ordering mechanism that created innovations constantly. Too much power concentrated in one ordered system becomes self-destructive in the universe. If such a system does not maintain that finely tuned balance between entropy and order, organization and chaos, it will be recycled through explosive supernovas or it will implode into a dark hole that will slowly evaporate.

Is there is a lesson here? It sounds familiar when Gary Hamel, management professor and author of the bestselling book *Leading the Revolution* says:

> *"...Every company must become an opportunity-seeking missile – where the guidance system homes in on what is possible. Not on what has already been accomplished. A brutal honesty about strategy decay and a commitment to creating new wealth are the foundation for strategy innovation."*[20]

I don't know about 'missile', but one thing is clear from watching the evolution of the universe: *harmony* between the creation of *chaos* – which is the starting point for innovation – and the creation of *order* inherent to innovation is a necessary condition for the continuity of any form of organization. Maintaining that harmony is the most challenging task for the future. Nature knew how to do it, right from the beginning. It's called evolutionary expansion: a multi-dimensional expansion that involves everything in the universe. It has nothing to do with any 'progress' that is measured with any one-dimensional yardstick, be it (monetary) wealth, corporate bottom line, or even 'equality'.

The missing Piece

"JJ, one last thing. For the most part, I've absorbed the stuff. I've now a better feel for what evolution is. Gravity is much clearer now, but I don't understand the 'expulsionary' forces. What drives the universe and everything in it, apart?"

You caught me, Alvas, right on the threshold to the next chapter. It is one of those many remaining mysteries. Part of it are the centrifugal forces of rotating bodies, but that explanation is incomplete. Scientists assume the existence of 'dark' energy that pushes the universe outward.[21] However, even more mysterious is gravity. The total mass of all observable matter and energy accounts for less than ten percent of the mass that would be necessary to account for the gravity that rules the universe. More than ninety percent of the mass we 'need' is missing! Most scientists assume the existence of 'dark' matter that together with the 'dark' energy, accounts for the missing mass, stuff that is out there, we just cannot observe it.

"Thanks, JJ. I must tell you what the most intriguing part is, at least to me, especially since I have been part of a few hundred thousand years of evolution."

Which is?

"If it continued for so long in the past, what reason could there be for evolution not to continue into the future?"

I cannot see any, Alvas, except for some catastrophe on a universal scale.

"Then why are people still ignoring it?"

"I can answer that, Alvas. Ignorance is the reason. The most powerful and dangerous of all possible reasons."

"You're a pessimist, Novare."

"You're naïve, Alvas."

Chapter 5: The Cosmic Domain

"As when the voyager sights a distant shore, we strain our eyes to catch the vision. Later we may more fully resolve its meaning. It changes in the mist; sometimes we seem to focus the substance of it, sometimes it is rather a vista leading on and on till we wonder whether aught [anything whatsoever] can be final."

Sir Arthur Eddington[1]

"I conclude from the existence of ... accidents of physics and astronomy that the universe is an unexpectedly hospitable place for living creatures to make their home in. Being a scientist, trained in the habits of thought and language of the twentieth century rather than the eighteenth, I do not claim that the architecture of the universe proves the existence of God. I claim only that the architecture of the universe is consistent with the hypothesis that mind plays an essential role in its functioning."

Freeman Dyson[2]

"There are three great frontiers in science: the very big, the very small, and the very complex. Cosmology involves them all."

Martin Rees[3]

Cosmic Look & Feel

Ten billion years have passed. After crossing the threshold into existence, we watched the universe from the moment when quarks engaged in a mating ritual that produced the energy that filled the universe with radiating energy. Only one in a billion survived to build trillions of atoms, all 10^{78} (A '1' with 78 zeros: the number of visible atoms in the universe) of them. We watched as atoms formed the clouds of proto-galaxies that produced stars like the raindrops that condense out of water vapor. Finally, we watched the formation of ensembles of stars into galaxies, and the supernovas of exploding stars that produced our home, the planet Earth, and the planets that we call our neighbors in the solar system

Looking at the Universe

Nothing was ever added and nothing was ever taken away from the initial burst of energy at the time of the Big Bang. The total amount of matter and energy remains the same, at least as far as we know. We consider the universe to be a closed system that expands continuously. Expansion is the hallmark of evolution, as the universe creates and shapes its own tapestry of space and time.

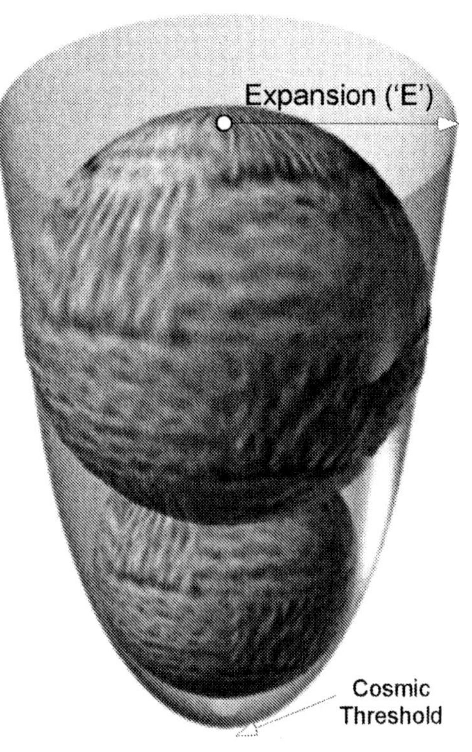

Illustration 15: The Universe, an expanding cocoon of existence.

History has a shape. It has the shape of an emerging cocoon that continuously opens up, as depicted in illustration 15. The tiny sphere at the bottom of figure represents the early universe after the Big Bang. We do not know what happened 'below' the threshold itself: the moment of the Big Bang. The spheres higher up represent the universe at later stages. The vertical direction in illustration 15 indicates the unfolding of history, and hence our sense of time. The universe can be imagined as a balloon that is constantly being inflated, symbolizing the expansion of the universe over the course of its evolution. Galaxies, stars, human beings, plains and automobiles can be imagined as existing on the *surface* of the sphere. This surface, just like the surface of a balloon, or the surface of the earth, has no boundaries, yet it is limited. This is consistent with the current scientific consensus of a boundary-less but finite universe. We used up all the dimensions available in a paper sketch, so we need our imagination to think of the surface as having *three* spatial dimensions.

"Pretzel alert! Even *I* cannot imagine *that* JJ."

Nobody can, Alvas. But you can try. The only solution is to build a digital virtual reality system in which you move in three spatial dimensions over a period of time, like our 'Einstein simulator', in which you can actually experience all four dimensions. Such a simulation should give you a better feel for the expanding universe.

"OK. As long as I am not the only one who can't imagine this thing."

Actually, Alvas, it is a four-dimensional surface. Remember that the sphere is constantly expanding. The vertical axis in the diagram is just symbolic for time. We need three spatial and one temporal dimension to locate any particular event.

"The only thing I can do is to get a feel for the whole thing, can't I?"

That's right. Only mathematicians have the ability to 'read', and thus, imagine, the universe, from their formulas.

In the last decade, scientists have found evidence for the acceleration of the expansion of the universe, hence the widening of the cocoon at the top. The cocoon itself is the envelope of all the states of the universe over the flow of its history. It symbolizes the course of evolution. Time and space can be imagined as a 4-dimensional tapestry, woven from the yarns of matter and energy, filling up the three spatial, and the one temporal dimensions of the surface of the sphere. As the universe expands, everything on -or rather in- its surface moves further apart, just like the images or letterings on our imagined balloon, thus stretching the tapestry.

This space-time continuum manifests itself as the gravity field that determines the movement of everything in it. Stars and planets and moons, asteroids and comets, complete solar systems and galaxies, and even the rays of light from stars follow paths that are shaped by the field of gravity. The shape of our universe is a self-organizing tapestry of energy and matter that constantly modifies itself as it keeps recycling its own resources, following the law of entropy. It is a self-perpetuating system that uses the two fundamental and opposite mechanisms of expulsion and attraction to continue its expansion and everlasting change. A critical look at the various scenes of cosmic performance shows us that space and time are not the only defining properties of this expansion. They only form the canvas on which the universe paints itself.

The radius "E" of the expanding sphere in illustration 15 is a measure of the expansion in space and time. This expansion of space and time are just the means of the universe to accommodate the innovative entities molded by evolution. In turn, the increasing variety of new entities, the emergence of synergetic properties, and the diversification of interrelationships between individual entities are instrumental in the expansion of the universe.

Cosmic Feeling

During the lifetimes of Ptolemy, the universe felt like a décor that seemed to be sculpted on a sphere rotating around the earth. The earth itself was the anchor of the universe; the rest was 'just' entourage that provided the earth with night and day.

"It still feels that way to me JJ."
"Only because you don't understand what is really going on, Alvas. It is time for you to open your mind. The universe is not what it feels like to you."
"Hear, hear! Isn't it you who has some serious *looking* to do? After all it is you who has to show me the way to new things."
Alvas is right Novare. You have a serious obligation to change his mind. After all, we now know that we are vulnerable specks on a tiny planet that spirals with incredible speed in a vast universe. Like the colored specks on the wings of a butterfly that whirls through a small patch of a meadow that we call the Milky Way. Interestingly, we are the only ones that can possibly channel the universe's vast energies, and if we want to survive in the end, we'll have to. Therefore, it is time to wake up, both of you.

Copernicus and Galilei started to change that feeling, when we started to understand that our planet is constantly on the move. Newton explained the movement and – in a way- gave us new peace of mind. Everything was moving all right, but the movement followed a precise set of laws and those would never change. The future, as well as the past, became predictable again. The people attending the Solvay conference in 1929 transformed our image of the small and the big in the cosmos in ways that our minds cannot grasp. Finally, Hubble discovered that the cosmos would never be the same again, and that it had never been the same before.

Our limited capacity for understanding might result in feelings of strangeness, volatility, flow, vulnerability, amazement, awe, insecurity, wonder, or disbelief. The universe doesn't have a marked pathway into the future. It will not stop moving, and it doesn't provide landmarks. We will have to design them ourselves. Our only guide is our evolutionary history.

It seems as if relativity might be just the first step in the transformation of the concept of time. No longer is it something absolute in which we can position the experiences we have. Maybe the ultimate perspective is one in which time is just a property of our experiences. Such a perspective would show us that beginning and end are only transitions between experiences, their sequence irrelevant. Aren't the spiritual dimensions of our experiences void of time? Is the total enjoyment of your favorite music just a 'passage' of time, or is it one total resonance of our being, body and mind, and time plays absolutely no role whatsoever, except when the experience is over? Eternity and an instant in time become identical when we try to open our minds to cosmic evolution, where change is the only phenomenon that we can experience. Forward and backward, long and short, blend together into one wave of sounds and sights and feelings that transcends time, that the poet Vaughan expressed so beautifully:

"I saw Eternity the other night,
Like a great ring of pure and endless light,
All calm, as it was bright;
And round beneath it, Times in hours, days, years,
Driven by the spheres
Like a vast shadow moved; in which the world
And all her train were hurled."[4]

A Universe of Information

Illustration 15 depicts the history of the universe. It is the history of the reality of the universe; it is the history of our perception of the universe, and the history of our understanding of the universe. The ultimate goal of science is to match the three histories and create a rendition that reflects 'true reality'. Until we solve the remaining mysteries, any rendition represents our knowledge about the universe. The cosmic domain is a universe of information that our minds turns into knowledge.

Spiraling Words

We want to look at the universe as an evolutionary process, - a 'natural algorithm' -, that continuously presents our senses with information, in the form of ever changing renditions that our senses can perceive. To do so, we need to define a few words that serve as our tools for such a description. These words, and the concepts they represent, are rendition, information, and knowledge.

The meaning of these words are interrelated; they are connected by threads that are circular if we consider them at a certain point in time and collapse the temporal dimension to zero. Time, however, transforms them into spirals that signify the flow of change. A complete turn along the spiral transforms the meaning of a word and we need another word to capture that new meaning. This process goes on until we arrive at the ultimate truth, a point at which the meaning of the words dissolve in the light of that ultimate truth, like the sun that dissolves the morning fog. Until that time, we'll have to live with the transient nature of the definitions that I'll explore in the next paragraph.

Words, Worlds, and Renditions

With modern computer applications, 'rendering' and 'rendition' became commonly used words. They refer to the 'painting' of our computer screens. When we download the latest pictures of Aunt Annie's vacation, a birthday card sent by a friend or an email from our boss, the software has to figure out what the incoming data are, and then transform them into a representation on the screen, which can be a graphical image, a block of text, a video clip or a sound bite. Sometimes, the software can even generate different renditions of the same data. Spreadsheet software for example, can render the same sales-figures as a table with text or a colored line graph or bar chart.

These are different *renditions* of the same underlying data. A rendition is a representation of any natural, artificial or conceptual phenomenon in our mind. Natural and artificial phenomena include such things as trees and bees, trains and planes, steaks and cakes, and paintings and plaintiffs. Conceptual phenomena include ideas and thoughts, emotions and feelings, theories and hypotheses. Renditions of conceptual phenomena can take the form of symphonies and paintings, mathematical formulas and philosophical deliberations, or graphs and diagrams.

Renditions are constructs of the mind. In the case of external phenomena, these renditions are fabricated from the raw data that enter our mind through our sensory organs. They are sense impressions, or *sense renditions* of the reality outside. Our mind then, processes these renditions and absorbs them into its overall way of understanding the world: its *mindset*. Whatever it is that the outside world presents to the senses, is *information* to the senses. The senses then produce their own rendition from this information. This rendition manifests the *knowledge* that the senses have about the outside world. This 'sensory' knowledge, in turn, is the *information* fed to our brain, which performs its own processing and derives its own meaning from the information, and thus produces the knowledge that a particular person has about the perceived reality.

The knowledge produced by an individual mind exclusively belongs to that person: that knowledge is subjective. To somebody else, or to a society of people, this individual knowledge takes, once again, the form of information, as that person communicates it. Sometimes, information reaches a 'final state' as knowledge. In science, for example, this happens when overwhelming evidence shows the validity of a new theory. In that case, we call it 'objective knowledge'. In every day life, this happens when a majority of people agrees on the meaning of certain renditions, or is coerced into such an agreement. In the first instance, we call it 'common sense' or 'consensus'. In the case of coercion, we call it 'the way things are done around here', or even 'brainwashing'. Information, and knowledge, are relative concepts, just as space and time. They are different renditions of the same thing. They change their status over time and between different mindsets. The renditions described thus far are the result of our sense perceptions. They are the result of the processing that our mind performs on our sensory input.

A second 'kind' of renditions are the things that our mind creates from the *inside out;* we called them conceptions. Some of the labels we use to describe mechanism that are at work to create conceptions are *imagination* and *intuition*. Contributors to the creation of conceptions are thoughts, memories and emotions that blend into a wave of mental activity that produces the final idea -or conception - seemingly 'out of the blue'. Conceptions are the things we can create in our mind, even when we shut ourselves off from the outside world.

"That's only possible when Novare is asleep JJ. Other than that, there is always some form of contact with the outside world. And sometimes, that's when I get my best ideas."

"I don't know about that, Alvas. But perceptions are different from conceptions. That's all Mr. J is saying."

That's right. The separation between external and internal renditions is itself a conceptual construct, and I realize that the distinction is a blurred one. It is a little bit like the distinction between the two of you, actually. It exists, yet it is blurred sometimes.

"Our roles are clear, JJ! I'm the creator of conceptions, Novare gives me perceptions."

"Actually Alvas, I am also the interface between perceptions and conceptions, and thus, the ultimate creator of both."

"Time for you to stop confusing the two, Novare."

"Time for you to stop confusing reality and illusion, Alvas."

Talking about blurred lines...

The renditions that we described thus far represent what things look like according to our own minds. But what do things *really* look like? This of course addresses the question that philosophers and scientists have been asking since humans can think. We want to be practical about this, by accepting that an outside reality does exist. A bridge is a structure we can drive on to cross a river that has flowing water in it and fishes that actually do swim in that water that jump out of the water to catch a real bait, and so on. They might or might not be Plato's 'shadows', but we assume that they exist, and are not mere illusions.

The water in the river exists of water molecules and in turn, the molecules exist of two atoms of hydrogen and one atom of oxygen each. Nature has its own way of rendering the oxygen atom and the hydrogen atom, the water molecule and finally, the river. At each level of the composition that we call 'a river', nature renders its components in *some way*, and the result of that process is some form of rendition. This points us exactly to what scientists do to figure out the world. They perform their work at the interface between the external world, and the internal world of the renditions of the mind. If we put a drop of water under the microscope, we find a rendition that the mind creates of the molecule with the help of the microscope and the human eye. Further down, we find a new rendition, that of the atom.

This is an elusive undertaking, since our own mind seems to be always in the way. The moment we try to imagine the 'real' nature of something, we form a mental rendition. If the microscope shows purple bacteria that live in the water of our river, we know that our mind creates the color from the wavelengths of the light that transmit the image to our brain. Whatever the mode of expression might be, I'll assume that nature creates a structure that has shape and texture. We 'know' what it looks, sounds, tastes, smells or feels like as part of the rendition that our mind creates. But we don't know what it *is*. The 'true' manifestation of such a structure, its existence without perception, is the rendition that nature produced.

Nature has only two things to play with: matter and energy. And even those are equivalent as Einstein discovered. The games that nature plays with these basic ingredients result in structures of great variety and varying complexity. The shape and the texture of these structures contribute to nature's rendition of the entity. We want to call these renditions *external renditions, perceptual renditions,* or *natural renditions,* whatever they might 'look' like in the 'eyes' of a visiting alien. With this definition, an atom is a natural rendition, just like a molecule, a star, or a fish. The creations of our mind then, are *internal renditions, conceptual renditions,* or *mental renditions.* After observation, the atom, the molecule, the star and the fish become mental renditions. We cannot perceive the space-time continuum that follows from the general theory of gravity, yet we have evidence that it exists. It has some external rendition, independent of our minds, at least, presumably.

Einstein's theory of relativity is a mental rendition of this natural phenomenon. Until we reach the final 'truth', science will continue to develop new mental renditions and match them with natural ones, and it will continue to peek deeper into these natural phenomena and modify existing mental renditions of these phenomena. This process of refining our mental renditions of natural phenomena will go on until we get to 'the bottom' of nature's way of creating its own renditions. When, at the end, our mental renditions and those of nature become identical, we will have found the ultimate truth. At that time, we do not need the word 'rendition' any longer, nor will we need this blurry distinction between the external world and the mental one. For the time being, we need the distinction to describe the relationships (and the discrepancies) between the physical world and what our mind makes of it. Einstein expressed the importance, and the mystery, of the relationships between and amongst perceptual and conceptual renditions in words that became some of his most quoted since.

> *"The very fact that the totality of our sense experiences is such that by means of thinking (operations with concepts, and the creation and use of definite functional relations between them, and the coordination of sense experiences to these concepts) it [the real world as we perceive it] can be put in order, this fact is one which leaves us in awe, but which we shall never understand. One may say 'the eternal mystery of the world is its comprehensibility'."*[5]

Books, movies and symphonies are renditions. Trees and airplanes, lamp-shades and dogs, woodstoves and roses, laughs and tears, TV-sets and Christmas trees, vapor escaping from a pot of boiling water and the steak on a dinner plate; they are all renditions -as we perceive them- of underlying objects or events. When we study a rendition, we might discover an underlying structure and produce a more refined rendition of the same phenomenon. When we put the water vapor under a microscope, we can watch the dance of molecules or even the position of single atoms. This might reveal more detail, or accuracy about the underlying process.

The result will be another rendition, maybe one that exposes the 'true nature' of the process. Further down yet, we might express the behavior of atoms using mathematical equations, thus experiencing yet another rendition of the process. An artist might produce a painting of the teapot with the emerging water vapor, or compose a tune to the rhythm of the emanating vapor. They are all renditions of vaporizing water.

All of them acquire meaning by a relationship to a concept. A rendition that is a manifestation of sensory information is an observation. Its association with a conceptual rendition makes it an experience. The boundaries between perceptual and conceptual renditions are not sharply drawn, and most of our human 'productions' are a combination of both. The boundary marks the interface between objectivity and subjectivity, and hence, the interface between scientific thinking and what Albert Einstein calls 'everyday thinking', when he says: 'The whole of science is nothing more than a refinement of everyday thinking.'[6]

The cosmic domain contains all the renditions that our minds create when we observe the universe: the moon, as in 'a moonlit night', the sun as in 'that gorgeous sunset' and the rings of Saturn as seen through a telescope. All of these are mental images that we create through direct perception. They constitute the real world of our daily lives.

This is the information, or the knowledge that we base our actions on. We use this information to send probes to Mars, build the space station, and send people to the Moon. We use it to create science fiction scenarios and ponder the future of humanity. The renditions from this information domain fuel the imagination of business entrepreneurs. They occupy the minds of religious leaders to explain the origin and destination of our existence. They occupy the minds of environmentalists as they wonder about the influence of the universe on our planet and vice versa. And they shape the political mindset that decides the direction in which to point our satellites: inward, towards the earth, or outward, towards the universe; and their purpose: defense, attack or exploration.

A Matter of Processing

The word algorithm goes back to 825 AD and Persian mathematician al Khuwa. Its most general definition is "the art of calculating with any species of notation."[7] This definition expresses the symbolic nature of information processing. We will use the idea of 'notation' in a liberal way and include anything - symbolic or physical - that represents information, in the same way the beads of an abacus present information, and where the manipulation of the beads is equivalent to the processing of that information. I will use 'algorithm' to mean any form of recipe or procedure.

Universal Algorithms

The recipe to create a fabulous apple-pie is an algorithm, and so are the instructions to assemble that new piece of furniture you bought at the store. The procedure to calculate the percentage of sales tax or VAT payable to the government is an algorithm, whether done by hand, a pocket calculator, or a PC. The process of building a house is an algorithm, the design of the house is an algorithm (even though we do not understand the creative portion of it), and the use of Computer Aided Design software to design the house requires computer algorithms to support the human design algorithm. Algorithms are generative procedures that produce - at some level of execution - perceivable results.

Algorithms express themselves as renditions, be it in the form of manual human actions, or in the form of mathematical formulas or a computer programming language. The results of the 'execution' of algorithms are also renditions, be they external or mental renditions. The finished apple pie, the finished house, or the image of the house on a computer-screen are renditions produced by executing algorithms. Likewise, the images, words, sounds or feelings that we form in our mind reflect the outcome of the execution of some form of algorithm.

Algorithms, as I define them here, don't need to be sequential like computer algorithms, and they don't need to have a preset design. If your child starts playing with Lego-blocks, it might or might not have a design in mind. The product however, will have the unmistakable characteristics of a Lego-construction. The algorithmic features of the Lego blocks limit its construction: the type of connectors on the blocks, the shape of the blocks etcetera. Similarly, the outcome is dependent on the algorithmic capabilities of the child who is doing the playing: its mental and its physical (dexterity) features.

The universe renders itself into perceivable phenomena. Our sense impressions of these phenomena are witness to their existence. The argument that our sense impressions are just mental renditions, and don't say anything about the true nature of reality does not change the fact that our minds do receive information, and that this information is processed in various steps. Therefore, the universe somehow carries that information, and transmits it - through our senses - to our minds. Does that mean that the universe is a gigantic computer? The universe certainly computes information, according to Seth Loyd[8], renowned for his work on quantum computing, although "The universe is not an electronic digital computer, it's not running some operating system, and it's not running Windows."

I already introduced the term 'natural algorithm' to discuss evolution's basic property of awareness. In the universe, information is equivalent to the properties of particles, energy, as well as higher-level entities, such as atoms, molecules and stars. Information is inherent in the spread and the collapse of the quantum wave, in the electric charge or the magnetic momentum of particles, and in the mass of celestial bodies that exercise influence on each other in the form of gravity. In the universe, we cannot observe a separation between the carrier of information and the information itself. If we accept that reality exists, even if 'nobody is watching', we must accept that the universe carries its own information, or that it is its own information. If we assume this to be true, the universe performs some amazing processing on its information.

First, it processes information to generate the patterns of behavior, performed by its inhabitants. These patterns are the 'recipes', or algorithms, that rule the motions in galaxies, the radioactive decay of atoms, the formation of stars, the nuclear reactions of particles, the tornadoes here on earth, and the movement of electrons in TV sets and in the feet of Geckos.

Second, it processes information to generate new entities with new, emerging properties. These new properties are the seeds for new, emerging algorithms. The universe has the capacity to process information in such a way that new algorithms can emerge from the results obtained from a previous generation of algorithms.

"Sounds like an artificial intelligence dream come true, Mr. J."

It does indeed, Novare. Thus far, we don't have the tools to mimic nature's creation of truly novel, emergent algorithms. Our current programming languages and operating systems aren't even self-correcting, as PC users will know from experience. Some hope (and some fear), that the right configuration of algorithms, or a high enough complexity *alone* can lead to such emergent phenomena.

"Do you agree with that expectation, JJ?"

No, Alvas.

"Why not?"

For now, let's just say that nature has a way to constantly reach out, into a void, stretching itself, and in the process change the constellation of the information structure - or algorithm - of emerging entities, and emerging 'tapestries': environments of interacting entities. Computers do not exhibit this self-organizing modification of algorithms. Human programmers can design algorithms that can modify existing algorithms, but that requires the predictability of the properties of the new - modified – algorithms. Randomness does not help either, because randomness without an algorithm that determines a direction is equivalent to arbitrariness. Computers need a recipe, or at least a definition of the solution space within which to deploy random variation.

"How do I reconcile this with your belief in Artificial Intelligence, Mr. J?"

Because I make a distinction between 'intelligence' and other properties and activities of the human mind, including consciousness, intuition and imagination. If we define intelligence simply as 'the capacity of solving problems', Artificial Intelligence has achieved major results, among which are programs that a lot of us use daily. For example, word-processing software and spreadsheets have reached problem-solving capabilities that exceed the abilities of a single human being. If, however, we define intelligence as problem-finding, even problem-creation, the story is a different one. In that regard computers have - in my opinion - not even the intelligence of bacteria.

"So the Turing test can be passed before 2029?"

As Turing has defined the test, I believe it will, Novare.

"Thanks for acknowledging my problem-creating capacity, Mr. J."

"You are right with *that* observation, JJ."

The first type of algorithms are recipes for the behavior of entities and their interactions at a given point in time, while the second type of algorithms relates to change over time. If we pick a certain point on the surface of the cocoon in illustration 15, the sphere that touches that point represents the universe at that specific point in time. The renditions that we find in that universe are expressions of the first type of algorithms. Let us call them 'lateral' (by lack of a better term). The second type of algorithm defines the relationship between the renditions of the universe between two points in time, and I will call them 'longitudinal' algorithms. The longitudinal algorithms 'describe' (at least in retrospect) the way in which evolution changed the universe in the direction of its expanding history. This expansion manifests itself by increasing diversity and increasing synergy, increased entropy and increased order, while maintaining a dynamic balance between order and entropy. While science provides us with rich set of theories describing or explaining many of the lateral algorithms, the longitudinal algorithms remain to be some of our biggest mysteries, especially at the thresholds of evolution.

Evolution itself is a continuous expression of such emerging algorithms. From an information-processing standpoint, this is a fascinating phenomenon, because it relates to the multi-dimensional expansionary character of evolution: the unfolding 'nature' of nature along the direction of its history, which is the direction of expansion. Somehow, nature manages to create new renditions in the form of new entities, and new relationships between entities. Then, the emerging information embedded in these new renditions combines, to form new algorithms that not only determine the emergent behavior, but that carry the 'seeds' for the continuation of evolutionary innovation. The evolution of the universe thus becomes a continuous, expanding spiral of new algorithms and new renditions that carry the seeds for new algorithms, and so forth. It is like the programmer's beloved recursion, but then 'inside out', 'inverse', and with expanding functionality. The laws of physics, including the thermodynamic laws, or the laws of quantum behavior, and the theory of relativity are mental renditions of some of the algorithms of the universe.

Universal Music

The universe plays a song with the harmony of a well-crafted symphony. Everything moves and radiates with perfect synchrony. Matter transforms into energy, and back to matter in tune with the demands of the composition. The only perceivable rhythm is a function of the expansion of its totality. The composition of its melodies is written by, and within, its own tapestry of space-time.

The individual notes manifest themselves as particles stringed together by energy, rendering themselves into the music of gravity, nuclear forces and electromagnetism. Similarly, large celestial bodies perform a ballet of movements, as tiny particles dance their strange dance as quantum waves, all of it choreographed by the symphony itself. As they render themselves, the music changes to accommodate the interaction of the players. In the orchestra of the universe, the players are the instruments, and they are the audience. All of the possible renditions are the components of the symphony as it manifests itself to us. In the universe, the symphony is the orchestra. The mysteries lie in its self-renewing composition that has the innate capacity for expansion.

Evolutionary Framework, Part 1

After watching the universe unfold over a period of fifteen billion years, we can assemble a framework that describes what evolution is. This framework is my distillation of the performance of cosmic evolution. It leads me to the following definition of evolution.

What is Evolution?

Evolution is a process of unfolding events that continuously transform knowledge (algorithms) encapsulated by the physical manifestations (renditions) of reality into novel renditions, in the direction of expansion.

Renditions are physical manifestations of reality in the universe as we can observe them with our human senses, oftentimes augmented by our technology (e.g. microscopes and telescopes): atoms, stars, our planet, the moon and supernovas are examples of such renditions. The properties of such renditions and the relationships they form between them constitute the algorithms of the universe. These algorithms express the knowledge embedded in the tapestry (renditions and environments) of the universe.

Illustrations 15 and 16 symbolize evolution as transformation and expansion. The two mechanisms that drive the process of evolution are the creation of chaos, and the creation of order. They are the two pillars on which the process of evolution rests.

The drivers of evolution are the creation of chaos and the creation of order.

Without either one of them, chaos or order, reality would fade away into nothingness, or collapse into a static and meaningless lump. The law of entropy governs the constant (re-) creation of chaos, while the creation of order is nature's continuous effort to (re-) organize itself. We do not have an accepted theory explaining the organizing principle behind the phenomenon of order, but the reality of the universe provides ample evidence for its existence. Order and chaos maintain a finely tuned balance, whereby chaos always stays 'ahead' of order, enough to provide adequate amounts of matter and energy to maintain existing order and to create new forms of order. Providing a constant surplus of chaos is what makes evolution continue. The maintenance of this harmony between chaos and order is evolution's most astonishing balancing act.

The continuation of evolution is dependent on the maintenance of harmony between chaos and order.

Chaos and order spark and fuel two processes that determine the way in which the universe evolves. These processes are innovation and communication. Over the course of its history, the universe combines and recombines the components of its own tapestry and forms innovative entities. No 'new' things appear 'out of the blue'; transformation, rearrangement, aggregation and addition - of existing substances - are the names of the evolutionary game. Innovative entities are forged from the chaotic presence of substances in the region of emergence. Quarks condensed out of the chaos of energy in a local field, and produced photons. One out of every billion combined with other ones in a chaotic field of particles and energy and formed protons and neutrons. Protons and neutrons combine into a variety of different nuclei. Nuclei and electrons form a range of different atoms with different properties, and so forth, all the way up to stars, galaxies and planets. Innovation feeds on chaos. That is why I equate chaos to opportunity. Chaos, through innovation, is opportunity.

For entities to organize themselves and form new, higher level constructs, they need to have the ability to communicate. In the physical world of the universe, this communication manifests itself through the physical and chemical properties of substances. These properties force substances to form relationships that are ruled by gravity, electromagnetism and nuclear forces. Communication is the pre-requisite for order, just as chaos is a pre-requisite for innovation. The processes of innovation and communication establish two types of entities.

Entities of the first type are those that can be considered as a single frame of reference in the time-space continuum, as subject to the rules of relativity. An airplane, the planet earth, an asteroid or a star are examples of this type of entities. I will refer to them as *individual* entities. The forces of the micro-world of particles and energy keep these entities together.

Entities of the second type are constellations of single entities, such as solar systems and galaxies. The forces of the cosmic macro-world, such as gravity and the mysterious expulsionary forces rule their relationships. I will refer to these larger constellations of single entities as *environments*, or *tapestries*. At its highest level of aggregation, we find the universe itself.

Evolution creates new, individual entities and environments through innovation, and establishes and maintains their integrity through communication.

The emergence of new individual entities change the environment in which they emerge. When stars condensed out of the pre-galactic clouds of dust, and gained a measure of independence, they changed the powers that ruled in that region of the universe, as the new star began to radiate its own energy and formed a single entity of concentrated mass. In turn, the changing 'fields' of power in the environment started to influence the behavior of other entities and changed the balance of order and chaos in the region, thus stimulating the innovative creation of new entities.

Creating order and creating chaos, while maintaining harmony between the two is what nature does. Innovation and communication are the means by which the universe evolves, and expands.

The Expanding Universe

In illustration 16, the vector 'E' (Expansion) symbolizes the expansion of the universe and is identical to the radius E of the sphere in illustration 15. In the *framework for evolution*, E represents the expansion of the universe that manifests itself in the following properties of the evolutionary products.

- *diversity*
- *versatility*
- *decision space*

As time goes on, the diversity, versatility, and the decision space of individual entities increase. Before I describe the nature of these properties, it should be noted that any measurement of their 'value' can only be expressed in the relation between two entities or between an entity and its environment.

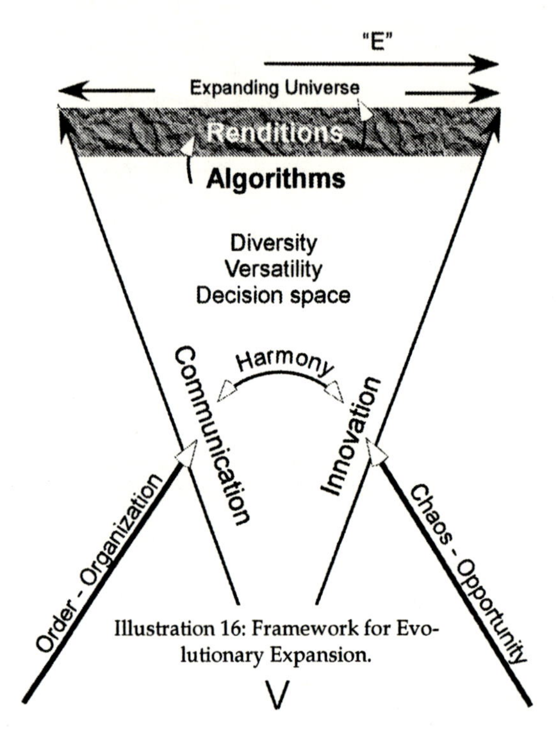

Illustration 16: Framework for Evolutionary Expansion.

Diversity. As obvious from our journey through cosmic evolution, nature creates an ever-increasing diversity of entities. Ninety-two atoms with different properties form the basis for the billions of different celestial bodies. Each body has a different configuration of atoms, and different relationships between these atoms (e.g. in the form of a different pressure or a different temperature). The properties of the components of the entity and the interactions (i.e. communication, relationships) between the components are the ingredients of the algorithm that renders the entity into the appearance that our mind creates from it.

Versatility. The properties of an entity determine its behavioral repertoire: the actions it is capable of performing, or the relationships it is able to maintain. I will call this behavioral repertoire the entity's versatility. As such, versatility is a measure for the entity's 'state of being'. As an example, the carbon atom possesses an enormous versatility to interact with other atoms and form a vast array of compounds. The carbon atom's versatility is the foundation for all organic chemistry, and of life itself. The versatility of an entity, independent of the nature of the entity (inorganic, organic, or even a human thought), is determined by three attributes that I will define as:

- *Awareness*
- *Processing*
- *Dexterity*

I have already defined awareness as the capacity to 'detect a presence'. We might think of it as a sensory phenomenon, or even an algorithmic operator. In the physical universe, this awareness is not some separate faculty possessed by an entity, but an innate characteristic embedded in the entity's property. The same is true for dexterity, defined by the Merriam-Webster Unabridged Dictionary as 'readiness and grace in physical activity' or 'readiness in the use or control of the mental powers'. I will use dexterity in both of these senses. In the universe, dexterity is the physical (or chemical) preparedness of an entity to manifest a certain behavior. This behavior is dependent on the innate properties of the entity. If awareness renders the 'input' to the entity, then dexterity renders the 'output' from the entity. The pathways followed by the entity, between input and output, constitute the algorithm that processes the entity's input information to determine the course of action (the entity's output). Some might argue whether the universe does 'process' information or not. When a particle with a positive electric charge 'meets' a particle with a negative charge, they will form a bond and 'stick' together. I am not implying the existence of some hidden, yet to be discovered, mechanism that processes the particles' electric charge information, but I do interpret their behavior as 'information processing'.

Decision space. This term is commonly used in mathematical optimization, where the total domain of possible solutions to a problem, given by a set of mathematical equations is called the solution space, or decision space. In contrast to mathematics, the formulation of such a 'problem' does not exist in the universe; at least not one that is known to us. At any particular point in the course of evolution, it is not clear 'what will happens next', or what the 'best' course of action is. With the increasing versatility of individual entities and - through the interactions of such entities - the emergence of more versatile and sophisticated environments, the number of 'options' available to individual entities increases, as do the number of options such entities are able to exercise.

Evolutionary expansion manifests itself through increased diversity, increased versatility, and an increased decision-space for individual entities. This, in my view, is the emerging framework for evolutionary expansion, as gleaned from the evolutionary performance of the universe itself. Cosmic evolution is not only expansion in the four dimensions of space and time, but also the expansion of the diversity and versatility of its inhabitants, and an expansion of the decision space offered by expanding environments. Evolution created time and space, and with it existence. The process of evolution relentlessly reached out into the unknown territory of entropy, or chaos, and equally relentlessly grasped elements from the unknown, creating new forms of organization, or order. If the Big Bang would have been the creation of the glass that optimists see as half full, then certainly that glass expanded in dramatic and unexpected ways. That is why I call it the expanding glass of opportunity.

A Stage for Life

Evolution created an orchestra of players. Now, at the end of ten billion years, it seems as if the universe was setting the table, expecting guests to enjoy lavish dishes of sights and sounds.

The universe, as turns out, was setting a stage, on which it could arrange the table for its guests. That stage is built by, and from, the players that performed its symphony. The players turn into a platform. The stage has no boundaries, is always changing, and has no anchor. Everything that constitutes the stage is different from every angle, and from every position in it. At every level of aggregation, the stage presents itself as coherent and independent frames of reference, just like the airplane on its journey to distant places or just going around the Earth. Seat 2A was a stable place, because its occupant is part of the same frame of reference. In his inspiring book *Consciousness in four Dimensions*, neuroscientist Richard Pico makes a strong case for relativity in every domain of evolution; in the cosmos, the domain of biological life, and in the domain of human consciousness. About relativity in the cosmic domain, Pico says:

> *"At all scales of organization that we may perceive, measure, and understand, the independent 4D [four-dimensional] referent frame essentially creates its own unique reality, relative to any other referent frame. Observations of one 4D referent system made from any number of other local systems in motion relative to it would all vary, and thus each observation would be unique to the particular system making the observations. Furthermore, none of the observations and measures would be exactly what was being measured within the referent system of observation! Thus the reality of the space and time dimensions, indivisible to each system, is relative to the point of view of the observer. What makes this vision so extraordinary insightful and revolutionary is that it is not a semantic nuance, not a metaphoric philosophical or mathematical conceptualization, but as real as anything we may understand."*[9]

"You seem to like Richard Pico's strong emphasis on relativity, Mr. J."

Yes, Novare. Especially his concept of *biological relativity,*[10] which we will discuss in later chapters. You see, Einstein provided the scientific foundation for relativity in the physical universe. Pico proposes a framework that expands the notion of relativity into the domains of biological life and human thought.

"Why is that important?"

We can judge the value of something only in relationship to something else, Novare. A single event, a single experience, even any phenomenon has no value in itself, but only in relation to another event, etc.; or in relationship to a larger context. A 'measure' of that value is only meaningful if taken from that relationship.

"When I watch CNN, I get a different impression, Mr. J."

How is that?

"The politicians, the Wall Street analysts, and many religious leaders give me impressions that sound like absolute truths. They seem to have all the solutions for all of the world's problems. They don't talk *relativity*. Good and bad, success and failure, virtue and evil, they have it all figured out. Like measuring obesity, they give you the yardsticks to measure them: of course you have to use *their* yardsticks."

The evolution of the universe as we have observed it, clearly shows the importance of the diversity of its individual species, the importance of the wholeness of its tapestry, and the importance of relativity to judge the position of individual entities in this tapestry.

"Excuse me JJ, but you talked about the 'look and feel' of the universe?"

What about it, Alvas?

"I imagine being there, all alone, and watch the universe before life ever appeared. I can't help myself but thinking that the universe is trying to tell me something. Know what I mean? For 10 billion years, the cosmos has been expanding and producing this enormous variety of entities. It feels like something new is in the making. Like it produced all the innovative technology possible, and was ready for something new."

"Alvas has a point, Mr. J, although it's easy talking in hindsight. For me, the universe was ready. It had, from some initial knowledge, produced renditions, and then these new renditions became the new algorithms - the new knowledge - that again produced new renditions, just as you have explained. In the last few billion years nothing truly new 'in kind' happened. That is why I would expect evolution to come up with something new that would create a new rendition from the knowledge encapsulated in its tapestry."

Thanks, Novare. Thanks Alvas. Let us proceed to see what evolution can tell us about its own continuation in the domain of biological life, and about the mechanisms, it uses to maintain its own expansion.

"I'm ready, JJ."

Chapter 6: *Act Two, "Making a Living"*

Illustration 17: His nose scarred and mane luxuriant, 'Chaca' may have been the only male in a lion pride on a small island in the Okavango Delta of Botswana. This picture is used with kind permission of nature photographers Gregory G. and Mary Beth Dimijian.[1]

"Our best theories are not only truer than common sense; they make far more sense than common sense does. We must take them seriously, not merely as pragmatic foundations for their respective fields but as explanations of the world."

David Deutsch[2]

"I will call a system that can act on its own behalf in an environment an autonomous agent. All free-living cells and organisms are autonomous agents"

Stuart Kauffmann[3]

Episode 5, "The Delivery of Life"

The Beginning of Life,
4 billion years before the awakening of knowledge.
4 billion years before today.

Prelude: What is Life?

Strudel's fur is a reddish-brown-blond mélange of fur, reminiscent of the colors of a freshly baked German apple strudel. Her constant companion Blue is an unusual patchwork of charcoal-grey and beige. The reason for his name is simple: he has blue eyes. Both cats are relaxing, only two feet apart, on this sunny Sunday afternoon. Not far away: the loosely connected remains of a mouse. That mouse has their full attention.

"Hey Blue, I met Boris out there, you know, where all the birds come to play. You know how smart he is. He claims that some things we play with are different from others."

"How so?"

He says that some things are *dead*, and others are *alive*."

"Strudel, you think too much. All we know is that *we* are alive, the rest is guesswork. Why does it matter anyway?"

"It matters, Blue, simply because I want to *know*. I think it has to do with the fact that some things play *back*, when you play with them, and others do not. Boris calls it *inter-action*. That mouse over there played *back* you see. It moved by itself!"

"You know better than that. Everything moves. All the things the Uprights give us, move too. One of the balls I had, moved itself all the way down the hill! It jumped, it flew, it rolled, and it turned. When I went after it, I couldn't find it, because it was hiding behind a tree! Everything moves, and you know it."

"But that mouse over there doesn't move"

"Just throw it in the air, and it will fly again"

"I still feel there is a difference, Blue. You see, when a mouse gets close to a little hole, you know, one we can't go through, it always goes in, and hides. Not this one though. I made it fly very close to a tiny hole, several times, and it did not go in. There is definitely a difference there. Sometimes you know where they move too, but mostly they are so unpredictable, and that makes for all the fun."

"Yeah, I had that happen to me too. Want to hear my theory?"

"*You* have a theory? Go ahead."

"When do you usually find a mouse to play with?"

"When they are eating, of course"

"There you go. I think there are things that take eat, and things that don't."

"Blue, what kind of a theory is that, everything has to take eats. If something doesn't take eat, it gets this pain inside, therefore, it takes eats."

"Still, remember that time when the Uprights left us inside and didn't give us eat? I tell you, I could not move as fast as I normally can. Hence, I think moving has to do with eat, therefore, like anything, that mouse needs eat."

"Hmm, I still believe it has to do with playing back. But maybe it's both to make for Boris' difference between dead and alive. See, he believes that life is not something that just happens. 'there is more to it than that', he says. It has to do with *autonomy*."

"Autonomy? What's *autonomy*?"

"Remember how things can fly, when the wind blows?"

"Sure. But I don't fly when the winds blow!"

"Exactly. That's because you have *autonomy*!"

"Come on, let's take some eat, you're just having a bad day. Jeeez, *autonomy? Inter-action*?"

Cleaning House

The earth has to be *just right* for life to evolve; the right position in the space-time continuum of the universe, the right attractive forces and the right position in relation to the sun for receiving the right amount of energy. It is going to need the right atmosphere, the right temperature, and all the right compounds in the correct mixture. Pressure and moisture in the atmosphere and in the surface, and the proper mechanical and electric forces: everything has to be just right for the recipe of life to be successful.

"The early earth is not a pleasant place at all Mr. J. The smell of rotten eggs is in the air. Fortunately, we can't smell it. The odor-generating system of our simulator is switched off. Budget cuts, you see."

"Alvas, your description is impossible! That smell is from hydrogen sulfide that emanates from the hot surface and the murky-brown waters of the hot planet. Sulfide and methane fill the ultra thin atmosphere that contains hardly any oxygen, and offers no protection against the penetrating radiation from the Sun. Intense bombardments by asteroids and meteorites continue to pound and alter the surface, the structure and the material contents of the earth. As the earth gains more mass, its increased gravity attracts even more dust, debris and meteorites."

"Be positive Novare. All those bombardments bring a host of new compounds that enrich the surface of our planet. Energy is available in abundance, can't you feel it? The whole spectrum of radiation, from ultraviolet to infrared, is coming from the sun."

"But radioactive rays hit the earth unfiltered, Alvas. No protective ozone layer. Only slowly, as the earth cools, heavier elements sink to its core and lighter elements float up. We can witness gigantic volcanoes continuously spewing substances into the atmosphere that fills up with hydrogen, carbon dioxide, nitrogen, methane, ammonia, and water vapor, but still no oxygen."

"But then, the rains begin. Torrential rains that fill streams that form rivers and lakes and oceans. An awesome sight.

Just a little over four billion years ago, the universe prepared a table for new guests that were about to appear. That table turned out to be our planet earth.[4]

How do we know how old the earth is?

Radioactive decay causes the transformation of certain radioactive elements into other elements. The time it takes (this time can vary from 700 million to more than 100 billion years!) for half of the element to transform is called the 'half life' of the element. Using this 'radioactive clock', we can determine the age of certain materials found on earth or in the universe. In Western Australia, zircon crystals encased in rock formations have been dated to 4.3 billion years ago, and the oldest rock formations, found in Northwestern Canada near Slave Lake, to 3.96 billion years ago. On their visit to the moon, Apollo astronauts brought back rock samples that are 4.5 billion years old.

Days on earth are only 10 hours long. All of the necessary elements for life, carbon, nitrogen, oxygen, and hydrogen are now present in the earth's surface in the form of compound molecules. After a few hundred million years, the earth becomes one vast ocean, with chains of islands of rock and volcanoes sticking out, and marking the presence of the consolidating lithosphere beneath.

New molecules form out of the prevailing gases and compounds of the earth's surface layer. Combinations of these molecules result in the synthesis of novel macromolecules. After many reiterations of recombining and rearranging their components, these macromolecules reach a biochemical complexity that can lead to the formation of ...*protocells.*

"That Mr. J is exactly what happened at some place on earth, in the mud of a pre-life shoreline, in the depths of an ocean, or in the murky waters bubbling to the surface ... That was 3.8 billion years ago."

Crossing the Threshold

"...but again, we need a digital simulator to dive into the murky waters of a deep mud-hole like the one in Yellowstone Park, as shown in illustration 18. I'll leave the controls of the simulator to Alvas, Mr. J."

All right, Novare.

Illustration 18: The first protocells might have been formed in the depths of bubbling pools like this. Picture is from one of the sulfuric mud-pools in Yellowstone Park, United States.[5]

In one of the muddy pools, we can watch the appearance of tiny bubbles. They look like microscopic versions of the bubbles made from soap and water, the ones we used to blow into the air and try to catch, as kids. Unlike the playfulness of the soap variety, the newly emerging bubbles are involved in a serious game: the struggle for life itself. Here is Richard Pico's hypothetical account of the events right at the threshold of life:

"The protocells that we watch seem to be fighting against the pace of the background thermodynamic dance, struggling to climb the collective activation hill against the pull of entropic gravity and to create a sustained rhythm of their own making. Ultimately, they fall into step and become subject to the background pulse of time, destined to decay to the lowest level of equilibrium, to randomness and disorder. This is the swansong of the protocell. However, this momentary dance of death is actually a prelude to the emergent shape of things to come. The first living cell was a semi closed membrane-bound system of interdependent chemical reactions whose structure

and function maintained an internal integrity, a nonequilibrium ordered existence, for a critical time..."[6]

"Pretty. *And* pretty difficult, JJ."

Chemical reactions take place between the compounds inside the bubble, Alvas, and materials are exchanged between the inside and the chaotic world outside, right through its semi-permeable membrane. The tiny bubble is fighting to sustain itself, and to define its independence by establishing an orderly interface with the outside world. Many bubbles lose the fight. First they seem to come to life, but then quickly lose their integrity. They fall apart and their components spread out through the murky environment. New bubbles form until, finally, one of them survives. That bubble is a cell, the basic unit of life. All living organisms ever to emerge on earth, including human beings, will develop from that cell. The first living organism inhabits the earth, almost four billion years before anybody could wonder about the mystery of its emergence, when:

Evolution crossed the Second Threshold, and delivered Life.

Qualities of Life

The first living cells were not a romantic species by our standards, but we would not be able to live without them. Today, one teaspoon of garden soil contains millions of them: they are the bacteria, the earliest living organisms, and still sustaining, (sometimes destroying) life today. The earliest ones are called prokaryotes, in short PKs. They do not have a cell nucleus (prokaryote means 'before nucleus', just some loose strands of RNA (Ribo-Nucleic Acid) in a bubble of protoplasm enclosed by a semi-permeable membrane. The strands of RNA enable them to make copies of themselves.

Oxygen is deadly to these early creatures. However, because of that they still fulfill a vital role today: in the food digesting function of the animal (and human) intestines, and in nature to produce the methane necessary to keep the oxygen levels in the atmosphere out of the danger zone of becoming explosive. Blue-green algae are tapestries of such tiny methane-producing creatures. Prokaryotes can perform a variety of functions. Some live on the oxygen-free compounds found in the early atmosphere, while others maintain themselves by using light to produce energy, or fabricate their own food from carbon dioxide.

An amazing feat was accomplished about 3.7 billion years ago when green and purple invented photosynthesis[7]. Using solar energy, they extract hydrogen from hydrogen sulfide gas and combine it with carbon dioxide to construct their own bodies, while producing sulfur waste. These early bacteria cannot use water as a source of hydrogen, because the released oxygen will burn them. In stead, they use hydrogen sulfide, to get safely at the hydrogen. Over time, a series of genetic mutations produce a new species, which can safely store the oxygen in more complex organic molecules (enzymes). Photosynthesis, the process by which prokaryotes learned to accomplish this feat, will change the atmosphere, and open new possibilities for life to evolve.[8]

During the late 1920's, British scientist John Haldane[9] and Russian scientist Aleksandr Oparin proposed the first serious scientific theories of the origin of life. Although they developed their theories independent of each other, they would become jointly known as the 'Oparin-Haldane' theory.[10] According to this theory, complex molecules randomly form new aggregates under the influence of ultraviolet light, until finally one such organism persisted, and was able to replicate. The theory did not attract much attention in those days, since no evidence could be provided. That changed a few decades later, when ...

In 1953, Stanley L. Miller, a graduate student and an eminent chemist and his teacher, Harold C. Urey, at the University of Chicago, conducted an experiment that would change the approach of scientific investigation into the origin of life. Miller took molecules of methane, ammonia, hydrogen, and water, which were believed to represent the major components of the early Earth's atmosphere and put them into a closed system. Next, he ran a continuous electric current through the system, to simulate lightning storms believed to be common on the early earth. By the end of one week, as much as 10-15% of the carbon formed organic compounds. Some were amino acids, the basic organic components of the proteins that make up every living organism! Miller's experiment showed that organic compounds such as amino acids, could be produced in laboratories. This enormous finding inspired a multitude of further experiments.[11] Miller-type of experiments have resulted in the production of 90 different amino-acids, which is interesting, because living organisms produce all the necessary proteins from *20 different amino-acids only.*

On April 25, 1953, James Watson and Francis Crick made a discovery that changed the future of medicine, and the future of life itself. They published their seminal article *Molecular Structure of Nucleic Acids: a Structure for Deoxyribose Nucleic Acid* [DNA].[12] in Volume 171 of the British journal Nature on April 25, 1953. Watson and Crick unraveled the structure of the DNA molecule, a molecule of incredible complexity that contains every bit of information necessary to build a living organism. We commonly call it the *blueprint* for life. The insert gives a brief overview of its structure and some of the related definitions and notions.

What is DNA, RNA and Genes?

All living organisms, from the simplest bacteria consisting of one single cell, to fungi, plants, animals and human beings, store their hereditary information in molecules of DNA (deoxyribonucleic acid) and RNA (ribonucleic acid). Some scientists believe that RNA might have existed *first* and that DNA developed from it.

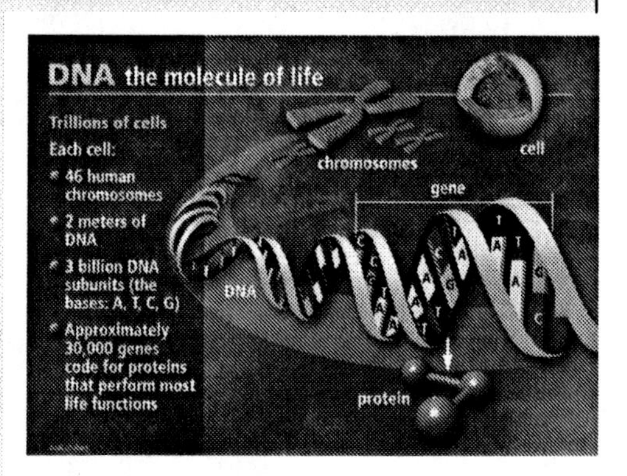

Illustration 19: Structure of DNA. Courtesy: U.S. Department of Energy Human Genome Program.[13]

The DNA molecule has a structure like a spiral staircase, a double helix. The *rails* on each side of the staircase are alternating molecules of sugar and phosphate. Each *tread* of the staircase consists of *two connected bases,* called a *base pair*. There are only four different bases: adenine, called 'A'; guanine, called 'G'; thymine, called 'T', and cytosine, called 'C'. A always joins with T, and C always joins with G. Because of this, one-half of each tread is predictable, and the duplication of DNA is made easy. When the staircase splits down the middle, the missing bases in each half are known, and can be rebuilt. One base together with the molecules of sugar and phosphate form a subunit. These subunits are called *nucleotides*. The four different bases result in four different nucleotides. Three adjacent nucleotides (as you walk up the staircase) form the instructions to produce one specific amino acid, out of the 20 different existing amino acids. This sequence of 3 nucleotides is called a *codon.* Amino acids are the building blocks for all 30,000 different proteins that make up the human body.[14] Cells are the fundamental working units of every living system. All the instructions needed to direct their activities are contained within the chemical DNA (deoxyribonucleic acid). DNA from all organisms is made up of the same chemical and physical components. The DNA sequence is the particular side-by-side arrangement of bases along the DNA strand (e.g., ATTCCGGA). This order spells out the exact instructions required to create a particular organism with its own unique traits. The genome is an organism's complete set of DNA. Genomes vary widely in size: the smallest known genome for a free-living organism (a bacterium) contains about 600,000 DNA base pairs, while human and mouse genomes have some 3 billion. Except for mature red blood cells, all human cells contain a complete genome. A strand of DNA is one long sequence of nucleotides. A strand of DNA, measuring 5 cm (2 inch) in length when stretched out, is tightly coiled around a core of protein and packed into a *chromosome*. Every human cell has 46 such chromosomes. If stretched out, the DNA stored in the human body would be enough to cover the distance from earth to the sun and back 500 times!

A *gene* is a (unique) sequence of codons, each codon consists of 3 nucleotides. This sequence determines the production of a specific protein that will be built from the particular amino acids as defined by the codons. RNA carries out the actual assembly of the proteins from amino acids. The genetic code, the way DNA codes the production of proteins was 'cracked' in 1965 by Gobind Koran and Marshall Nierenberg.

The number of human genes had always been estimated to be 50,000 - 100,000. The Genome project -decoding all human DNA- resulted (finished in 1991) in a surprise: the final count is 34,000.

Some scientists believe that the first DNA or RNA molecules formed as amino acids and attached themselves to crystalline matter such as clay or ice. Others believe that these giant molecules only form within the first tiny capsules that protect them from being dissolved. Such tiny capsules are called liposomes, meaning 'fat bodies'. They are tiny hollow spheres (like Richard Pico's proto-cells), like microscopic soap bubbles that form, whenever lipid molecules are in contact with water.

One group of tiny creatures, the viruses, is still a mystery, and a threat, today. We do not know whether to call them 'alive' or not. They are loose strands of genetic material that can only replicate by invading a living cell and using that cell's replicating mechanism: they have a code, but not the machinery for reproduction. When they invade another cell, to use the cell's reproductive mechanism, they can change or kill that cell in the process.

On September 28, 1969, two months after the landing of Apollo 11 on the moon, fragments of a meteorite fell in and around the small town of Murchison, Victoria (about 100 km N of Melbourne) in Australia. Local residents collected fragments soon after the fall, minimizing the chances of contamination. This meteorite transformed our ideas about organic material in the Universe, because the meteorite turned out to contain a wide variety of organic compounds, including amino acids. It raised the odds that such extraterrestrial material might play a role in the origin of life.[15]

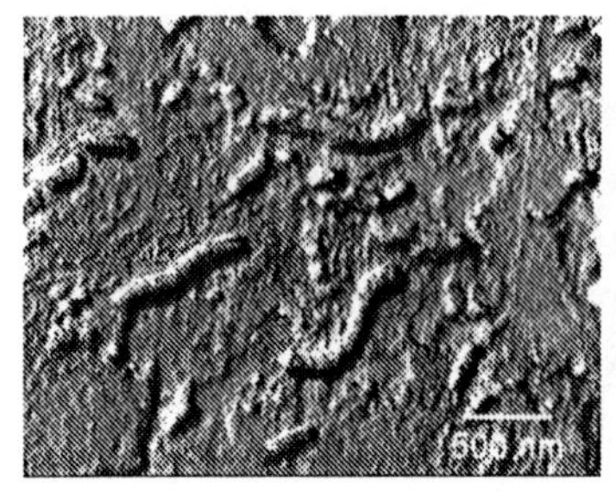

Illustration 20: Fragment of ALH 84001. This high-resolution transmission electron microscope image is of a case, or replica, from a chip of ALH84001. It shows the outline of what are believed to be possible microscopic fossils of bacteria-like organisms. The tubular features in this image are less than a micrometer in size, or about 1/500th the diameter of a human hair. Courtesy NASA[16]

In 1996, a meteor was found in Allan Hills, Antarctica. It bears the imaginative name 'ALH84001', and carries enough evidence to conclude that it was once part of the planet Mars. The meteor, which is 4.5 billion years old, fell to the earth 13,000 years ago, and possibly contains evidence of life on Mars. Inside the meteor, along tiny cracks, scientists found evidence of what many believe to be ancient bacteria (see illustration 20).

Several theories (or speculations) assume that life has been imported to our planet, either by the impact of meteorites or asteroids, or even –as Francis Crick suggested- by an alien spaceship. Most scientists consider these theories to be not very likely explanations for the emergence of life on earth. We know today that the basic chemical components, and some of the organic compounds that are vital for life, also exist in outer space, hidden in meteorites, asteroids and planets. We do not know where the first cells formed; our description of the emerging cell in a mud-hole is just one possible scenario. Some scientists assume that life emerged deep beneath the surface of the ocean, in deep-sea hydrothermal vents, that were first discovered in 1979. Hot gaseous compounds with temperatures exceeding 572 degrees Fahrenheit percolate through these vents from the center of the earth, and scientists discovered ecosystems containing worms, crabs, fish and bacteria surviving in this hostile environment.[17]

All options remain open. We do not know *where* or *how* the first cell was formed. We don't even know whether the first living entity was indeed a cell (see insert). Despite all the successes in manufacturing organic compounds in the laboratory – the Miller type of experiments - we are unable to build a single living cell. The simple fact remains that we cannot build life, at least not to this day.

Life before the first cell?

An important new research avenue has opened with the discovery that certain molecules made of RNA, called ribozymes, can act as catalysts in modern cells. It previously had been thought that only proteins could serve as the catalysts required to carry out specific biochemical functions. Thus, in the early prebiotic world, RNA molecules could have been 'autocatalytic'--that is, they could have replicated themselves well before there were any protein catalysts (called enzymes). Laboratory experiments demonstrate that replicating autocatalytic RNA molecules undergo spontaneous changes and that the variants of RNA molecules with the greatest autocatalytic activity come to prevail in their environments. Some scientists favor the hypothesis that there was an early 'RNA world', and they are testing models that lead from RNA to the synthesis of simple DNA and protein molecules. These assemblages of molecules eventually could have become packaged within membranes, thus making up "protocells"--early versions of very simple cells. Some scientists favor the existence of such RNA before the first cell emerged.[18]

For scientists involved in the study of the origin of life the question no longer is if life has originated from inorganic matter, but which of the many pathways nature choose to accomplish the task. Dr. David L. Abel is Program Director of the life emergence project. This non-profit organization offers a prize of US$ 1 million for the scientific theory that sheds light on the origin of life. The prize is called the 'The Origin-of-Life Prize'.[19] The search for the origin of life continues.

"Are the first living cells inferior living beings, Mr. J? Their versatility is certainly less than that of the spider or the whale, and even less compared to the versatility of human beings, isn't it?"

I am glad you bring that up Novare. 'Quality of life' can only be judged against the total tapestry of a given environment of living organisms. There is no scale to measure an 'amount' of life. Some organisms might be more versatile than others, allowing them to adapt to more complex environments, but that doesn't imply that they possess a larger 'amount of life.

"When you are asleep Novare, life still goes on, even though you have a hard time believing that. I can tell because I'm the one that keeps *you* alive. Life is life, no matter what life *is*. Bacteria are as much alive as are jellyfish, dolphins, chimpanzees or human beings, asleep or awake. Easy!"

We can only look at the 'quality of life' of a living creature in its relationship to other living creatures, and its relationship to its total environment. No creature, from the first prokaryote to the most complex plant or animal, has 'more life' than any other living creature. There is no absolute standard for 'life'. The quality of life is relative. The behavior of a living organism relates to its environment as a planet relates to its solar system, and the solar system to the cosmos. The behavior of a planet is part of the interplay of the cosmos. Likewise, the behavior of a living organism is part of the interplay with its surrounding creatures. In Richard Pico's words:

"Life is a frame of reference, it is an emergent organization of organic processes, it is an irreducible system, and it is the most general name for the property of the entire field of cellular behavior. Therefore, we must explicitly state before moving on that in a seemingly paradoxical way, life has no inherent units of measure, no specific quantity, and absolutely no force of action back on the cell or background environment. There is no vital force of life, no élan vital. No cell has more life than another; no organism is less alive than another."[20]

Every living cell and all the living creatures yet to emerge, every butterfly, every rose, your beloved dog or cat, and every human being represent their own biological referent system, and every single one of them lives a biological existence that has its own 'quality of life'. Richard Pico calls this the 'relativity' of biological life. Every organism lives by the rhythm of its own baton: a metronome powered by the clock of the individual's metabolism.

Autonomy

In his book What is Life, Erwin Schrödinger said, back in 1944:

"What is the characteristic feature of life? When is a piece of matter said to be alive? When it goes on 'doing something', moving, exchanging material with its environment, and so forth, and that for a much longer period than we would expect an inanimate piece of matter to 'keep going' under similar circumstances."[21]

After all, our cats Strudl, Blue, and Boris were not that far off with their analysis. They intuitively felt that life had to do with their ability for autonomous behavior: they had options to move about, and gained an unprecedented independence from the forces that rule the universe. While the cats just wonder about the peculiarities of behavior of living versus dead things, Schrödinger seeks, like many after him, to explain this behavior in terms of the underlying dynamics of ultimately inert particles. As quantum behavior of electrons is responsible for the stability of atoms, it might also explain the amazing stability of the genetic code, give or take a few favorable mutations, over a period of hundreds of millions of years. Schrödinger explains how quantum effects can certainly play a role in the explanation of the stability of small atomic or molecular constructs, like the nucleus of a single cell, but that different and yet unknown laws govern the ordered behavior of macro-scale organisms.

"For it is simply a fact of observation that the guiding principle in every cell is embodied in a single atomic association existing only in one copy (or sometimes two) – and a fact of observation that it results in producing events which are a paragon of orderliness. Whether we find it astonishing or whether we find it quite plausible that a small but highly organized group of atoms be capable of acting in this manner, the situation is unprecedented, it is unknown anywhere else except in living matter."[21]

We know with certainty, that subjecting an organism to very low temperatures will cause the organism's death. At even lower temperatures, approaching absolute zero, it will disintegrate. Lowering the temperature is equivalent to the slowing down of any moving particle within the structure, which is equivalent to lowering the energy of those particles. At a certain point on the temperature scale, particles start to behave "predictably," as if they are tired of playing "hide and seek." It is a known fact of science that, at room temperate, entropy plays "an astonishingly insignificant role"[23] in many chemical reactions.

Schrödinger and many others today, believe that we will find the laws of physics that govern the behavior of living organisms. Others try to find explanations beyond the realm of the physical world, or claim that our notion of the physical world needs an expansion or transformation of perspective in order to include and explain the phenomenon called "life."

Intuitively we know that living organisms are different from inanimate, inorganic materials. Scientifically, we know that life means a new form of order and that a living organism has to – literally - feed on the chaos of the background environment to maintain that order. There is no scientific agreement as to what life is, and how and when it began. There is a common understanding that the cell is the most basic unit of life, however this is not certain either. More or less loose strands of DNA or RNA could have been the first 'living' molecular structures.

No matter which theory will prove to be the right one, scientists agree that life had a *beginning*, just as the universe had a beginning that was marked by the Big Bang. At some point in time, somewhere on earth, the first living organism was delivered. *For the duration of 11 billion years, nothing, absolutely nothing could act outside of the thermodynamic tapestry of the universe.*

Particles, atoms, molecules, stars, planets, solar systems and galaxies: nothing could escape the forces that rule the cosmos. The nuclear forces, the electromagnetic fields and the gravity fields determined the behavior of everything in it.

With life, something unprecedented happened. A structure built from cosmic matter and energy *broke loose from the thermodynamic tapestry of the universe*. From the time-space continuum of the universe a creature formed, and broke the chains that anchored its components to the weave-world of matter and energy, like the majestic appearance of a dolphin breaking through the surface of the ocean. This new creature organized itself and established its own boundaries – its own interface - to the outside world. Against the odds of entropy, life created its *own rules of existence. It became autonomous.* This might be the most fundamental characteristic of life: the *thermodynamic independence* from the cosmic background, and its capacity to successfully maintain its own structural and functional integrity – to *maintain its own order in the face of increasing entropy* - to secure its newfound *autonomy.*

Just as the Big Bang marks the beginning of *our* universe, the first living organism marks the beginning of *our* life. Life as we know it on the planet earth. No matter what bio-chemical pathways in the world of biochemistry led to the emergence of life, a new domain of existence was entered. We call it the domain of life, or *the life domain.* As the gate opened to this world of new possibilities, another threshold was crossed. We will call that threshold the *life threshold.*

DNA, the Algorithm of Life

"JJ?"

You have a question about DNA, I bet.

"Easy guess, since it's in the title of this paragraph. But I do think, I know what you're after."

You do?

"When you talked about the algorithmic nature of the universe, I had a hard time understanding. Then, when you talked about DNA-molecules being part of a 'blueprint', I got an idea."

Novare seems agitated. "Well, I'll be ...! Who put that idea together, Alvas?"

"Big deal! OK. You did put it together, but I gave you the pieces."

May I just hear the idea?

"Here it is, Mr. J. With the emergence of DNA, nature accomplished something incredible. It separated the design specifications for the 'production' of an organism from the organism itself! So, not only did nature produce the autonomy of living organisms, but it also produced an autonomous set of design specifications, and rendered them in the form of DNA molecules."

Indeed Novare. As we have seen, DNA molecules are grouped into genes, and all the genes form the blueprint. We call that totality of genes the genotype of the organism. The organism itself, its appearance, its structure and its functionality we call the organism's phenotype.

"That is not, what I'm after Mr. J. What I find so fascinating is that nature has created an algorithm, nothing more and nothing less. It has created an algorithm that encapsulates all the information necessary to define an organism's blueprint. Isn't that the true brilliance of nature's accomplishment?"

Alvas seems impatient. His thoughts seem to be at the point of exploding.

"You see, once Novare put that together, something else fell into place."

Which is?

"Can't you see? Genotype and phenotype: how does nature get from one to the other? Through bio-chemical procedures that involves DNA, RNA, and a host of compounds, processes and organelles present within each cell. Whatever the precise nature of those mechanisms, nature had created the algorithm that defines the whole process, from genotype to phenotype!"

Thanks for your thoughts, both of you.

From the earliest moment onward, life has a feature that nothing that came before, ever had. Nature had isolated the information it needs to create a living organism. It had separated design from finished product. It had found a way to distinguish between the parameters that define the organism and the rules for producing a living organism from those parameters, a distinction that many software developers still struggle with today. Nature had provided autonomy to the blueprint of life. The human genome project, genetic engineering, Dolly the sheep, and now, Eve the cloned human being (as of March 15, 2003, the validity of the claim has not been verified), are part of a new reality because of it. Nature had invented the algorithm for the production of life.

Episode 6, "Living Earth"

2 - 3 billion years after the beginning of life,
2 - 1 billion years before the awakening of knowledge.
2 - 1 billion years before today.

Infecting the Planet

"PKs are now everywhere, Mr. J. They have formed complete tapestries in some cases, but when they do, all the cells in the tapestries are the same. Hardly any differentiation has taken place."

"The PKs first had to saturate the planet, JJ, before something new could happen. It feels as if the earth had to be totally infected with life for new surprises to occur, if you pardon the expression. It reminds me of the universe, when it was halfway through its preparation for life to occur. By then, it had filled itself with all the stuff it needed, and it had spread it evenly. "

"I'll have to rephrase Alvas' simplistic description, as usual. Maybe the surface of the planet had to be recycled enough to create a high state of entropy in vast areas of its surface, to create the opportunity for innovation to happen. Anyway, that is what it looks like to me. Radioactive radiation has practically disappeared as the virtual Geiger indicator on my virtual digital dashboard indicates. New continents have formed, and vast shallow lakes emerged, in which the sunlight can penetrate. Bacteria produce enormous amounts of oxygen, first in the water, then, after all the iron in the water has rusted, the oxygen escapes into the atmosphere. Life became the agent of change for the whole planet."

"The air and the water are clean. The skies are the eternal blue of the 'big sky' of Montana, matched by the azure of the water. The beauty is overwhelming, JJ, but I do miss the colors and patterns of flowers, trees, butterflies, and birds. None of them exist yet."

Then, the impossible happens. Under the clear skies, we witness a phenomenon that is unique in the history of biological evolution. One that will determine the future of all life.[24] Different bacteria start to cooperate. As biologist, Lynn Margulis would observe in the 1970's:

"Many living bacteria work together, exchanging biological services that help them to move about, or make food, or even reproduce. Some species live inside the walls of another, using its food but returning useful molecules. Sometimes more than two species collaborate in this barter economy."[25]

"That is how a new cell emerged, Mr. J. Its composition is an amalgamation of different PKs, loose strands of DNA, and other compounds. Nature had discovered the synergetic power of mergers and acquisitions, so to speak, as mimicked in the corporate world today."

"Just like the corporate world? Aren't their mergers done to decrease diversity, to serve the 'economy of scale'? They look more like the unicellular tapestries of PKs to me, JJ. 'Cost cutting' is the label for an elimination of diversity. Evolution in reverse, if you ask me."

"Nobody asked you, Alvas. Please, learn to wait for your turn, and keep your strange associations to yourself."

All right you two. Let us have a closer look at this new cell.

The new cell is a eukaryote, or EK, meaning 'true nucleus'. This feature, a nucleus with its own enclosing membrane is one of the many prominent new mechanisms of the EK. This nucleus contains the DNA, folded into more complex chromosomes than the PKs. The EK contains many new organelles, little (previously independent creatures) structures, such as the mitochondria that can produce and store chemical energy.[26] Originally, these mitochondria were bacteria. As Richard Dawkins puts it:

"Two billion years ago, the remote ancestors of mitochondria were free-living bacteria. Together with other bacteria of different kinds, they took up residence inside larger cells. The resulting community of ("prokaryotic") bacteria became the large ("eukaryotic") cell we call our own. Each one of us is a community of a hundred trillion mutually dependent eukaryotic cells. Each one of those cells is a community of thousands of specially-tamed bacteria, entirely enclosed within the cell, where they multiply as bacteria will. It has been calculated that if all the mitochondria in a single human body were laid end to end, they would girdle the earth not once but two thousand times."[27]

After their appearance, EKs diversified explosively, through variations in the composition of their DNA, and through symbiotic[28] processes that gave rise to them in the first place. Anywhere between 36 to 80 different phyla emerged, according to current estimates, although the knowledge about their evolvement is still very patchy. However, we do know that they are the building blocks of every living organism to ever emerge. Also, we know that they have the versatility to take on any function and structure that living organisms require. They are the cells in our bones, our heart, our kidneys, our skin, our lungs, and our brains.

Right after their appearance, they produced many innovations on their way to perfection. Two of these accomplishments stand out. One will determine the structure and function of every living organism; the other one will determine the future of evolution itself. The first accomplishment is multi-cellularity:

"EKs made a further and revolutionary jump. Differences in their genetic program enabled them to build bodies that used several cells instead of one, cells with a range of different structures and functions but which could still chemically communicate with each other. This was the turning point that set life expanding out of the microscopic to exploit new resources by building

complex structures that would range in size from mosses to sequoias, aphids to dinosaurs."[29]

A host of multicellular organisms will enter the scene in the coming billion years. Algae, fungi, and animal life forms, already existing in unicellular form, now appear as multicellular organisms. Multicellularity will be accomplished repeatedly by nature, usually as the attempt for a division of labor, at least so the theory goes. Ultimately, this leads to the formation of the three great kingdoms in Nature that we know today: the Plants, the Animals, and the Fungi.

The second accomplishment is the *invention of sex*. One billion years have past since the first EK emerged. Another billion years to go, until evolution would face another threshold.

"Sex, JJ?"

Yes, Alvas, I know you have a special interest in it. But that will have to wait until later.

The principle is simple, yet incredibly effective. Eukaryotes developed, through symbiosis and the occasional mutations of DNA due to copying errors, the capacity to form special reproductive cells, called gametes. The parent cell contributes a randomly selected half of its own genetic material from each pair of chromosomes to the gametes. In this way, no two gametes are the same. Each time that the gametes of two parents combine to create a zygote (the zygote is the fertilized cell that will grow into a new organism) they share a different package of information that has been reshuffled and redealt.[30] Scientists assume that the burst of new forms of life, and the mechanism of sex are related. Why did nature invent sex in the first place? Sex could be nature's attempt to avoid the deadly visit by viruses. By continuously varying genetic codes - according to this theory - viruses would get 'confused' about the cell's identity, giving them less chance for an organized attack, thus increasing the chance for survival of the organism.

"I can think of better reasons for the appearance of sex, JJ."

That doesn't surprise me at all Alvas. But evolution had not yet invented certain 'side effects' you are undoubtedly referring to. The emergence of sex happened a billion years ago, way before your time. Your role in these things didn't start for another few hundred million years.

"Oh. I was just thinking."

"*That* is what Alvas calls thinking, Mr. J. Thinking! Alvas, let me give you a lesson about sex. A cell can pinch itself, so that it simply splits in two identical parts. Before the pinching, it duplicates its own

DNA, so each copy is identical. No sex involved here. It is called *mitosis* and happens every day to every cell. That is simply the way organisms grow. When two different cells rub up to each other and slip some DNA to each other, a *new* DNA pattern is created, hence a new type of cell. That, dear Alvas is sex: the mixture of DNA to create a new combination, and a new *individual*, or in the case of bacteria, a new *strain*."

"How utterly unromantic. Sounds like bioengineering. I always realized, it was all about the side effects. Makes me wonder though. Will biotechnology expose the means to an end as the end itself? Would make my job a lot easier, that's for sure. However, from my experience there is more to sex than that. But thanks anyway Novare."

With the emergence of the EKs, nature is ready. The algorithms for life are in place. A method for the creation of multicellular organisms has been tested. Sex has been invented. As Ernst Mayr puts it,

"The origin of the eukaryotes was arguably the most important event in the whole history of life on Earth. It made the origin of all the more complex organisms, plants, fungi, and animals possible. Nucleated cells, sexual reproduction, meiosis, and all the other unique properties of the more advanced multicellular organisms are achievements of the descendants of the first eukaryotes."[31]

Algorithms and Renditions, part 1

"I'd like to return to the subject of sex, JJ."

"Alvas!"

"Just keep quiet, Novare. Hear me out, JJ. You talked about the separation of 'parameters and rules', remember?"

Sure.

"If that hadn't happened, sex would have been impossible, right?"

"That *does* make sense, Alvas. Nature devised a universal standard to code and store the information of the blueprint. Therefore, pieces of information can be 'mixed and matched' to form new combinations. Nature's universal production mechanism can still 'read' the new combination and make perfect sense of it. I'm proud of you Alvas."

"Thanks for the compliment. My point is a different one, though."

"Oh?"

"Sex works on the blueprint of organisms, right?"

"Right."

"Blueprints reflect the design of something new. They reflect ideas and concepts for a new organism, right?"

"Sure."

"Therefore, sex is the recombination of existing ideas?"

"You can see it that way, if you want to."

"Now I have a question for you, JJ."

Shoot, Alvas.

"Why do people organize 'brainstorming sessions?"

They do that to produce new ideas, for a variety of purposes. Why do you ask?

"Simple. When they do that, they are involved in sex. Therefore, brainstorming is mental sex."

"Alvas! Your associations simply don't make sense."

Maybe this one does, Novare. You might want to think about it.

In cosmic evolution, matter and energy are the parameters for the design; and they are the components of the final product. Information and the carrier of that information are one. The algorithmic components are the forces and properties that produce the space-time continuum. The algorithm is the rendition. Sex, the rearrangement of design properties, is the reorganization of matter and energy in the universe, as nuclear forces, electromagnetism and gravity forge the new arrangements.

Life separated algorithm from rendition, information from manifestation. Now, nature could design new blueprints by rearranging information, before an organism ever saw the light of day. Now, sex became the design of new entities, instead of their direct production. Production itself became a gentler and separate task, performed by the bio-chemical processes that lead from the genotype to the phenotype. Information and manifestation, algorithms and renditions, became the dancers in nature's way of innovation.

With the appearance, and the perfection, of the eukaryotes, nature established the ground-rules for the evolution of life: the algorithm for biological innovation. This algorithm continued the 'tradition' established by the universe, the tradition of creating diversity and synergy, but with a salient difference. Now, nature had the ability to create innovative 'products' at the design level as well. Symbiosis remained to be a powerful way for innovation by blending finalized renditions (merging actual organisms). However, the way of the future was the recombination of design elements – the genes – to create novel blueprints; a new set of parameters that the universal production algorithm could 'understand', and render into fully-grown organisms.

As Darwin discovered, nature accomplishes this by a process carrying the fearful name of 'natural selection', a process that features two distinct and separate mechanisms. Together, these mechanisms determine the course of biological evolution; they are the core elements of the algorithm for the evolution of life.

The first mechanism produces variations of the genotype of an individual organism, either by random mutation of particular genes, or by creating gametes containing a random selection of the organism's genes.

The second mechanism is selection. Variation leads to a new genetic blueprint, which nature renders into an organism that possesses a new structure and new functionality: the phenotype of the new organism. During the development from genotype to phenotype, and during its adult life, the organism is subject to the influences of the environment in which it develops and acts. If its 'feature-set' is such that it can successfully interact with its environment, the organism will survive, thrive, and eventually produce offspring. If its genotype produced a creature that is unable to handle the challenges of its particular environment, it will not be successful; it might be 'sick', it might die, or it might simply be unable to produce offspring. The populations of organisms with features that are beneficial to handle the challenges of their environment will survive and thrive, and the less fortunate ones will vanish.

"Excuse me, Mr. J."

Yes, Novare.

"I wonder if there is any way of predicting the survivability of a new organism, like a 'recipe for success'?"

Thanks for reminding me Novare. The simple answer is 'no'. Nature performs its design task for new organisms on the level of the genotype, while the 'testing' of the new organism takes place in the 'real world' of interacting living creatures, and living creatures interacting with the inorganic world of the planet and the cosmos. There is no 'formula' for success in nature, except the versatility of new organisms. Increased versatility helps the organism to adapt to different environments. But remember that every feature, every mode of behavior of a living organism has only value in its relationship to other organisms, or in its behavioral expressions within the environment it lives in.

"Sounds to me like modern times, JJ."

How so?

"Years ago Xerox invented the color copier, a truly breakthrough invention, as I recall. However, hardly anybody bought one, because 'the market' wasn't ready for it. For one thing, there wasn't enough colored material to make copies from!"

Good thought, Alvas. There is an important difference, though.

"Alvas doesn't realize that nature had only one way to carry out 'what if' scenarios: it had to actually *produce* the reality of a new design, in order to test it."

"Judging by the quality of today's software, nothing changed. Real users are the ones that test it, or so it seems."

"At least Alvas, today, a lot of new ideas can be tested by computer simulation. The behavior of a new airplane will be thoroughly tested for stability, strength, and reliability, before it ever leaves the ground, even before the first wing is actually constructed. That is the difference: biological evolution can't do that."

That's right Novare, even if, sometimes, a new biological design will not survive because it is incompatible with the bio-chemical environment that has to render it into the final product. The separation of design and final product, the separation of design parameters and the rendering algorithms, is the most important achievement of nature at this stage of evolution.

The planet earth is getting ready. Everything is in place, tested, and ready to go: the mechanism to store and copy blueprints, a new way to develop innovative blueprints from existing ones, and a factory to render blueprints into organisms. The mechanisms to grow its own structure by replication of cells has been tested, and the machinery to feed and sustain itself - so that its own order could be maintained against the forces of increasing entropy - has proven to be robust and reliable. The autonomy of living organisms is firmly established. The interfaces with life and the inorganic world of the planet and the cosmos have shown to be successful. That is when the cold sets in.

Changes in the orbit and the tilt of the earth might have been instrumental to the climatic changes. Life might have depleted the atmosphere of too much carbon dioxide, and diminished the greenhouse effect of its protective layer. Whatever its causes: the earth was freezing over. The biggest ice age in the history of the planet had started, around 800 million years ago. After that, the planet would put the creative powers of life's inventions to work and produce the most colorful scenery in the universe, at least, on this side of the universe.

Episode 7, "Quest for Harmony"

500 - 65 million years before the awakening of knowledge,
500 - 65 million years before today,
3½ - 3.935 billion years after the beginning of life.

Chaos & Opportunity

Are you ready to describe the new world, Novare?

"JJ, I believe it is my turn. After all, I appeared in the world first. Granted, it was the most primitive form imaginable, but from *a history of the unconscious*[32], I know that it all began with the Jellyfish that had just appeared around this time, at about 600 million years ago. Jellyfish developed a simple ring of neurons that coordinate their swimming motion."

"Go ahead, Alvas, steal my show."

"OK, we'll do it together. There is nothing unscientific about Jellyfish, Novare. They evolved from the Anemones that managed to break loose from the chains that anchored them to the bottom of the sea. Jellyfish possess a ring of neurons that coordinate their behavior, which is quite elaborate. If a jellyfish has ever stung you, you'll know what they are capable of. Anyway, here is what the earth looks like these days: The sheets of ice have melted, and glaciers have receded, leaving new rivers and lakes. High up in the atmosphere, the ozone layer has been created by splitting oxygen molecules and attaching each oxygen twin to a remaining oxygen molecule, protecting the surface from destructive UV rays. Your turn, Novare."

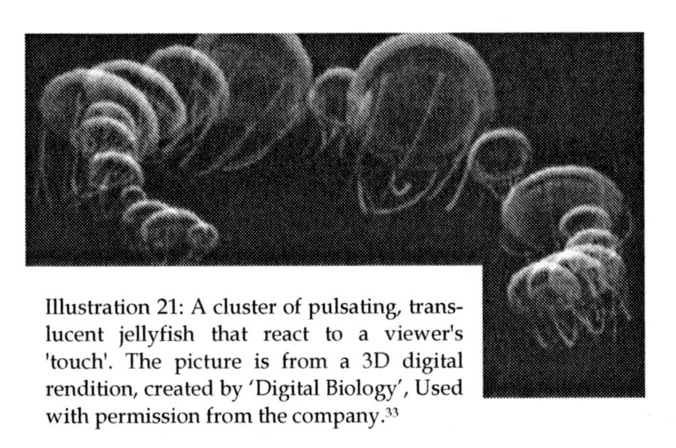

Illustration 21: A cluster of pulsating, translucent jellyfish that react to a viewer's 'touch'. The picture is from a 3D digital rendition, created by 'Digital Biology', Used with permission from the company.[33]

"The Anemones, mentioned by Alvas, represent the first multicellular species that emerge from the colonies of highly evolved eukaryotes that are stuck to live their life attached to rock. Eventually, some of them manage to detach and survive. Later, they learn to move, and develop into fish. Where are we now, Alvas?"

"My digital time-map shows that we are approaching the beginning of the Cambrian period, at about 530 million years ago."

Alvas is at the controls of the simulator, and has taken a vantage point high above the rotating earth. While the earth rotates at a higher simulator speed, we get a good overview of the changing landmasses and oceans on the planet. Since the start of torrential rains, now billions of years ago, landmasses appeared and started drifting. like toy-boats in a bathtub. Over the ages, we can watch them collide, merge, and create mountain ranges. Then, they break up and drift apart again.

"Let's have a closer look at the earth. Can you stop it, Alvas?

"Let me halt the simulation. Here you go."

"The earth looks quite different from today. 'Laurentia' is located right on the equator at ninety degrees of its current position. Today Laurentia is known as North America. Siberia is to the southeast, enjoying a subtropics climate, and Northern Europe is even further to the south. Most of the other modern continental masses are merged into one super continent called Gondwana, extending from high northern to high southern latitudes. South China is positioned off Gondwana's Western coast in tropic regions of the planet. You can proceed Alvas."[34]

"I'll change the speed of the simulation to 100 million years per minute, so that we can cover the time-period of this chapter in a few minutes. Look at that, Novare."

Starting 530 million years ago, we can watch the landmasses drifting towards each other. North America is rotating counter-clock wise, Siberia and Northern Europe are floating north, Gondwana drifts south and west, and South China is breaking up and drifting north. Barely a minute passed: 100 million years went by."

"Please hold the simulation, Alvas. What's that change in color, Mr. J?"

That is snow and ice, Novare. Glaciers and ice caps are covering the land and the seas. It is another ice age. It happened 440 million years ago. The seas are full of life by now, and the first plants, fungi and insects just begin to flourish on land. As a result of this ice age, half of marine and land species are extinguished. It will take nature 25 million years to recover. You can proceed, Alvas.

"OK. But wait. Looks like another one is coming, JJ."

"Another disaster?"

"Indeed Novare."

Less than a minute after the previous one, apparent climatic changes cause another extinction of half of all existing species. During the recovery from the previous one, fishlike creatures left the seas and ventured ashore. Some of them, the amphibians, managed to live in the water and on land, maintaining a double life (amphibious means 'double-lived'). Half of them disappear as well. It takes nature 30 million years to recover from this one.

"You forget to mention the positive side of this, JJ. These climatic changes, as well as the continuous change of continents and oceans, act like a giant recycling machinery. Inorganic materials, and organic compounds break down, redistribute and new soil compositions, as well as atmospheric compositions emerge. New chaos is new opportunity, remember?"

Excellent point, Alvas. Show us what happens next.

"We are now at 365 million years ago, right at the end of the Devonian period. That's what my time-map says."

"Look, Alvas, we can now start to recognize the shapes of the continents as we know today. Except that, they are lumped together. Gondwana breaks apart, and big chunks become the African and south American continents. South America, North America, and Africa are very close together, while Europe, Siberia and China are off to the Northeast. Other parts of Gondwana are drifting South and East, and will become Australia and Antarctica. India is still lingering in the Southern hemisphere... Why are you stopping the earth?"

"We paused for the previous disasters. My monitor indicates another one, although I don't see glaciers, JJ?"

We do not know what caused this one, Alvas. Scientists assume that it was a domino effect: One species in the food chain failed to adjust, and the others following in the wake of this failure. Climatic changes might have played a role also. It is the largest extinction ever. Ninety-five percent of all species disappeared, flora and fauna alike. The earth will be hit by another extinction, the fifth one in the history of the planet, just 37 million years later, at around 208 million years ago. Half of the recovered and new species will disappear in that one.

"That one just passed by, JJ, we're three minutes into the trip. It is now the beginning of the Jurassic period, yes indeed, of 'Jurassic park' movie fame."

"We cannot see them from here, Alvas. But they are there all right. Because the continents are joined together, they can spread all over the planet. The dinosaurs populate the earth. That's why we would find their fossils all over the world, a few hundred million years later."

"The Jurassic period has ended now. The time-map indicates '130 million years ago'. The continents start to drift apart again. North and South America detach from Africa, leaving space for a new and familiar ocean to begin filling up: the Atlantic Ocean. The European countries, Russia, China, Japan, Australia, and Antarctica are on their way to their current positions. Greenland is emerging from the ocean.

Really makes you wonder, JJ: where are the continents headed *next*? Anyway, my dashboard now shows 65 million years ago."

"It's time to get down to earth again, Alvas. Too much time has passed. The dinosaurs have appeared, and are about to disappear again."

"Just hang on Novare. Can you see that asteroid, just to your left?"

"Wow! It hit the earth!"

"Isn't that frightening, Novare? The asteroid hit the earth near the Mexican Yucatan peninsula. The dust and debris blocks the earth from our view and keeps the planet from receiving any sunlight."

Back on earth, the dinosaurs are gone. All of them. Brachiosaurus, Hadrosaurus, Protoceratops, Stegosaurus, and all the others: gone. The infamous Tyrannosaurus Rex and Velociraptors, and their flying cousins: the Pterosaurs, gone. They were too big, consumed too much food; they were too much of a threat to other living creatures; they had to go.

While we watched the dance of continents from the comfort of the simulator, dramatic things had happened on the surface and in the waters of the planet. Out of the chaos of five major extinctions, nature created new inventions that again would change the rules and raise the bar for yet more innovations to come.

Illustration 22: A large body of scientific evidence supports the hypothesis that a major asteroid or comet impact occurred in the Caribbean region at the boundary of the Cretaceous and Tertiary periods in Earth's geologic history. Such an impact is suspected to be responsible for the mass extinction of many floral and faunal species, including the dinosaurs. Artist: Don Davis. Courtesy NASA. http://impact.arc.nasa.gov/

Innovations

The Cambrian explosion[33] right was the biggest explosion of life nature has ever seen; it was the most productive era of biological evolution. Within a few million years, nature developed all the basic body plans for every future organism.

Soft-bodied and shell-bodied species split and proceeded on different paths. The shelled creatures, producing lobsters, go in one evolutionary direction as the arthropods,[35] and later produce the insects and spiders that we know today. The soft-bodied species go in a different direction, producing the worms, and later, the animals with backbone such as fish, our early ancestors, and later yet, human beings. The *internal* skeleton was invented.

Along the way, tiny external tentacles, used to propagate the organism through water, or grab food, seem to turn inward to be used to send signals from cell to cell. An internal communication system emerges: the nervous system was invented.

The nervous system facilitates the construction of more complex multicellular and multi-functional organisms. Some cell-groups develop into muscles, the beginning of backbone, mouths and guts. Squid and cuttlefish appear.

A concentration of nerve bundles at the head-end of animals become nerve centers and thus, the brain is invented. Light sensitive spots near this new brain evolve into eyes, and a host of other sensory organs evolves over the course of evolutionary time: nature invented a vast array of sensory organs.

With new sensory organs appearing, nature redesigned the interface of its creatures. It totally changed the way living organisms can perceive the world, a world that consists of matter and radiating energy. Nature devised a way to transform the renditions of the universe into more subtle and differentiated perceptions that increased the versatility of living organisms. Organisms developed a way to render the pressure of moving particles of air into something we know as sound. The organism didn't need any 'loudspeakers' to accomplish the task. It created a group of cells that was able to sense variations in the pressure of the airwaves, and translated it into something the creature would experience as sound.

One of the most amazing senses is vision. Living creatures found a way to measure the impact of photons of light, and transform it to something that they experience as an image. Even more astonishing is the fact, that many animals, as the human being, developed the capacity to transform the two-dimensional photon bombardments on two separate eyes into a three-dimensional image. Animals equipped this way, could perceive 3D-space!

Using the properties of new structures of matter and energy, nature used all the tricks in its arsenal to create new mechanisms to communicate with the outside world, to perfect its interface, and to increase its capacity for awareness of its environment.

> "There might be some lessons for modern computer-interface designers here, Mr. J?"
>
> "I'd say yes. Just imagine, we use a mouse to point at our computer-screens, and we still use a keyboard to speak to it. Ten thousand years from now, judged by those devices, human beings would be thought of as creatures with two fingers on one 'hand', and eighty on the other. Of course, Macintosh users would confuse the issue, since they need only one finger to manipulate the mouse."
>
> On a more serious note, your point is a good one, Novare. The future of computer interfaces will be the elimination of devices, or at least the development of less obvious, less obtrusive devices.
>
> "What about the invention of the nervous system, and the brain, Mr. J?"
>
> The invention and the evolution of a nervous system, a brain, and sensory organs went hand in hand. It would be no use to receive and transform the bombardments of photons of light without a way to process them. To process them, they had to be transported to the place in the body where that processing takes place, hence, a brain and the nervous system.

With the advancement of these new features, living creatures were on their way to develop complex information processing systems.

Information Processing

The universe creates renditions. The 'substance' of these renditions is matter and energy. The features of this substance allow plants and animals to communicate. Such features include the speed, spin, or electric charge of particles, and the frequency, wavelength and amplitude of radiation of energy, be it in visible form, or otherwise. Nature found ways to use all of these properties of the universe. Living creatures live by them. No interaction with their environment would be possible without an awareness of this environment, just as their inert ancestor entities. However, the autonomy of living creatures required more advanced mechanisms to sense an environment that changed as they moved through it, and that changed as time went by.

The renditions of the universe - the way the universe manifests its properties – now becomes the information for the sensory organs, and an information capability is needed to process this information into something the living organism can 'deal with'. With the advance of multi-cellular organisms, the demand for information processing increased for another reason.

The organism had to coordinate the functions of an organization of cells and organs that became increasingly complex in structure, and in functionality. The bio-chemical processes inherent in the activities on the level of the cell, and the direct communication between cells, no longer suffice to manage and control the organism: the 'society' of cells that make up the organisms needs a separate network to connect all of them; a network that is able to transport information from, to, and between cells. Hence the nervous system, and to control the new network, the brain.

Nature had achieved another break-through innovation: the separation of information processing and information distribution on the one hand, and the organic 'devices' that carried out certain functions, on the other. Through this separation, it increased its 'plug and play' capability for the future of evolution. When new organs and functionality became available, the organism was better prepared to integrate it into its already established 'standard' of information processing. Not Microsoft, neither Apple, nor IBM invented 'plug and play'. Nature did.

The universe provides information. Nature learned to process that information and render it into something that biological creatures can 'live with', in a very literal sense. The individual nervous system was not the only network that nature developed. Another network established itself because of the increased capability of organisms to process sensory input information. This is the first wireless network in the world, and it is almost as old as the world itself: the network that connects all the living creatures in a given environment. Creatures start to communicate 'from a distance'. After billions of years of close contact, dictated by chemical communication, the new sensory organs allowed awareness from a distance through smell, vision, sound and other means. We still don't know the secrets of some of the communication mechanisms that organisms use. Whales seem to communicate their presence over large distances, but we don't know how.

Versatility

All of these inventions took place in the water of the seas; their perfection would continue in the water, and on land. The water in the oceans that face the moon is 'pulled up' by gravity. The gravitational and rotational interplay between the earth and the moon generates the tides.

These tides carry, and leave, mats of algae and fungi on the beaches. Some dried out and died; some developed protective mechanisms and survived. Worms and arthropods are the first animals to go ashore. Contrary to the algae and the fungi, they already possess a protective coat that allows them to stay moist inside.

Molds, mushrooms and yeasts, belonging to the kingdom of the fungi, learned to digest minerals by excreting enzymes that dissolve the rock they live on. Some of the fungi merge with algae and produce the lichen. The winds carry the tiny spores of algae and fungi and disperse them all over the planet. Plants might be the result of symbiosis of fungi and early animals, because their cells are more complex than the cells of either animals or fungi, and have both of their features. Whatever the ultimate story is, somewhere on the planet the first plant starts growing on land, and develops protective fibers to keep the moisture inside. They adapt to living in the open air, take hold, grow and multiply, wherever enough moisture is available.

This invasion of new life on the earth's surface accelerates the production of an enormous biomass. A thick coat of soil forms through continuous biological recycling, and provides the opportunity for new species to exploit its diversity; plants develop deeper roots, and larger trees can emerge. Animals develop mechanisms to reproduce in the absence of water; ovaries evolve, producing eggs that become embryos after fertilization by pollen. These embryos develop in the protective shield of their own bodies.

Other sea animals find their way ashore, and the first amphibians appear. Some of them develop more muscular fins to walk on land; and they exchange their gill slits for air breathing lungs. Eventually fins become legs, while a second pair of legs develops over time. We can still observe this process in today's frogs and salamanders. After their eggs hatch in water, the swimming tadpoles have to grow legs to come ashore.

Some of the amphibians evolve into pre-reptilians that produce eggs, containing enough food and offering the protection of a shell that can hatch on land. These pre-reptilians split into two lines. The 'synapsids', that would develop into the mammals, and ultimately humans, and the reptiles, which in turn split into three branches: the turtles, the snakes and lizards, and the archosauromorphs (meaning "ruling lizard forms") which include the dinosaurs and pterosaurs (the "flying dinosaurs"), the crocodiles and the birds.

By the time of the extinction of the dinosaurs, nature had devised an astonishing variety of instruments for organisms to act in, and act upon, the world around them. Front legs and rear legs; tentacles and wings and beaks; claws, tongues, stings, hair, and the beginning of hands and feet; hoofs, jaws, teeth, horns and tusks; eyes, heads, noses, ears, and the beginning of sound producing organs; lungs, hearts, livers and kidneys; just a sample of the accomplishments of nature. They determined the behavioral repertoire of the animals. The increasing physical dexterity, together with sensory abilities and the animals' processing power, made up the animal's versatility. This versatility will be put to the test when a new creature enters the ecosystem into which it is born. If it can interact successfully with its new environment, it will survive and thrive. In that case it will change the environment by its presence and its interactions with it. If the new creature does not have the necessary tools to cope with the environment, it will perish, sooner or later.

Successful organisms change the environment. Bacteria started this process by turning the surface into organic soil, and the atmosphere into a protective layer. Plants, fungi, and animals continue that work, feeding on the chaos in their environment, and continuously recycling that chaos to maintain the order of their own bodies and leaving different compounds behind, ready for other organisms to repeat the cycle.

Over long periods of time, the versatility of organisms increases, thus diversifying the spectrum of interactions that occur between the organism and the ecosystem it participates in. The contributions to its own environment raises the bar for other creatures that might appear and participate. Evolution is mercyless in this regard: innovative creatures raise the bar, which breeds new innovations, which raises the bar again.

Illustration 23: A small family of white-faced capuchin monkeys stops for a curious glance at human intruders in the tropical cloud forest of Monteverde, Costa Rica[36]

Harmony, Part 4

The dinosaurs are gone. Many smaller animals were unable to face the light of day when they were around. The competition had been unfair because the dinosaurs had a monopoly on daylight. The explosion of debris and dust from the asteroid blocked the sunlight for thousands of years, and eliminated the massive animal.

In the dark shadows of the night, small mammals are waiting in the wings, waiting to come out of their nocturnal existence. Small primates, our ancestors, have emerged on earth and wait for their change to shine, away from the threat of the dinosaurs. They came out of their burrows and caves[37] and out of the trees. They would learn to live by daylight. The chaos created by the asteroid (or by whatever caused the extinction of the dinosaurs) provided the opportunity nature needed to once again put pressure on the process of innovation. For evolution, chaos is opportunity, opportunity stimulates innovation and innovation means the creation of new synergies that increase diversity, while maintaining harmony through the processes of variation and selection.

Episode 8, "Matter that Minds"

60 – 8 million years before the awakening of knowledge,
60 – 8 million years before today,
4 billion years after the beginning of life.

Time for our Life

In 1758, Swedish botanist Karl Linne, better known as Linnaeus, listed the various orders of the mammals. Naturally, this was done from the static worldview of the eighteenth century. He coined the name primates (meaning 'first in rank') for the group of animals that included the human being. The primates shared some of the same features that distinguish human beings from the rest of the animal world. They had hands, that could hold things (and, sometimes with their feet), with thumbs that opposed the forefinger, and fingers with flattened nails instead of claws. They had novel skulls, and unique teeth and bone structures. Importantly, they had a longer gestation period, and vision with binocular vision. Finally, they had larger brains.[38] The functionality of the brain lead Linnaeus to call human beings 'Homo sapiens', meaning 'wise man', "a classification that has come to seem more like a challenge than a scientific description," according to Gould.[39]

"That hits home, Mr. J."

You have to realize that Stephen Jay Gould was (he died during 2002) a great advocate for the random nature of biological evolution; he does not accept any notion of 'direction'. Human beings are merely *different* form other biological creatures, more so than being *better*.

"So, in fact, he stressed the relativity of the 'value' of creatures. This value can only be measured against the creature's ecological environment. Quite similar to what you called 'relativity of life', and similar to Richard Pico's 'biological relativity', right?"

You could put it that way, Novare. However, something happened that sets human beings apart in ways that go beyond biology. In fact, Gould's *The Book of Life* expresses this by saying "it goes beyond the scope of science, a human invention, to account for all the elements that make up the measurable body and immeasurable mind of humankind."[40]

"*Wise* man, after all?"

'Wisdom' is a concept, which itself is loaded with relativity, Novare. Usually, it relates to knowledge relative to a specific environment, or situation, at a given point in time. Knowledge related to changes over a longer period is more elusive, because that knowledge will itself invoke changes in unpredictable ways. Maybe true wisdom is something that relates to change itself.

"In that case we might not be very wise at all."

"Pretty obvious to me, JJ. As I said before, just watch CNN."

CNN?

"TV is one continuous display of different 'wisdoms' in the world, JJ. The wisdom of the political left is anarchy according to the right, and the wisdom of the right is stupidity in the eyes of the left, or vice versa. Has to do with the relative nature of knowledge. To me, wisdom is to realize the relativity of knowledge."

"Alvas! Such a difficult concept, coming from *you*! I find that amusing. After all, I'm the observer of change, of relationships, and of relativity itself. Usually you see everything against the absolute back-

Illustration 24: Chimpanzee 'Washoe'. Full Name: Washoe Pan Satyrus. Washoe was named for Washoe County, Nevada, where she spent her early childhood. Washoe is a Native American word from the Washoe tribe meaning "people." Pan Satyrus is the old classification used for chimpanzees. Washoe's Birth (estimated) December 1965. Courtesy of Central Washington University's Chimpanzee and Human Communication Institute (CHCI)[41]

ground of your own thinking."

"The thing is, Novare, that I have been *practicing* relativity for millions of years. Ample of time to realize that my existence depends on my adaptability to changing environments. When the environment changes, I have to change as well. That's how I survived many storms. That is wisdom. I have to admit that I learned the words wisdom and relativity from you. Shows you who the wise one is, Novare. At least as measured relative to what I'm up against."

"Just a friendly reminder, Alvas. Without me you would still be climbing trees."

"What's wrong with that?"

Wisdom might, or might not, be reserved for human minds, but with the emergence of the more elaborate brains of the primates, nature was on its way to create matter that minds. That matter came in the form of a more elaborate brain.

The evolutionary history, from the primates to the appearance of our ancestors, the great apes, is patchy and might change with new discoveries and testing methods in the future. Three families split off from the early primates, starting at about 34 million years ago. One is the now extinct family, the Proconsulidea, and one is the family of the Gibbons; they are alive and well today. The third family diverged at least 17 million years ago, and includes the Orangutans, the Gorillas, and the Chimpanzees. The orangutans split off at about 15 million years ago, and the gorillas about 11 million years ago.

"Are we descendants, then, of the Chimpanzee, JJ?"

That is a common conclusion, Alvas, however it is wrong. The chimpanzees and human beings have *common ancestors*. Since about 8 million years ago, when the human and the chimpanzees diverted from one another, we share more than 98.8 percent of our genome with the chimpanzees.

"What a difference one percent makes!"

"Don't get cocky, Alvas. After all, you share more with them than I do."

When Novare and Alvas start arguing about their history, their differences and their similarities, the time of human life, and the human *mind*, must be approaching.

Diversity

Three trends characterize the evolution of mammals in these last 65 million years, the period that takes us from the primates to the great apes; and then, to Homo Sapiens. The first trend is the diversification into a large number of species and body-forms, the second is the trend towards larger sizes and the third is the increase in brain-size.[42]

After nature had spread the enormous variety of genotypes all over the planet during the Cambrian period, the continents had started to drift apart again. Now isolated from their 'siblings', these genotypes started to diverge on different continents, increasing the diversity of different animals. Climatic changes influenced this trend to diversity. Slowly, but continuously, the planet changed from a warm, wet world covered with tropical rain forests to a colder world with distinct seasons, that varied in time, and with longitude and latitude. Species were forced to adapt to the new environments that varied widely in temperature, moisture, available chemical compounds, and flora.

The second trend, toward increased body size, contributed to an even greater pressure for diversification. The bigger the animal, the less body-surface it has in relation to its body mass, the less loss of heat it will suffer, and hence, the less food it needs to keep up its temperature.

The drawback of bigger size is gravity. The bigger the animal is, the more it will have to adjust its body-plan to support and balance its own weight: It would be hard for elephants to survive with the body plan of a bird. In this way gravity contributed to even greater diversification of the mammals. The reduced influence of gravity under water might be one of the reasons for a lesser variation in body-plans in fish.

The general trend towards a greater brain size seems to have stabilized at about 37 million years ago. The great exception was the apes and the human lineages that continued to increase their brain size over time. The growth in brain size to the 1350 grams of human beings, however, started only a few million years ago.

Algorithms and Renditions, Part 2

Since the Jellyfish, the nervous system and the brain grew in functionality and complexity in animals. Since the bacteria and the multicellular EKs, fish acquired fins, gills, and tails that had to move in a coordinated fashion. Reptiles, dinos, and mammals developed legs, hearts, heads, and wings that had to be controlled. More sophisticated internal and external organs evolved to sustain the animal's life and increase its versatility to deal with the environment it was born into, or was forced to move into after birth. The animal might face a different climate from the one its genotype expected.

More importantly, the animals had long left the state of direct biochemical interaction with the environment. As animals increased their autonomy to move about more freely, in the seas or on land (or both for the amphibians), sensory organs emerged allowing the animal to perceive its environment. The environment acts on the body of the animal in the form of particles or the radiation of energy. As we remember, matter and energy is all the universe provides for any body, including animal bodies, to 'sense'. The sensory organs had to be able to 'pick up' the properties of particles, like the pressure of water molecules in the sea or the speed of molecules of air in the atmosphere, or the properties of quanta of energy (e.g. photons of light), like the frequency and the intensity of their impact.

The emerging sensory organs needed to be equipped with the mechanisms to render the incoming information into something the animal could *understand and interpret*. This forced nature to develop *algorithms*.

First, nature needed to develop the algorithms that transform the information 'hitting' the sensory organ into a rendition the animal can 'read', not unlike a digital video camera, that transforms the impact of photons of light on the CCD-chip into an array of colored pixels on its digital screen, so that we can view it. Since animals, like us, do not have a TV screen in their heads, nature had to invent something that gives them the experience of an image, a 'virtual TV-screen, so to speak. Light has only its frequency (wavelength) and its amplitude to differentiate the different types of waves, therefore nature had to invent ways to represent these differences to the animal. Nature developed a way to do this in the form of colors. Indeed, the beautiful blossoms in the spring are illusions created by the brains of higher animals and human beings! Nature invented equally ingenious solutions to create the 'illusions' of sound, smell, touch, taste, and a plethora of other sensors to render the world to the animal. Many senses are still not understood today. How dolphins and whales find their way around oceans, how the pacific salmons find their way back to their birthplace after traveling thousands of miles over a period of up to five years, or how ospreys find their way back to their nests every year, just to name a few, remains unsolved.

The *second* type of algorithms nature had to develop, are the algorithms that integrate all of the different sensory renditions into one coherent 'picture' of the world surrounding the animal. This then, is the mental rendition the animal creates as its 'comprehension' of the world around it. From it, the animal develops strategies for action.

These *strategies* are the *third* type of algorithms. They will result in the activation of the functions of selected organs of the animal: a lion lies down in the tall grass to hide its position from its prey, a female dear produces the smell to attract a mating partner, or a wolf pup starts playing. These algorithms link the renditions created by the sensory organs to the activation of the right organs.

The *fourth* type of algorithms are those that control the function of specific organs, like the coordinated movement of jaws or beaks, the deployment of the sound organ that makes a parrot speak, or the production of adrenaline to provide a burst of energy.

"Sounds like a computer to me, JJ."

Your observation is a valid one, Alvas. Not just as a metaphor, but also as a real information processing system, embedded in the living organism. The first type of algorithms compare to the input devices of a computer, like a video camera, a scanner, a keyboard, a microphone for voice input, a mouse to locate a position on the screen, or a joystick for flight simulation.

Illustration 25: The National Advanced Driving Simulator at The University of Iowa. The capsule in the picture can hold a variety of cabs of real cars. It simulates the forces on the car when driving on the road. Inside the capsule, an array of video projectors creates a 360-degree view of real landscapes surrounding the vehicle. This simulator has been used to successfully execute a substantial driving safety study. Courtesy of The University of Iowa.[43]

Commercial systems like flight simulators or car-simulators (see illustration 25) have elaborate sensory input devices. These devices measure the driver's actions and the car's reactions, like the movements of the steering wheel, pressure on the foot pedals, and so forth. Every device requires its own algorithm to render the input into a form the computer can understand, just like the sensory organs of animals. The same is true for the fourth type of algorithms that control the behavior of specific organs. In the case of the car these 'organs' are devices like fuel injectors, brakes and so on, and a computer needs algorithms to control printers, hard drives, CD recorders, or monitors. Readers who had to install such a computer device, might know that the computer asks for a 'driver' to handle the device.

"What about the other algorithms, Mr. J?"

We will turn to them in the next paragraph, Alvas. Nature developed the algorithms mentioned here alongside the organs themselves. As evolution proceeded, the genetic code was altered such that the structure and the functions of new organs developed simultaneously. This means that the algorithms embedded in the DNA and the RNA had to be modified such that both -structure and function- were rendered at the same time.

"That structure-function is the animal's phenotype then?"

"Precisely"

"Can I say then that the phenotype is the technology the animal uses to express its genotype?"

Well put, Novare.

Operating System & Applications

An organism is a complex network of communicating organs. That network is the animal's nervous system. The device drivers for localized internal and external organs posses their own information processing capabilities that inform each other by sending signals over the elaborate network of the nervous system. In the individual organism this nervous system is 'hard-wired' through long fibers, called axons that have connectors called 'synapses' that transmit the signals to neurons. These neurons are specialized body cells (and descendant of the eukaryotes, as all body cells); they are tiny information processors by themselves, and relay the information to a host of other neurons within the nervous system, and on to the central processing unit: the brain.

The second type of algorithms that we mentioned above, the ones that coordinate and 'make sense' of all the incoming signals, are part of the core of the *operating system* of the animal. These algorithms manage all other algorithms that control the functions of all organs. This organic operating system is embedded in the structure-functionality of the nervous system and the brain, and the physical, chemical, and bio-chemical processes that take place in -and between- its organic compounds. Some scientists, most notably Roger Penrose[44] assume that quantum phenomena might play an information processing role at the lowest (nucleus of the neuron) level of these processes.

Like the operating system of a computer, the 'organic operating system' of the animal manages all of the functions of the animal. With every evolutionary innovation, the operating system changed accordingly, and automatically.

"I truly wish that were the case with my PC, Mr. J. I have to upgrade the operating system almost yearly to be able to make use of new technology. Still, I get that 'blue screen of death', especially when I forgot to save my work. Does that happen in nature?"

Yes, it does. We might compare the symptoms of an infection or other organic malfunctions as such a screen. Fortunately, in many cases we have the 'reset buttons' of medicine or surgery to cure the problem. At one point however, all living organisms face the final 'screen of death': death itself. Back to your problems with upgrades: nature automatically takes care of that.

The organic operating system, unlike Windows ('95, '98, XP, etc.) or MACOS (Apple Macintosh operating system) that operates your PC, is a distributed operating system. Every neuron in the animal body represents a processing unit, and the operating system consists of every algorithm controlling all of the billions of neurons in the body. In fact, it is a networked hierarchy of networked algorithms, embedded in the organic matter of the animal nervous system and brain.

With every evolutionary step, the behavioral repertoire of the animal changes. The changes in the animal's phenotype will provide it with new physical capabilities, and the operating system needs to provide the algorithms to use the new capability. We referred to these types of algorithms as the third type of algorithms. They might be compared to the application software on a PC. Examples on our PC's are spreadsheet of word processing software, software to edit the pictures from your digital camera, the web browser or the software to play videogames.

In nature, many such applications become part of the operating system. When sea animals left the water to live on land, they had to change their gills into lungs. Not a bad thing to have the new 'application' for breathing embedded in the organic operating system, away from the 'conscious' control of the animal.

Many algorithms for innovative patterns of external behavior were embedded in the animal's operating system. They became 'habits': automatic responses of the animal to certain stimuli. The newborn mammal immediately finds the mother's nipple for nourishment. For fish, swimming is a natural habit. A newborn deer can stand up right away and 'knows' how to walk. Nature embedded a growing multitude of such habitual modes of behavior into the organic operating system.

As animals increased the versatility of their behavior, more and more application algorithms were imbedded in the species' operating system. The brains of such animals processed the input information coming from its external and internal environments. Then, the brain chooses an application algorithm that seemed appropriate, the algorithm took over, and the animal acted accordingly. Everything was done automatically; the animal had no choice in the matter.

As the environment itself became more complex, caused by the appearance of more sophisticated animals and varying climatic conditions, the animal needed greater flexibility to adapt its behavior to different situations in the environment. Nature needed to give animals a surplus in processing power, *and* it needed to provide the animal with a faculty to use that processing power. That is exactly what nature did.

Natural Programmers

"It was just a matter of learning, JJ."

How did you learn to learn?

"At some point, I forgot when, I discovered that I didn't have to take a right turn if I didn't want to. To instruct the body to do so, I mean. I could choose. Mind you, the choice was still dictated by the circumstances, but there was a flickering awareness of having options. Sometimes, it was as if Novare was trying to interfere with the algorithms of my operating system. As if she had something new to tell me. She made me consider things I had never considered before. Soon after the discovery of options I started to experiment. That's how I started to learn."

I guess, something like that must have happened along the way, Alvas.

After that, the road was easy, yet full of challenges. Wrong moves could lead to injury or death of the animal. However learning was born. Along the road of evolution, the surplus in processing power and memory increased. The animal could now learn, by experimenting, and from the examples of their parents, siblings, and their peers. They learned to translate those experiences into algorithms, and to render those algorithms into bio-chemical patterns in their brains and nervous systems. By the time the primates, and later, the great apes arrived, they were able to program their brains with algorithms that gave them the power to program their own applications.

In that way, they learned the rules for social behavior within the group they belonged to, and they learned to build strategies for hunting, gathering food, play or attracting the opposite sex. Nature made yet another distinction in the way organisms function: it made the distinction between the operating system and application, automatic and learned behavior.

Animals became natural programmers.

Chapter 7: The Life Domain

Illustration 26: Picture used with kind permission of nature photographers Gregory G. and Mary Beth Dimijian, see http://www.dimijianimages.com[1]

"How could we hesitate even for a moment about the evolutionary origins of the layer of life on the earth?"

Teilhard de Chardin[2]

The Tapestry of Life

Evolution composed the universe - a tapestry of space and time - in one sweeping symphony that lasted ten billion years. Energy played the music and created local concentrations of matter that became the players in one vast ballet of atoms, molecules, stars, planets and galaxies. The music and the players themselves choreographed the ballet using the properties of mass, energy, and nuclear, electromagnetic and gravitational forces. Nobody was swinging the baton in that evolutionary period. The only absolute 'dimension' was change, with the direction of a multi-facetted expansion. As we have seen, we can observe this expansion not only from the increase of space occupied by the universe, or the increase in entropy, but also through the continuous innovation of its entities, and the increasing diversity of these entities.

Guests at the Table

As it turned out, the universe was not just the self-perpetuating ballet of celestial bodies to the tune of the symphony of swirling energies. Rather, its players formed the stage of something new to arrive. In one corner of this stage, a planet was prepared to create the most amazing innovation yet. The position of the planet earth in the Milky Way galaxy turned out to be just right, and the planet turned out to have the right physical and chemical composition for this innovation to emerge. The planet was ready for the formation of life. The universe had created a stage and set a table - the planet earth - that provided all the ingredients for new 'guests' to arrive. In addition, arriving they did, with the survival of the first living cell.

"Guests, Mr. J?"

"Of course Novare. *No body* lives forever. Every living organism is a *temporary* composition of matter and energy. After a while, every organism succumbs to the recycling effect of the law of entropy. We all fall apart at the end, and die, hence we are guests in the universe."

"Thanks Alvas. That helps clarify my question: why are living organisms guests, when they are built from the stuff of the universe, and therefore an integral part of that universe?"

I have two reasons to call life a 'guest', Novare. First, although we do not know what life *is*, we know – as I discussed earlier – that life gave its organisms an increasing independence from the thermodynamic mechanisms that rule the cosmos. The enormity of the effects that living organisms had (and have) on this part of the universe – our planet – led me to differentiate them from inert objects in the universe and call them 'guests'.

"I find that rather dangerous, Mr. J. If we would consider ourselves to be an integral part of the planet and the universe, we might treat our 'table' differently."

Therein lays the second reason for using the word 'guests', Novare. As Alvas pointed out correctly, all living organisms are, just like anything else in the universe, temporary configurations of matter and energy. However, along with the increased complexity of living organisms came increased vulnerability. With that, new interdependencies emerged between the universe and living organisms. The 'guests' became – albeit in new ways - dependent on their host for their survival.

"Is that why we say 'life is precious', JJ?"

Many people use that phrase to point at the temporary nature of their lives. You are right, life is precious, not only because of its limited existence, but because of the increased freedom enjoyed by living organisms. It simply meant new opportunities for innovative activities. After all, if the first cell had never appeared we wouldn't be here to talk about it.

"It still confuses me, Mr. J. On the one hand, we are made from the stuff of the universe; on the other hand, we are 'independent 'guests'."

Let us compare a living organism to a candle, Novare. The candle is composed from the same basic ingredients as everything else in the universe. Life then, is the heat that *lights* the candle, and that keeps it lit; we don't know what that heat is. Unlike the spark that lights the candle, however, life is not an additive to the organic compounds of a living organism. Rather, life is an emergent property of the biochemical structure-function of an organism that crosses some threshold of interactive processing that is still unknown to us. We can synthetically produce some of the compounds of the candle that are vital to life, like carbohydrates, amino acids and nucleotides ever since the first successful attempts by Stanley Miller at the University of Chicago in the 1950s. Nobody however, has been able to create life in a test tube thus far. We can build the candle, but we cannot find the spark that lights it. Until we find out what the nature of that heat is, I want to use the word guests to point out a unique difference.

"As it seems to me Mr. J, autonomy within the environment and interaction with the environment are the most important properties of life."

I agree, Novare.

"JJ, I'm not sure I like the idea of the *heat,* because it is the same heat that *burns* the candle."

"That's following the law of entropy, Alvas. Burning the candle is just recycling order into chaos."

"How utterly unromantic, Novare. But I guess you are right. Is that what people mean when they say 'life is a bitch'?"

Cocoon of the Living Universe

Life is an extension of the universe. What we have dubbed 'the domain of life' is built from, and built upon, the universe. Life cannot be maintained separate from the universe. This does not only relate to the obvious dependence on food and the energy from the sun. Internally, the organism's structure is maintained by the same rules that govern the integrity of atoms and molecules.

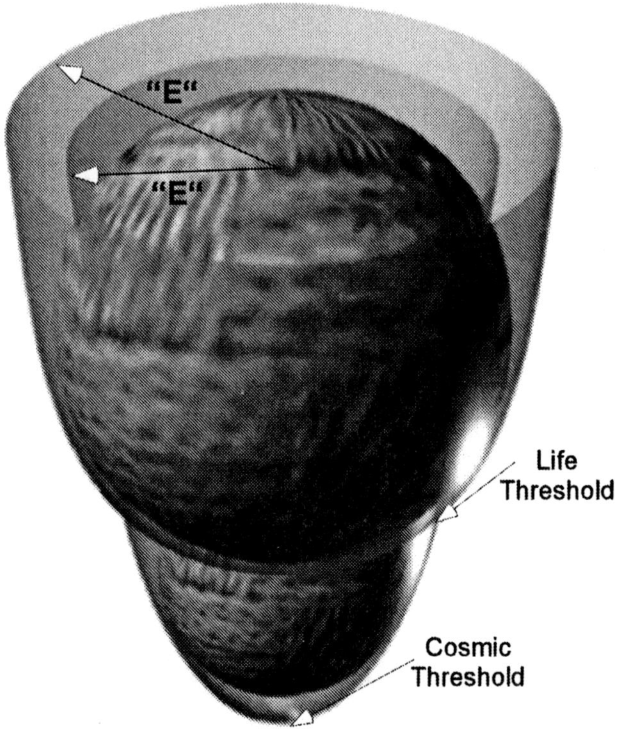

Illustration 27: Expanding, living universe. The cocoon continues to open up.

Its inner frontier, at the level of elementary particles, is the micro-world of quantum behavior. Equally important, communication between living organisms and between an organism and the inorganic world is impossible without the cosmic energy moving between them. Without light, we cannot see, and without waves of air, we cannot hear, and so forth. We can now expand our 'history of the universe' as depicted in illustration 27 and include the domain of life.

At the threshold to the domain of life, because of the ways life accelerates the creation of innovative organisms, evolutionary expansion starts to accelerate. On top of the expansion of the cosmic time-space continuum itself, life expands its own domain of existence. In illustration 27, the larger sphere symbolizes the addition of life to the smaller sphere of the universe. The discontinuity at the threshold to life marks the appearance of the first living organism. This discontinuity might disappear if, and when, the emergence of life can be explained as a continuity from inorganic matter to the living cell.

With the appearance of life, the universe starts to transform inorganic matter into living organisms. This happens on our planet and maybe on other planets in outer space as well. Is that what the universe is doing? Is the long-term future one in which the atoms of the cosmos will be transformed into life?

All 10^{78} atoms in the universe can only contribute to the existence of one entity at a time. One atom might contribute to the movement of one of my fingers hitting the keyboard of my computer today, and the same atom might contribute to the snow next winter, but it cannot do both at the same time. The larger sphere in illustration 27 represents the – now living - universe at a certain point in time. We may remember that the actual universe is contained in a three-dimensional space represented by the surface of the sphere. Thus far, life has only been found on the planet earth, although several organic compounds vital to life have been found in outer space. Projects, like the DARWIN project mentioned earlier, are designed to continue the search for life in the universe.

"Excuse me, Mr. J?"

Yes Novare.

"Will the atoms in the universe be transformed into life in the long run?"

There is no established scientific position on that issue, Novare. Frank Tipler, the scientist I mentioned before, envisions such a transformation as a necessary step before the universe reaches the ultimate point Omega. In an article on his website[3] he says, "Life must eventually engulf the entire universe and control it." It is one of several basic claims he makes for his theory about the way the universe will evolve in the future.

"Do you agree?"

Alvas is eager to respond. "I can answer that Novare. To me that transformation seems natural. After all, the universe had time enough to try out different scenarios, apart from evolving living organisms. However, evolution did come up with life as the next step to create innovative entities. On this planet it is simply what happened."

"I rather see the appearance of life in terms of knowledge, Alvas. I see it as a new rendition of existing knowledge. These renditions took the form of ever more sophisticated organic technology that helped living organisms to execute on the encapsulated knowledge."

An important point Novare. As I have discussed before, I see evolution as a series of transformations of algorithms (knowledge) into renditions and renditions into algorithms. Genes, encoded in molecules of DNA, are the carriers of that algorithm and the organism itself is a rendition - the expression - of that algorithm. As soon as the genes find the right environment, that environment (a zygote with he right conditions) will replicate the genes and start to construct the organism. The *adult* organism then, will exhibit a behavioral repertoire that reflects the knowledge embedded in the genotype.

In *The Fabric of Reality*, David Deutsch assumes that "we can infer the 'intention' of genes to render an environment that will replicate them, from Darwin's theory of evolution."[4] Whether intentionally or not, genes that don't render their knowledge into an organism, or genes that render an unsuccessful one, will not produce offspring. A 'successful' organism is one that proves to be able to adapt to its environment and therefore replicates the genome, and therefore propagates the knowledge embedded in it. As Deutsch puts it:

"...what the phenomenon of life is really about is knowledge. We can give a definition of adaptation directly in terms of knowledge; an entity is adapted to its niche if it embodies knowledge that causes the niche to keep that knowledge in existence."[5]

"Is the expansion of knowledge the key to the mystery of crossing the threshold to the domain of life, Mr. J?"

It seems that way, Novare. However, I don't see it as some form of linear progression from the cosmic domain to the domain of life. That is why I call it a 'threshold'. What happened at that threshold has nothing to do with some mysterious 'substance' or 'unknown energy'. Rather, I believe it to be some kind of 'algorithmic leap'. Something that changed the rules of programming, like a new instruction set, or a new programming language that can produce algorithms that are new in kind, rather than only affecting efficiency and the like.

"We are not on a straight path from here to Teilhard's or Tipler's Omega points then, JJ? Novare tells me that many scientists are able to look beyond any possible future thresholds."

"Alvas, you take everything I tell you so literally. Scientists are not better than anybody else, including Mr. J, in predicting the future. Most people try to extrapolate their current worldview: scientific, economic or political worldviews alike. They have to explain the *unknown from the known*, otherwise they risk their reputation. With the Big Bang, the universe crossed a threshold into the existence, as we know it. With the appearance of life, a second threshold was crossed. We are about to witness, in the next chapter, the crossing of a third threshold. What entity in the universe could 'look' across the threshold to the domain of life? Which living organism can look across the threshold to human thought and consciousness? Strudl and Blue were stuck in their discussion of life, as we humans still are. Sure, we have concluded that knowledge is the essence of biological life, but the same can be said about knowledge and the cosmos. We can try to glimpse into the future and look for possible new renditions or algorithms, but who can predict the nature of another threshold, if one were to emerge? And who could look across such a possible threshold, when we haven't yet figured out the previous ones?"

Thanks for you thoughts, Novare. Your observations about knowledge and the expression of that knowledge are a core ingredient of the perspective that I describe in this book. The millions of living creatures, from the first bacteria to the primates and the great apes, are witnesses to the unfolding and the expansion of knowledge, and the expression of that knowledge through their diverse behavioral repertoires. Let us now turn to look at the evolutionary expansion and the framework of such expansion as I have proposed in chapter 5. After the emergence and evolution of life, we observe an accelerated increase in the radius 'E' in Illustration 27 above.

Evolutionary Framework, Part 2

The crossing of the threshold to the domain of life marked the beginning of dramatic changes in the way nature deals with knowledge and renditions, as we have already observed over the course of four billion years of biological evolution.

First, nature separated the knowledge about a living organism from the rendition of that organism: genotype and phenotype. The genotype, the blueprint for the organism, includes the knowledge to manufacture of the organism and the knowledge for the behavioral repertoire of the organism. Nature rendered this 'design knowledge' in the physical form of ordered sequences of DNA molecules supported by RNA molecules, and various other biological mechanisms embodied in every cell. Together, these constitute the algorithms *and* the biochemical 'computer' that produces the 'finished' organism *and* maintains the organism over the lifetime of its existence.

Second, from the first multicellular organisms (such as the anemones and the jellyfish) onward, nature rendered the internal communication algorithms into a separate mechanism: the nervous system and the brain (as the executive function of the nervous system). Every living organism is a network of cooperating components, where the network itself has its own physical manifestation in the form of the nervous system (including the brain). In the cosmic domain, this network consists of the fields (e.g. gravity) and forces (e.g. nuclear forces) that forge the relationships between individual entities. In the domain of life, the internal communication network evolves its own physical structure and function: nature separates the rendition of the organism itself and the rendition of the communications network, which embodies the algorithms for the organism's behavioral repertoire.

Third, the nervous system and the brain evolved into a *programmable* mechanism. Along their evolutionary path, animals started to learn by the actions they performed, thus forming new and repeatable habits, or algorithms, in their brains. Now, they were able to 'replay' the behavior, by executing the algorithms, acquired genetically over hundreds of millions of years, they could also learn and repeat new strategies. In the domain of life, nature performed its acts of variation on the level of its blueprints stored in the genetic code. Hence, nature created new concepts through variation of existing features of existing concepts. Then, nature constructed organisms by transforming those concepts into real physical organisms, and tested them in the physical environment of the real world.

What is Evolution?

This separation of concept (blueprint for a physical system) and the living creature (the physical system itself) occurred at the threshold of the cosmic domain and the domain of life. The continuity between the two lies in the expansion of knowledge, and the expansion of the physical manifestation of this knowledge. My initial definition of evolution, as given in chapter 5, applies to the continuation of evolution into the domain of life:

> *Evolution is a process of unfolding events that continuously transform knowledge (algorithms), encapsulated by the physical manifestations (renditions) of reality, into novel renditions, in the direction of expansion.*

Throughout its journey, in the cosmic domain and the domain of life, evolution can be viewed as the execution of algorithms that produce physical systems that exhibit emergent algorithms. By 'systems' I mean both, 'tightly' connected physical entities (I called these individual entities), as well as 'loosely' connected individual entities (called environments). In the cosmic domain, the physical forces of the micro-world of the cosmos maintain the coherence of individual entities, while the forces of the macro-world maintain the coherence of environments. In the domain of life, the coherence of environments continues to be maintained by physical forces, while bio-chemical forces maintain the coherence of individual entities. I have to note here that quantum theory blurs (or even eliminates) the traditional boundary between physics and chemistry. The meaning of 'tightly' and 'loosely' connected is, of course, relative to the observer. An observer the size of an atom will view a living organism as loosely connected, whereas an observer the size of a galaxy might view the planet earth as one 'tightly connected entity', all six billion human beings included. Therefore, the distinction has the quality of relativity.

In the domain of life, *renditions* of individual entities (tightly connected) are the millions of different species, and the trillions of variations amongst individual creatures within the populations of those species. Bacteria, mushrooms, oak trees, oranges, wolf, giraffes, cats and human beings are examples of such individual renditions. The internal communication between the parts of an individual organism takes place through bio-chemical processes within the organism. It has a well-defined boundary and interface with the outside world.

A pack of wolves in their hunting territory, the oranges on an orange tree, a community of human beings, and the fish in a lake are examples of renditions of environments. All of the entities, alive or not, in a particular environment from a 'tapestry' of 'loosely' connected individual renditions, and the well-being of each is dependent upon the harmonious interactions between the participating entities. Communication between living creatures takes place through the interfaces of the particular organisms. Eyes, ears and all other sensory organs that make up this interface make use of all of the physical properties of particles and energy of the cosmic domain. The resulting patterns of behavior of living creatures, and the resulting interactions and relationships between them, constitute the *algorithms* of the domain of life. These algorithms harness the knowledge embedded in the tapestry of life. Combined with the algorithms of the cosmic domain, they form the algorithms of the universe.

Illustration 28 symbolizes the expansion of the universe into the domain of life. This expansion does not assume the addition of a new substance to the mix. It does symbolize the acceleration of the expansion of knowledge - the expansion of algorithmic 'power' that life added to the universe. As was the case in the cosmic domain, the two mechanisms that drive the process of evolution are the creation of chaos, and the creation of order. The formation and the maintenance of the order of living organisms are possible only through the background chaos on which the organism feeds. Order and chaos continue to be the two pillars on which the process of evolution rests. Its continuation is dependent on the maintenance of a finely tuned balance - the harmony - between the two.

The chaos present in the systems and the interactions that constitute the environment for living organisms stimulate the creation of new order and the maintenance of existing order. Random mutation of genes, random encounters of mating partners, and the availability of various foods that keep the creatures that participate in an environment alive, are the opportunities for nature to foster communication and innovation. Over the course of its biological history, nature combines and re-combines the components of its genetic knowledge to form innovative entities. No 'new' things appear 'out of the blue'; transformation, rearrangement, aggregation and addition - of existing knowledge and technology - are the names of the evolutionary game. Innovative entities are forged from the chaotic presence of randomly selected algorithmic pieces of knowledge in the region of emergence. Innovation feeds on chaos. That is why I equated chaos to opportunity. The creation of new order is only possible through communication. Without communication, there are no interactions, no innovation, no evolution, and no creation of novel living organisms.

In the domain of life, the tools for communications evolved from the biochemical interactions between organic compounds to the visual, vocal - and so on - interactions made possible by the emergence and refinement of sensory organs. In evolved living organisms, nature invented systems for the 'remote sensing' of the environment by making use of the properties (information, knowledge) of energy and particles that transmit reality (at least a certain interpretation of it) to the living creature. The sensory organs depend on the physical properties that we already know from our discussions of the cosmic domain. Biochemical processes rule the communication within the individual organism, while sensory organs facilitate the communication between organisms. Evolution creates new, individual entities and environments through innovation and maintains their integrity through communication.

The domain of life is the expansion of the cosmic domain; the creation of order and the creation of chaos, while maintaining harmony between the two, is what nature continues to do. Innovation and communication continue to be the means by which the universe evolves, and expands.

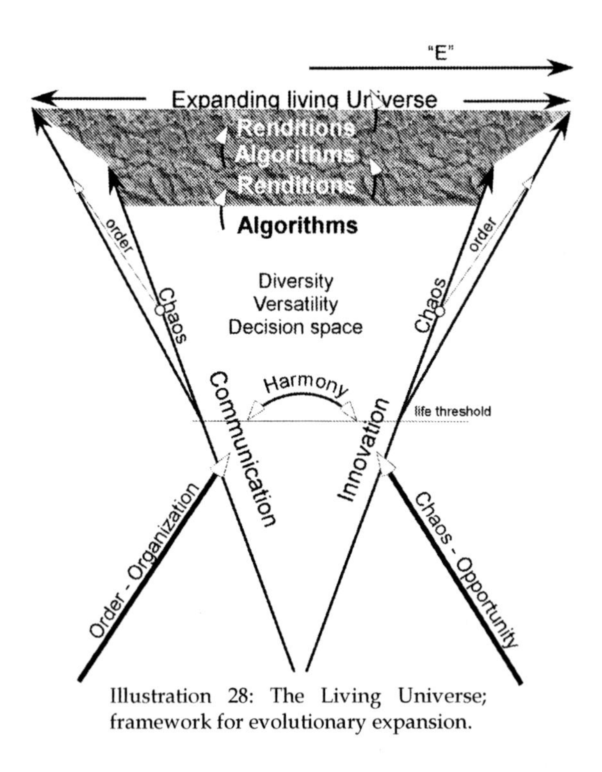

Illustration 28: The Living Universe; framework for evolutionary expansion.

The Expanding Living Universe

In illustration 28, the vector 'E' – again - symbolizes Expansion, and is identical to the radius 'E' of the sphere in illustration 27, and represents the continuous expansion of the vector 'E' in the cosmic domain (illustration 16). The continued evolutionary expansion in the domain of life exhibits the properties of diversity, versatility and decision space; they are the same expansionary properties found in the cosmic domain. Together, they determine the length of the expansionary vector 'E'.

Diversity. Today, most biologists classify living organisms into the five kingdoms of life: the kingdom of the monera (the bacteria, or prokaryotic organisms), the kingdom of the protista (a collection of single-celled eukaryotes and algae), the kingdom of the fungi (some multicellular eukaryotic organisms and single celled yeasts), the kingdom of the plants and the kingdom of the animals.[6] Each kingdom subdivides into several phyla, and each phylum has a number of levels of subdivisions. Today, about 1.8

million different species have been described in the animal kingdom alone. The grand total of animal species is estimated to be at least 5 to 10 million, according to Mayr.[7] The diversity in the world of living organisms is obvious from these numbers alone. If we consider the variations within a species, diversity takes on yet another dimension. I am sure that you'll agree that your chocolate Labrador dog is different from mine, and that other higher animals differ in 'character' or personality. Even flowers and plants, based on the same genetic codes, differ in quality dependent on many properties of their respective environments. If we include the different sets of unique physical features of the 6 billion human beings on our planet, the domain of life features an amazing diversity if we compare it to the tiny size of the terrestrial neighborhood that it occupies in the universe.

Versatility. From the single-celled prokaryotes to the dolphins, and from the worms to the chimpanzees, living creatures increased their versatility in all three of its contributing components, awareness, processing capability and dexterity.

Nature developed a plethora of sensory mechanisms that make organisms aware of the characteristics of the environment they had to operate in. In the domain of life awareness expanded from the mere capacity to 'detect a presence', as I defined it, to the faculties that allowed animals to generate an ever more complete 'picture' of the environment through vision, sound, smell, taste, the measurement of temperature, the sense of gravity that allowed them to maintain equilibrium, just to name a few. From the vastly improved sensory input information, the animal's brain - with its increased processing power - was now able to generate a complete mental rendition of the environment. With this evolutionary development, nature invented something that we usually contribute to modern computers or video - games. With the invention of mental renditions of reality however, nature invented virtual reality. The 'gadgets' that nature developed is the eye with its lenses, retina and optic nerve, or the ear with its drum, cochlea and various other components, and so forth. Nature created a virtual reality in the animal's brain as a means to live its life in the real world. Against the backdrop of this virtual reality, the animal's brain can now execute a behavioral strategy (or behavioral algorithm) and propel the animal into action. Nerve pulses are sent to various organs and the organs are put to work: a golden eagle spreads its wings, or a whale breaks the surface of the ocean. Evolution created organic 'tools', for the animal to manipulate its environment, of increasing functionality, precision and grace. The webs of spiders, the nests of birds, the dams of beavers, the jumps of squirrels, and banana peeling of a monkey are witness to the ever increasing dexterity, physically and mentality, of the animal world.

"JJ?"

Yes, Alvas.

"I must conclude that with humans, mental dexterity sometimes gets ahead of physical dexterity?"

How so?

"I just remember the hurt that you felt the other day when you hit your thumb with that hammer. You *thought* you could do the job, but you couldn't!"

Thanks for reminding me, Alvas.

"You're welcome."

His smile runs from ear to ear.

"As far as information processing goes, you did alright, Mr. J."

You too, Novare?

"I just want to put the focus on the body's performance as a sophisticated information processing system, that's all. Your physical *clumsiness* was too painful as it was. No need to rub it in."

Both of them smile. They agree.

Decision space. Armored with improved versatility, the animal (as well as other living organisms) was able to adapt to differences presented by the environment. Changes in temperature, varying atmospheric conditions, or a changing chemical composition in the seas, could be handled by the creature with increasing ease. The repertoire of strategies and actions available to the animal expanded as its versatility increased: more options for behavior within a certain environment, and more *possible* environments to survive in, entered the realm of its decision space.

Evolutionary expansion manifests itself through increased diversity, increased versatility, and an increased decision-space for individual creatures.

Once again, evolution reached out beyond the domain of the knowable, into he unknown of entropy, or chaos. Once again, nature grasped elements from that chaos and formed new patterns of existence, created novel knowledge, and novel physical entities that harnessed and expressed that knowledge. In doing so, once again, evolution expanded the glass of opportunity in unexpected ways.

Composing Nature

While the universe continued the performance of its ever-expanding ballet to the melodies of its own symphony, a new type of player emerged: the living organism. From the first living bacterium onward, the new players had to play in tune with the symphony of the universe. If new organisms were not able to play in harmony, they would not be able to maintain their own performance and fade away into the background of inorganic chaos:

they would die. With appearance of life, nature separated the composition of the symphony from the performance itself. Strands of DNA contain the information of the composition to be played, while the organism itself became the instrument that played the genetic composition through the algorithms of its own bio-chemical processes.

Then, with the appearance of multi-cellular organisms, nature began to create a separate system for playing the instrument: the nervous system and brain. Composition, instrument and player of the instrument, all embedded in separate organic systems, yet integrated into one coherent living organism, that grows from a zygote to a full grown organism that participates in the performance of the - now living - universe, contributing to the cosmic ballet.

Along an evolutionary history of billions of years, nature as a whole remained to be the composer of the symphony and the choreographer of the ballet. This continued until the versatility of living creatures crossed a certain threshold (whatever that threshold may be, if such a threshold, like 'complexity' exists) and afforded the living creature *options* for its behavior (to vary its tune in within the composition of the symphony). Living animals were now able to 'play' variations on the theme dictated by its genes. Some animals developed enough versatility to communicate these new variations from generation to generation, by the 'languages' of the expressive mechanisms nature had given them. Sights, smells and sounds were among the important means for parents to 'program' the brains of their offspring, to teach them the habits they needed to do well in the environment they were born into. Indeed, animals had become natural programmers, of their own biological 'executive computers' and those of their offspring.

Almost fifteen billion years, or about 99.9 percent of evolution have passed. The Big Bang had started a process of rendering initial information (or 'algorithms', as I called it) into physical entities that we, humans, can perceive and comprehend, thus turning that information into (our) knowledge. As this process continued, new physical entities expanded the existing tapestry that we call our universe. As history progressed, the amount of available information increased, while expanding its physical rendition.

With life, information was rendered into the separate mechanism of genetic 'programming', and the creation of new organisms took place by recombination of the information (or knowledge, or algorithms) embedded in different genetic programs. Sex became the label for the process of creating new information, new knowledge, and new living creatures to execute on that new knowledge. Living creatures translated this new knowledge in new modes of behavior, changing or expanding their existing behavioral repertoire. As an integral part of the expansion of their versatility, nature invented new technologies to support the expanded behavioral repertoire of its creatures. From the tiny flagellas that propelled bacteria forward, in fluid environments to the versatile hands of the primates that could perform the most intricate operations.

At this stage of the evolution, with the appearance of the great apes, both, the knowledge embedded in the genetic algorithms of the organisms, and the technology to deploy that knowledge had advanced to the create the most versatile animals, living in the most varied environment, yet. What would be next, if anything? More sophisticated technology, expanded knowledge, or both?

And so nature composes its music. At each stage, the symphony expands. The orchestra consists of all the players on the planet Earth, and beyond. The animals, the plants, the fungi and the bacteria. The chemicals and organic compounds in the oceans, streams, rivers and ponds. The minerals in rock formations, boiling up from the earth, enriched the growing soil. The winds, rainstorms, lightning, tornados, hurricanes, snow and ice. The freezing, moving and melting of glaciers. The relentless sun and asteroids that affect the planet's surface. All of them are the players, and play their own instruments. If a single one of them stops playing, the symphony, if ever so slightly, changes.

The individual part of some players is hard to change. The sun, for example, is part of the fabric of the time-space continuum of the universe; the modulation of its music, its position in the orchestra is dictated by fields of gravity. The intensity of its music depends on the supply of its own fuel dictating its nuclear reactions. The musical part of our earthly atmosphere is written by its own composition, the temperature and moisture differences around the globe, and the local topography of the terrain it blankets. The populations of living organisms play their own part, written in their genetic code: their contribution to the symphony.

Who is listening to this symphony? The audience is made up of the musicians themselves. They change their music, as they go, not unlike an improvising Jazz orchestra. Some of the players do not follow the rules of change. They might be bound by the sheet music of their design, or the limitations of their instrument. They might play their music in the shadow of another instrument, so that their dissonance stays undetected. Nature has its own unique way of dealing with unharmonious players. It is an astonishing display of democracy. The players are the instruments, and the players are the audience. All of them decide together as one audience.

The audience is called *nature*. Players that produce the dissonant tunes will soon stop playing. They will be unable to find the energy to sustain themselves in the orchestra, in the form of food, in the form of compatible mates, or in the form of lacking interactive abilities. Unable to maintain their own well-being, they will be unable to reproduce, and die as a species.

The usual description for this phenomenon of successful participation of a new species in its environment is 'survival of the fittest'. A better description, one that gets rid of the connotation of 'strongest' might be 'survival of the fitting'. It is not the player or the instrument producing the loudest music that survives, but the player that produces its tunes in harmony with the overall symphony. Harmony is the player's destiny, while the expanding symphony widened the spectrum of allowable variations on the overall theme.

Déjà Vu

"There is something I keep wondering about, Mr. J."

What is that, Novare?

"Nature always builds upon itself, correct? I mean, nature never starts from scratch, but always uses what is already there?"

Correct.

"If nature produces a new and successful mechanism, will that mechanism be kept for all time to come?"

Alvas is annoyed. "Of course Novare. I am living proof of that. I have been regulating blood pressure, heart rate and the production of hormones for millions of years. I controlled the movements of primates from the beginning, and I made wolf-pups learn to play since they could *learn*. All of the algorithms, as JJ calls them, and all of the organic technology was already in place, and kept through the ages. Nature just improved upon them, increased my processing power, and refined the technology. Life was exciting, you see. Over many generations, improvements came along that kept me on my toes constantly. Until you came along, Novare. That is when the trouble began!"

"Things are not that simple, Alvas. Sometimes nature replaces a mechanism with a new technology. Fish provide a good example. They needed gills to extract oxygen from the surrounding water. That technology wouldn't do you much good. When amphibians evolved from the fish, and left the water, they had to get their oxygen from the air in the atmosphere. They had to develop lungs, and discard the gills."

"Even gills are not lost, Novare. When an embryo starts growing during pregnancy, it clearly shows gills at a certain stage. Nature doesn't forget, Novare. It keeps track of its own history. Especially when it comes to old algorithms. In extreme situations that involve fear or anxiety, I have to fall back on old algorithms to react to the situation. In those cases it's *Déjà Vu* to me: I have to fall back on things I learned hundreds of thousands, or even millions of years ago."

"Who is right, Mr. J?"
Both of you are, Novare.

On the one hand, as Steven Gould puts it: "Our human bodies record a sequence billions of years longer than human culture. Nested like painted Russian dolls are features that we share with more and more organisms the deeper we probe."[8] On the other hand, nature sometimes makes drastic changes, such as the change from the gills of fish to the lungs of mammals.

Evolution builds upon itself. And it does so in extremely gradual ways. Indeed, this is process of adaptation can even be observed in ontogeny, the biological development of an individual organism from embryo onward. The early-stage embryos of humans are extremely similar to those of other mammals like dogs, cows, or mice, and even to those of earlier ancestors like reptiles, amphibians and fish. Sometimes characteristics of earlier species emerge at certain stages of the development of the embryo. Later, these features are transformed into the structure-function that is needed by the specific individual under 'construction'. All bird and mammal embryos develop gill slits, just like fish.

In many respects, ontogeny recapitulates ancestral conditions, thus repeating part of the evolution of the species in each one of us. Readers involved in software development might recognize this phenomenon, when older pieces of code are modified to suit new functions. Later these 'old features' might shine through in new versions of the software, sometimes even causing havoc. The 'blue screen of death' of the windows operating system might involve this phenomenon of ancestral features coming to the surface in new environments where they are no longer useful.

The great apes, and human beings, share the mechanism of DNA with every living thing on earth. The mechanisms for coding, replication and engineering life are the same since almost four billion years, from bacteria to human beings. Since a billion years, from the world of the fungi onward, we share the chromosomes and the aerobic (using oxygen) metabolism. We share tissues and the intake and digestion of external food with all species in the animal kingdom, starting 560 million years ago. From the fish (365 million years ago) onward, we share an internal skeleton (in contrast to the external skeleton of the insect world). From the amphibians (320 million years ago) onward, we share four limbs with every other species in the animal world. We share the mechanism for internal fertilization with all animals from the birds and reptiles (310 million years ago) onward. Like all mammals (emerging 65 million years ago), we share hair, while we share fingernails with all the primates (5 million years ago). Our human bodies are a selection, and an aggregation, of features that developed over billions of years, like a set of painted Russian dolls, layer upon layer, each feature added to older ones that already had proven their value.[9]

"Why is this important Mr. J?"

"She still doesn't get it, JJ."

She doesn't get *what*, Alvas?

"Evolution might have replaced and changed technology, but I still have to deal with, and sometimes rely on old algorithms that were developed a long time ago. Sometimes it makes me feel like I deal with an onion, JJ. There are layers and layers of algorithms that manage all of the functions under my control. Sometimes I get confused as to which one to use!"

"I get it, Mr. J. Alvas is talking about the distinction between rational and emotional behavior. I have to admit: sometimes he does not follow my instructions at all, but acts in such strange ways! It makes me wonder where he gets the strategies for such behavior."

Alvas may be right, Novare. Many strategies, or algorithms, for behavioral patterns were developed over the course of evolution as they became vital for the animal to cope with specific environments. As animals expanded their versatility, many such algorithms, or fragments of them, remained to be part of the deeper layers of the organic operating system of the animal. These algorithms might be activated, under certain circumstances, and control the behavior of the animal.

"Finally, JJ! Just look at the world today. Where else does murder, rape, torture, or war, come from? Ignorance, JJ, is to disregard the algorithms that make us human. It is falling back on old algorithms that had good use in the past, but not any longer. That's how I feel, JJ."

"Am I finally getting some recognition here, Alvas?"

"You got that right, Novare."

It seems to me that it is time to finish our journey. Let us see what happened after the appearance of the great apes, when Novare made her first appearance.

Chapter 8: *Act Three, "Creating Knowledge"*

Illustration 29: Rendition of 'The Thinker', sculpture by Auguste Rodin[1]

"The real voyage of discovery does not consist of seeking new landscapes, but in having new eyes."

Marcel Proust

"The artist is the man in any field, scientific or humanistic, who grasps the implications of his actions and of new knowledge in his own time. He is the man of integral awareness."

Marshall McLuhan[2]

Episode 9, "The Awakening of Thought"

The beginning of knowledge,
100,000 years before today,
4 billion years after the delivery of life,
15 billion years after the birth of existence.

Arr's Waves

Arr is alone. He is not supposed to join the others. His duty, as it is on most days, is to watch the camp.

"Time for the water," he thinks.

He makes his way to the river, avoiding the sharp brushes that line his path. No shadows float on the blue water: it mirrors a cloudless sky. He enjoys sitting in the sand of the riverbank, and watch his reflection in the wide, quiet stream that meanders through the valley.

"Arr," he said to the reflection in the water. That's what the others call him. Arr is his name. "Does the reflection have a different name?"

As he wonders about this, the reflection of a *Bo* appears right next to his. The Bo is similar to him, smaller yet stronger, but, to be sure, not as smart. They move in funny ways on arms and legs, although he has to admit, they are quick at moving through the trees. He had to hide fruits from them, but otherwise they were harmless. He turns around abruptly, the Bo shrieks and runs away. He watches the Bo leave and smiles. The Bo's moves always make him smile. Work will come later. Now, he has to create waves, and catch them.

He tosses a stone in the water and marvels at the disturbance of his own reflection. He follows the waves as they spread out, until finally, when they reach the sand, he can catch them.

Arr can feel it: something is different today. He ponders the waves much longer than usual. Until his eyes catch something that never entered his mind before. The waves spread out in *every* direction! He intensely watches the perfect shape of the waves when it happens. His hand is moving deliberately, yet unconsciously, as if moved by magic, through the sand. The feeling in his stomach intensifies.

He looks at the pattern that his hand creates in the sand, and then looks back at the expanding waves. They were the same! At first, he doesn't move. It is absolutely astonishing. He could make waves! How did he do that? But then, his own waves did not move. He had to *make* them move. He gets up quickly, looks around, finds a stick, and starts carving in the sand, more circles, each one around the other, one more, and another one, and yet another one. Exhilarated, he starts running. He has to find the others, and share his discovery. Show them how sand can make waves. Arr grasped the shape of what would become known as *the circle*.

"What's the big deal, JJ? I give 'circle' performances all the time."

I know Alvas. Arr was just the first living creature to ever produce the geometry of a circle in his mind. He was the first biological creature to produce something profoundly new, something that never existed before. Those circles in the sand were the first ones in the history of the universe to be drawn by a living being. Arr's eyes looked at a wave, and his brain created a new rendition of that wave, rather than a memory of the wave itself. Arr created a thought that had its own meaning, its own existence, independent of the origin of the thought. His mind was able to maintain that thought, and store it as an entity that no longer had the identity of the wave. He had created a new piece of knowledge about the world. From the information of the water, Arr's mind developed a rendition, and from that rendition, his mind derived a conceptual entity, a new piece of information. To him, it was a new piece of knowledge. From that knowledge, he created a new rendition: *a circle in the sand.* Arr's brain was able to understand the shape of those waves. He produced a thought that captured the essence of the wave's geometry. Then, not only could he *hold that thought*, he could reproduce it, quite independent of the wave phenomenon itself. He could produce a very *different rendition*, right there in the sand. And Alvas, I have to remind you, it is Novare who played the key role here.

"You are kidding, right? I gave her the images to work with. It is *I* looking for similarities in my memory. It is *I* who puts the new renditions in memory. It is *me* who knows the way around brains!"

"*Me, me, me! Typical* Alvas, *always* seeking recognition. No wonder, people get sick. Clearly, I came up with Arr's circle. That's what I do: come up with *new stuff*."

"That's exactly what makes people sick. If you leave them alone, without your 'new' stuff, they are happy as can be. I can sense it. Change is OK, if it comes from me."

Please guys, we are trying to figure out what happened to Arr. After all, this was the most important moment in the history of thinking.

"It was, Mr. J, because I finally gained my independence, you see. Before Arr's time, Alvas had to handle most things by himself. I was totally dependent on him. Just played a limited role, kind of a small window between him and the outside world."

"Oh boy, did she ever have a limited attention span. Whenever she saw something, I had to react instantaneously. On the spot decision making; no time to contemplate anything."

"With Arr, however, it was different. You see, the story didn't end with the wave, JJ."

It didn't?

"Novare first had to acknowledge that the reflection in the water was Arr's. That was a big thing in itself, although we had done similar things before. Then she had to make him aware of the shape and the dynamics of those circles. Now, she had to get him to recognize those circles independent of the water, lift them out of the water, so to speak. That was huge. First, when she gave them to me, I didn't know what to do with them, but after a few tries, she kept insisting on those darn circles, I got the picture, and put 'm away as a new thing. Of course, I didn't know they were circles. That came much later. Novare and I worked as a real team on that one, I'd say."

"Don't brag, Alvas, you had a hard time with it. I had to try so many times to give you the shape of those curves, it wasn't even funny. However, the circle was the easy part. The project was a first of its kind, because of what came later."

"After Alvas made him draw the circles in the sand, I had to make clear to him, what he just did. When he knew, he got very excited. Arr's mate hated water and avoided it like the plague, so he could never explain the waves to her. After drawing the circles in the sand, however, he could. He tried hard, by drawing circles in the mud and speaking about water, gesturing his heart out. For a while, she thought the water had put a spell on her beloved one, until one night, while sitting in front of the fire, she understood. She asked Arr to draw a bigger circle, a real big one, so that the fire would be right in the middle, and he did. She got all the folks she could find and gathered them, while Arr looked on with growing curiosity. She motioned everyone to take a place right outside the edge of the circle. From this time on, they would always sit around a fire in this way, because they discovered that everybody got the same amount of heat from the fire. She admired her mate for inventing such an ingenious scheme, which the water had taught him."

"That was my idea, JJ. Once I got the picture – of the circle that is -, I could communicate that picture by drawing circles in the sand. It took a while for Arr's mate to understand, but that is Novare's shortcoming. Once she got the picture, she could do her own thing with it."

"Once Alvas had labored for so long enough to 'get the picture', I quickly created a new interpretation of the circle for her."

The importance of his discovery

"What happened after the appearance of the great apes, JJ? Maybe that helps explain Arr's new ability."

"JJ, I am sick and tired of always getting the blame for her shortcomings. She even doesn't understand what I do, all the things I know. After all, the likes of me have been around for so much longer than her. I keep things running around here. She is just icing on the cake. A thin layer at that, if you ask me."

"I'll let that pass, Alvas. I wonder what happened between the tree-climbing apes and the appearance of Homo sapiens. What made my independence possible?"

Out of Africa

We left the story of biological evolution with the appearance of the great apes. That was almost eight million years ago. How did we get from the great apes, like the chimpanzees and the Gorillas, living in trees, with brains of 400 - 500 grams, to Arr, and all of us, with a brain triple that size at around 1350 grams that can discover quantum behavior, DNA, and create space shuttles, symphonies, movies, videogames, poetry and remotely controlled warplanes? To bridge this gap of millions of years, we will follow Ernst Mayr's view of the story, which, according to him, is still full of speculations:

> *"The resulting picture is entirely based on inferences and any part of it may be refuted at any time. But developing a cohesive story is far more instructive than merely compiling a list of unconnected facts."*[3]

Three major ecological changes marked the road of evolution from the Chimpanzee to the human that we know as *Homo sapiens*. This road began 5-8 million years ago in Africa as the *Rainforest Stage*, when chimpanzees spent most of their lives in trees and there was plenty of plant-food. Somehow, a population split off, and moved to the areas of lesser tree density surrounding the rainforests.

Supposedly, the African rainforests at the time were surrounded by Tree Savannas, hence the name *Tree Savanna Stage* for the evolution of the *Australopithecines*, the species that moved away from the rainforests. They were enormously successful and appeared wherever there were tree savannas in Africa. Trees were further apart, and they had to learn bipedal motion, although there were always plenty of trees to escape from predators such as lions, cheetahs or wild dogs.

Fossils are found, indicating several species of Australopithecines, ranging in age from 2.4 - 3.8 million years ago.

Since they still needed their hands for life in the trees, the young born had to hang on to their mothers; the adult females could not carry them in their hands. This meant that the babies could not be born prematurely like humans. They had to be completely ready for the struggle of life, and 'stand on their own feet' from the get go! This implied that the possession of a bigger brain, as humans have, was impossible: the human brain needs to develop after birth, which requires an incomplete skull that can partly collapse during birth. Evolutionary change of the female anatomy would be incompatible with the mechanism required for bipedal life; hence the only solution was a 'collapsible skull' combined with 'premature birth'. The Australopithecines hardly changes during the 1.4 million years of their tree savanna existence. Is the Australopithecus an ape or does it belongs to the genus Homo?

This debate went on for the better part of the 20th century. The Australopithecus has about the same brain size, a little over 400 grams, as the chimpanzee, they did not produce any tools, and although bipedal, did not evolve any other significant changes towards the features characteristic for the Homo species. The debate concluded with a classification as ape. The most important evolutionary stage from animal to human therefore was the development from the ape like Australopithecus to the first species, classified as *Homo*, for the evidence shows, that Arr's ancestors had been tree-climbing, yet bipedal, small-brained, plant-eating Australopithecines.

In a treeless environment, the Australopithecus would be defenseless against predators, and that is exactly what happened when Africa grew more arid during the ice age starting 2.5 million years ago. We do not know what happened during this *Bush Savannas Stage*, but we do know that it was the most important step towards humanity. Somehow, some Australopithecine population(s) managed to survive the environments with fewer trees, maybe using campfires, or primitive stone weapons as defensive tools. At the end of this stage, the primitive bimetallism of the tree-climbing Australopithecus had evolved into the terrestrial bimetallism of *Homo*.

A new species, *Homo erectus*, had appeared, featuring twice the brain-size of their ancestors, around 825 grams, smaller teeth, shorter arms, longer legs, and smaller anatomical differences between the sexes.

Now the story becomes truly confusing, because the evidence is fragmentary, and scientific interpretations vary widely. In short, it goes like this. Home erectus appeared, evidently, in Eastern Africa, but is assumed to exist earlier in Java and China. A different species however, *Homo Rudolfensis* (I wonder about the names, too), appeared 'out of the blue' at the same time, in the same region of Eastern Africa, featuring about 700 - 900 grams of brain-mass, even shorter arms and longer legs, and they used stone-tools.

However this might be, the fossil record shows, that Homo erectus was very successful for the next million years or so. 'Homo' spread out geographically, and evolved towards the characteristics of Homo sapiens, as their brains 'grew' to over 1000 grams, which took incredible anatomical changes, as briefly described. A newborn chimpanzee has the same state of maturation as a 17-month-old human infant, as far as body-mobility and independence are concerned. Now we know why: the price for a better brain is a premature birth that requires nourishment by parents for a longer period. Not a bad deal, really. Our human babies will double their brain size in the first year after birth. It took evolution a few million years to figure this out, without having to widen the birth canal, which would be incompatible with bipedal behavior for beings with an upright posture.

In August of 2002, the fossils of two Neanderthal humans were found, a female and a child, in the Neander valley near Düsseldorf, Germany. It was in the same valley, that the first Neanderthal fossil was found in 1856, hence the name 'Neanderthals', who flourished between 250,000 and 30,000 years ago, when the smarter Homo sapiens replaced them, all over the world. A fossil record showing gradual, slow evolvement toward Homo sapiens accounts for the million or so years between Homo erectus and the Neanderthals. However, it was the population coming out of Africa that won the final battle for dominance. Homo Sapiens, descendents of the 'African version' of *Homo Erectus,* is believed to have come from sub-Saharan Africa, where they originated 150,000 to 200,000 years ago, and became the dominant species in the genus Homo, the so-called hominoids, about 100,000 years ago. They, the ones we call Cro-Magnons, invaded Western Europe about 35,000 years ago. The disappearance of the European Neanderthal took several thousand years, and we don't know how that has been accomplished. The genetic split between the species seems to have occurred 465,000 years ago, according to recent DNA analysis. At about 50,000 to 60,000 years ago, Homo sapiens reached Australia, and later, 30,000 years ago, they reached eastern Asia. Some estimate the arrival of Homo sapiens in America at 12,000 years ago, while there is some evidence for an arrival as early as 50,000 years ago, crossing the landmass (isthmus) that connected the continents at that time.

Arrière's Brush

"Ils ont été ici."

> Éliette Brunel cried out, again, and again, "They were here!" In that instant they began searching all of the walls with great attention. They discovered hundreds of paintings and engravings. It was Sunday, December 18, 1994, when Jean-Marie Chauvet led his two friends, Éliette Brunel and Christian Hillaire, on the Cirque d'Estre toward the cliffs. A faint air current emanating from a small opening at the end of a small cave had attracted his attention... [4]

This anecdote marks the discovery of the cave of Chauvet-Pont-D'arc, at one of the meanders that the Ardèche River has carved out of the Massif Central in the south of France since the tertiary era. In 1995, the works of pre-historic art, drawings of rhinoceroses, lions, bears, and a many other animals, were dated to be approximately 31,000 years old, thereby representing the earliest expressions of art yet. Éliette assumed correctly that the Cro-Magnons, the European version of Homo sapiens had been in the cave and created the paintings. The humans had arrived in the region less than a

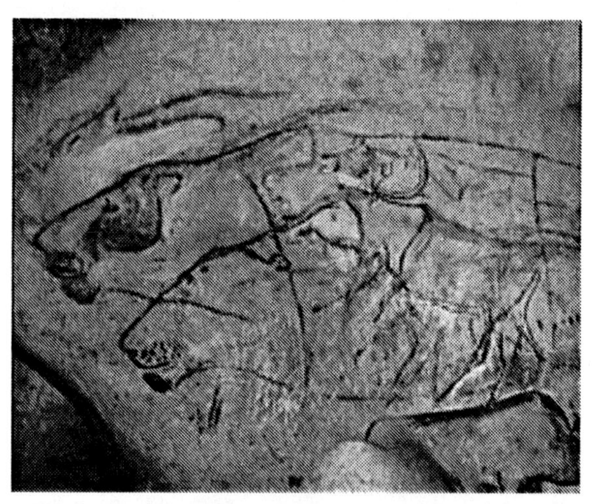

Illustration 30: Painting from the Chauvet Pont D'arc cave, estimated to be 31,000 years old. Three big lions have been drawn on top of earlier figures. The fine engravings in the foreground are cave bear scratches, which shows that the bears came back to the cave after the people had made their drawings. Picture used with permission of French Ministry of Culture and Communication, Regional Directorate for Cultural Affairs - Rhône-Alpes region - Regional department of archaeology.

few thousand years before.

> "The painter of these drawings must have been quite a character, Mr. J. A lonely artist that might be considered strange. I mean, everybody must have been busy just surviving, gathering food, taking care of the young, or going through rituals, all that. The others might have

considered him *backward*. He didn't join the rest and just smeared colored mud on the walls. Probably, the others did not realize, that he was actually ahead of them, being creative I mean."

"I got it! Alvas smiled from ear to ear. The French word for backward is Arrière. Therefore, the painter's name could have been just like that: Arrière. *Arr*ière could be *Arr*'s descendant!"

Nice twist, Alvas. Maybe that is even how the word originated. For all I know, it's as good an explanation as any. Anyway, between Arr and Arrière there seems to be an increase in 'mind-power'. To go from circles to paintings is real progress. Maybe Arrière was considered to be ahead of his time, and not backward at all.

"Alvas is dreaming. The Arr - Arrière connection is exactly the kind of thing he comes up with when I am trying to sleep."

"That's called imagination, Novare, not dreaming. It's what Arr needed to create his circle, and what Arrière needed to create his drawing. Scientists need it before any hypothesis can even be developed, and artists need it before the creation of any work of art. The hypothesis, or the artwork, is the rendition of that imagination. Both need it, and I provide it. If you choose to call that dreaming, fine."

"No painting was ever created, and $E=MC^2$ never discovered, before I came on board. The conclusion is a simple one, in all modesty."

"Nothing new has ever been created out of thin air, Novare. Anything new is created from combinations and interpretations that already exist. Who stores and maintains all the existing information? Who does the combining? Who presents the results to you? The conclusion is a simple one as well, with similar modesty."

Let me ask both of you: could Arr or Arrière have accomplished what they did, if either of you had been absent?

"No, Mr. J."

"No, JJ, although I ..."

Thanks.

Les jeux sont faits: Knowledge was born, creativity was born, imagination was born, and humanity was born.

Evolution crossed the third threshold, and awakened knowledge.

At some point in time, between the appearance of Homo sapiens - 150,000 to 200,000 years ago - and Arrière's cave paintings 31,000 years ago, Arr made his discovery and his mind created an image that that obtained meaning independent of the context in which it was created. A threshold was crossed. A threshold that had to be crossed if thought had to ever become re-productive, and allow us to create new and original manifestations of ideas, that became independent of the mechanism that produced them. At this threshold, we passed through a gate: the gate to the domain of knowledge. It was evolution's third passage through a gate that would change the rules for existence. This one was the gate that would lead to the first conscious human love, and the first conscious human war, two activities that join to form the perimeter of human endeavor.

Independent Thinking

Something new had arrived, something that had never ever happened before in the history of the universe. In the domain of the cosmos, no planet, star or galaxy had managed to do this. In the domain of life, no plant, insect, bird, fish, reptile, or mammal had ever done it. Arr's mind could capture a sensory input, and the next one, and the next one, and on and on. His mind could weave these three-dimensional images together into a complete tapestry of the experience of the waves in the water. The time between each image was determined by the time it took the neurons in the brain to propagate his sensory signals to all the brain regions involved in the process. That was the brains own frequency of operation, determined by its bio-chemical processes. Arr had no influence over that.

He could just stand there and watch, nothing forced or urged him to take any action, like diving into the water to catch a fish, or run away for fear of the Bo, or fear of the water itself. His mind was just creating a thought; a thought that lasted after he turned away from the water. It had been propagated through neurons and axons, until it became finally stored as a four-dimensional information-construct in the organic matter of his brain.

Arr's thought became a four-dimensional information construct that existed independent of the thermodynamic forces of nature, and independent of the biochemical processes that produced it. Thought had gained its own autonomy.

Arr conceptualized the waves, and derived essential properties from it. Then he created a completely new rendition from the information he had derived. He now possessed a new piece of knowledge: circles. From now on, they existed independent of waves, water, or rivers. All that happened in his mind, which is the master of the human brain.

The brain, neurons and synapses.

The brain contains about 100 billion neurons. They are the central processing units of the brain that help determine what we can possibly think. A multitude of incoming signals influences the chemical composition of the neuron. In turn, that chemical composition determines the potential of that neuron to send out a signal to other neurons. The output signal propagates along the axon of each neuron (see figure 31).

The axons vary in length that can be over a meter. At various places along the way, they can split, producing branches that can lead to various places in the brain, or the body, on their way to other neurons. The end of each branch has a little bumper-like structure with a concave "face," called a *synapse* that ends right in front of another bumper-like structure, this time with a convex face, called a bouton (French for "button"). These button-like structures are found on the neuron-body itself, or on *dendrites,* that extend like tiny bushes from the neuron body. Each neuron has an average of 1000 of incoming synaptic connections, giving a total of 100 trillion synapses. There is a tiny gap between each synapse and its bouton, called the *synaptic cleft.* Only *neurotransmitter chemicals,* can bridge this gap, and transmit a message between the now connected neurons. Some neurotransmitters are stimulatory in nature, and will enhance the potential for the receiving neuron to "fire," while some are inhibitory, thus lowering the firing potential of the receiving neuron.

The combination of all messages coming in to a neuron will then add up to a total potential (or probability) for that neuron to initiate the transmission of a signal. The totality of neurons and synapses form an enormous network of processing nodes, and connections between those nodes, called a neural network. From a computer-programming point of view, this seems like a relatively simple (at least in principle) computational scheme to simulate on a computer. Hence the attempts by the artificial Intelligence community to try just that: simulate neural networks to build, in software, intelligent systems, that can "understand" certain things, are self-learning, and can come up with intelligent solutions that cannot be achieved otherwise, just like the human brain. This is the "classical" picture of brain behavior. The assumption underlying this model is that neural processing and synaptic messages follow computational rules that can be modeled on a computer.

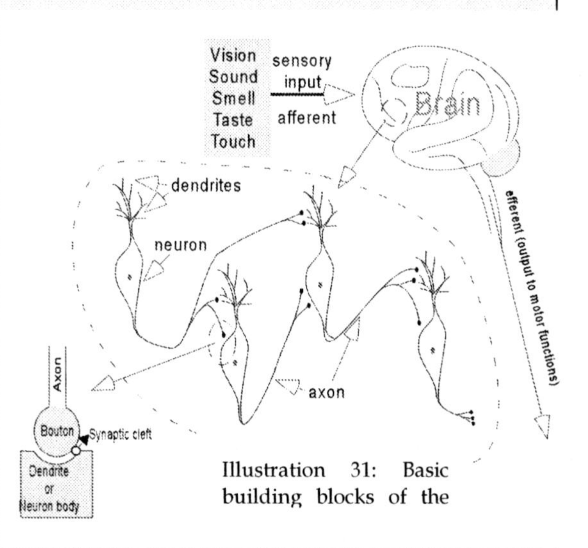

Illustration 31: Basic building blocks of the

The assumption by some AI experts is that, if the complexity of such a network surpasses a certain threshold, it would produce new and unexpected solutions, thus mimicking human imagination or creativity. In this view, only two things would be needed. One is a way to capture the synaptic network of the brain, the second is a computer that is powerful enough (speed and memory), to produce results in a minimum of time. Furthermore, when computers are faster than the relatively slow human brain (it is likely that computers will be faster in the very near future), and have more memory capacity than the human brain (which they will have), then computers would be smarter than humans. If the 'synaptic network' view of the way the brain works is correct, then predictions like the ones made by some scientists, most notably Ray Kurzweil[69], seem to be a fair assessment of what could happen in the future when machines take over from humans and leave them behind to wonder what in the world happened. The view developed in this book goes beyond 'machines taking over', or 'the end of the dominance of the human mind'. My view is a more hopeful, and a more challenging one, because we will have to create the future ourselves. No machine is going to do it for us. *We are that machine.*

The creation of autonomous thought is uniquely human. With it, a new information space has been entered. A space that is filled with all the thoughts ever thought, and with every experience ever experienced in human lives. Thoughts produced from the sensory signals entering our minds from the external environment, while we interact with all of nature and with other human beings, the information produced from the internal environment of our bodily functions, emotions, feelings, and motivations, and the thoughts produced by re-combining and associating all of the thoughts already present, thus creating novel thought products from within the existing information space. Something new has arrived, and we do not know what that 'something' is.

"I need to speak with you, Mr. J. In private, please."

That never ever happened before. What is the matter, Novare?

"I have to make a confession."

I wonder what that could be. Please, go ahead.

"Remember how I did not really like my name *Nowhere*?" Then I said I like it because - to me - it means '*Now - Here*'."

Well, what about it?

"I made that up, the now - here part, I mean."

You mean, that was not the reason you accepted the name?

"Indeed, it wasn't. Now, that the autonomy of human thought is out in the open, I want to give you the real reason. It is simple, really. In Latin, *Nova* is the feminine form of *Novus*, which means 'new', and *Res* means 'thing'. Therefore, I thought of 'Novare' to be 'derived from '*New Thing*', which in fact I am. I liked that a lot, but did not want to upset Alvas. Therefore, I did not mention it. Sorry for keeping that from you."

Apologies accepted, Novare. All of us live with the illusions of our little lies.

"Thanks for keeping Alvas away"

You're welcome.

We commonly call this 'new thing', that we consider exclusively human, consciousness, self-consciousness, or self-awareness. Several serious theories try to explain the phenomenon of consciousness, such as the theories by Penrose, Dennett, Crick, Searle, Edelman, and Damasio, just to name a few who's work I refer to on this book. I will follow the theory of 'biological relativity', as proposed by Richard Pico, because it focuses on the independence of thought as an explanation of the phenomenon of consciousness. Following Richard Pico's theory, I use the 'independence of thought' as the primary property that distinguishes human beings from the rest of nature, because it captures, in my view, the true essence and impact of human thinking.

Central Processing Unit

Pico's model of consciousness rests upon the existence of a Prefrontal Integration Module (PIM) in the human brain. The justification for this hypothetical module in the brain is based on neuro-anatomical, physiological, and behavioral evidence.[5] The function of this PIM marks the threshold for the emergence of exclusively human properties. The PIM is the device that can be viewed as the gate to the domain of human knowledge. Pico explains the term 'PIM' as follows:

"The construction of this term [Prefrontal Integration Module] reflects the neocortical region of interest (prefrontal), indicates the general biophysical computational function (integration), and honors the generally accepted interlaced columnar design of the neocortex (module). Since we cannot fully define the exact dimensions, cellular components, synaptic patterns, informational content, or specific biomathematical operations performed in the PIM, it must ultimately be viewed as a heuristic construct. The PIM, as envisioned, is a repeating structure within each hemisphere of the prefrontal area in which afferent fibers, conveying several informational products, topographically converge. It is a living, structured, multimodal information space. Indeed, it is a location where the other true multimodal representations, constructed in parietal, temporal, and frontal cortices and in the hippocampal system, may be further transformed into even higher-order representations."[6]

PIMs are (hypothetical) cellular constructs in the prefrontal neocortex of the brain, located right behind, and upwards from our eyes. It consists of one or more cylindrical neuronal structures, each maybe several hundred microns in diameter. An "executive" neurological brain component receives the information from internal and external sensory worlds. All the waves of information reach their highest possible form of integration within this module. A PIM contains a complete three-dimensional map of a momentary experience, with all the properties of the 'snapshot' of that moment in time. It may obtain these properties from outside the body, like smell and taste, or from within, like blood pressure or feelings. This map can be a momentary merger of an immediate experience and a memorized one, from the recent or distant past.

Many, if not most of the functional regions of the brain and the nervous system may be involved in the pre-processing and transmission of information. The PIM is the last stage where all information converges and blends into one snapshot of the world. Pico suggests a cycle time of 200 milliseconds (based on brain experiments) for the development of one complete snapshot, a process that takes place in every mammalian brain. Computations in the PIM result in an efferent activity: an output signal that propagates through the axons to other parts of the brain that might, in turn, evoke signals to certain parts of the body. As a result, a horse might stamp its hoof, a bird might start singing, or a parrot might say "hello."

Crossing the Threshold

The internal and external behavior in animals is determined below the level of the assumed PIMs. algorithms formed, and changed, in the brain and the nervous system together with the organic technology of the animal as it became available in the hundreds of millions of years of biological evolution. In mammals however, the emergence of PIMs might be responsible for increasingly more sophisticated algorithms, resulting in more complex behavioral patterns, because they added a layer of decision-making. Information from various sources is now subjected to an additional level of processing. Information that is more diverse is further blended, compared and weighted, thus producing an inhibiting or stimulating influence on behavior. This influence is exercised at the biochemically determined frequency in the millisecond range of the brain's cycle time.

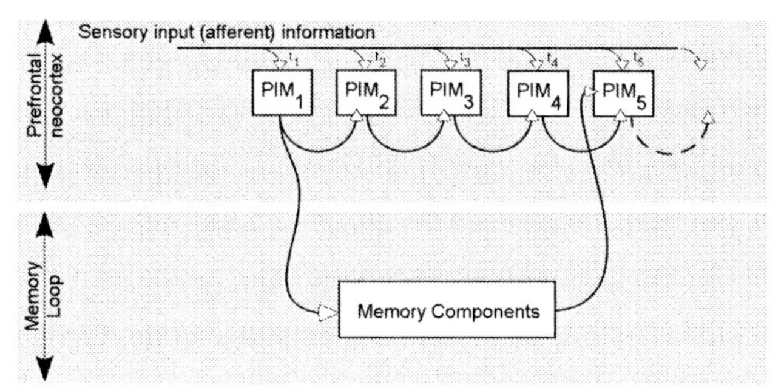

Illustration 32: Structure of the Prefrontal Integration Modules according to Richard Pico[7].

Pico imagines the existence of a series of these PIMs, as pictured in illustration 32. In every mammal brain, including the human brain, the output from each PIM is relayed to many brain regions, including the next PIM. Each PIM in the series might have slightly different characteristics, such as an emphasis on specific sensory inputs, e.g. visual, or slight variations in their computational functionality. Each PIM receives its input signals from a multitude of sources, computes fresh output signals from the incoming information and sends them on their way to connected neurons in the brain.

In Pico's model, at each moment in time, one particular PIM is the *dominant module* that receives the primary afferent signals from the sensory system. Therefore, this dominant PIM represents the *now* in real world background time. At the same time, this dominant PIM receives the output signal wave from the previous PIM, representing the previous moment in time (i.e. 200ms earlier). While this is going on, the current, or a previous afferent input might have triggered (initiated an association with) a specific memory from the recent or historic past. Seeing a particular bird feeder in a particular tree might trigger the memory of a lurking cat in the bird's brain. All of these information waves reach the *dominant PIM representing the now* for processing.

We are now standing right in front of the threshold that separates the animal's mental world and the domain of human thought, the gate that links the domain of life and the domain of knowledge. Where is this boundary? What small step makes the difference that changed the face of the earth?

This is the moment of *confusion*. As it turns out, it also a moment of *fusion*: the fusion of sense impressions (from the outside world and the inside world of the functioning organism) or memory constructs (stored sense impressions of past moments) *in time*. This is a new 'kind' of time, created by the brain's own operational rhythm.

Monkeys at the Gate

It is a fascinating experience: the woman on the stage, with her monkey called Rufus. Rufus is smart, very smart. Every time his master claps her hands, Rufus claps too. She can also tell him to clap, and he will. He can give kisses, she can make him stand on his arms and forehead, and he can even give the finger in his nastier moments. And he can count. When she speaks any number, between one and five, he will repeat it by clapping his hands as many times. He can even add. When she says: "2+3," he will clap five times. Many months, even years, of loving, caring training went by to teach Rufus these routines. Can Rufus think? Does he have human-like intelligence? The answer is no, and no.

Rufus has an overall healthy brain, a good memory, and a good set of functioning PIMs. Every gesture she makes, every word she speaks, is the specific informational constructs that Rufus' eyes and ears carry to his brain. From there they proceed to relay stations in the brain. An association is made to his memory, his perceptional information and his memory constructs are related to his dominant PIM, and an output is created to his body to initiate the reaction he is supposed to give.

This presentation of what actually happens is an over-simplification of the process that goes on in the billions of synaptic connections in the brain. We have to keep in mind, that the mechanism of PIMs is a hypothetical one. The important point is that Rufus is not thinking the way humans do. He does not have a clue, how much 2+3 is, he doesn't know what his clapping means, and he doesn't have a notion of "4" as a quantity of something. You might be able to teach him to pick out four people from the audience, and then clap his hands four times. Whatever he learns will totally depend on an immediate sensory input, and an association with a memory to retrieve.

If anything in the immediately perceived environment, externally (e.g. a known word spoken by somebody else), or internally (hunger, a memory of her teacher clapping), resembles closely the stimulus used to teach him (and to get his reward), Rufus might exhibit the learned behavior. In the absence of such a context, Rufus will be his own self: a monkey called Rufus. Close to 99% of the genetic code of the great apes, the chimpanzee, the gorilla, and the orangutan are identical to the human genome. Yet, over millions of years of evolution, and after hundreds of years of training by humans, they remained unchanged. Never has one been observed to draw circles in the sand, because of an association with the waves created by a stone thrown into a pond.

Rufus proceeds through his life from moment to moment, directed by the information that his brain derives from his external and internal sense organs, including his memory, at any given point in his life. The mental reality, created by his brain, guides his actions in the real world, just as our own mental reality guides our own actions.

Rufus however, has no way to share his mental reality with other living creatures, other than through his behavior. His short mental attention span prevents him to form mental renditions that lead a life of their own, independent of the moment-to-moment functioning of his bio-chemical apparatus. His brain has not crossed that threshold of functionality that would allow him to create more elaborate thoughts, 'look' at them and express them by other means than mere physical behavior. Rufus is unable to fuse different snapshots together to form an informational construct that is independent from the chains of immediate bio-chemical action. Rufus has to execute algorithms that are triggered by the information content of one PIM, because in the next execution cycle of his brain, this information might be lost, or it might have changed. His brain does not have the faculty to weave together a four-dimensional scene from the snapshots (with a maximum of three dimensions) that his senses deliver: his 'window' on the world around him, and the memory-world within his brain, is a limited one. His brain is not able to lift the information above the immediate bio-chemical processes that determine his behavior. Like a dolphin leaping out of the water, but forced to return to its natural environment of the water, the informational constructs in Rufus' brain are bound to return to the bio-chemical environment of his physical brain. Higher animals, like Rufus, might exhibit 'leaps' of thought, like the graceful jumps of dolphins and whales, but they are mere attempts, ultimately unsuccessful, to cross a threshold: the threshold to the independence of thought; the threshold to the domain of knowledge.

Minds that Matter

The One Percent Difference

What did that one percent difference in the genetic code, between Rufus and a human being, accomplish? In Pico's model, when the efferent information wave leaves a PIM on its way to the next PIM, it contains information about the 'snapshot' generated by the sending PIM. In the animal brain, the information content of this snapshot decays on its way to the next PIM. The resolution of the snapshot might deteriorate or the signal strength might weaken: information gets lost. Whatever the character of that information loss, the snapshot is lost to the next PIM, unless the information content exceeds a critical threshold level. This is, according to Pico, where we find the essential difference between the animal and the human brain. If the order contained in the efferent information exceeds that critical threshold, the receiving, now dominant PIM, can fuse the 'now' snapshot with the most recent one to a sequence in time. The time that links both snapshots together into a four-dimensional scene, is the brain's own internal biochemical cycle time of 200 milliseconds. In Pico's words:

> *"The integrated maximal 4D [four-dimensional] sensory representation created within a dominant PIM functioning above the critical flicker/fusion threshold at each millisecond computational moment is a thought. Each operational cycle of these PIMs thus produce a thought, and this thought is encoded in one or more efferent output projections. The individual thought product thus contains a novel 3D informational structure and the inseparable, internally defined temporal dimension."*[8]

The ability to connect snapshots to form a scene makes all the difference. In the mammalian brain, the snapshots in different PIMs do not connect to form a scene, because the information contained in the efferent signal wave is below the threshold necessary for the dominant PIM to successfully detect its informational content, and to fuse the snapshot to the current one with the 'glue' of the brain's internal time. Mammals might commit individual snapshots to memory, but they do not form coherent information constructs, tied together by the brain's temporal rhythm. Like the separate musical notes scribbled on a piece of paper, they don't add up to a tune that transcends the notation on the paper. The mammalian brain resembles modern day computers that can only execute one instruction at a time. Like animals, computers do not have an overview of their own algorithmic behavior. Although software can perform the most elaborate calculations, and produce results of mind-bending complexity, it never has a 'thought' that encompasses more than the content of its own CPU (Central Processing Unit) at a given point in time.

In his book *The Executive Brain*, neuro-psychologist Elkhonon Goldberg explores the distinctively human and complex functionality of the frontal lobes in the human brain as the latest evolutionary accomplishment achieved by nature:

> *"The frontal lobes are the latest achievement in the evolution of the nervous system; it is only in human beings (and great apes, to some extent) that they reach so great a development. They were also, by a curious parallel, the last parts of the brain to be recognized as important ... they were called the 'silent lobes'. 'Silent' they might be, for they lack the easily identifiable functions of the more primitive parts of the cerebral cortex, the sensory and motor areas, for example, but they are overwhelmingly important. They are crucial for all higher-order purposeful behavior – identifying the objective, project-*

ing the goal, forging plans to reach it, organizing the means by which such plans can be carried out, monitoring and judging the consequences to see that all is accomplished as intended."[9]

Both scientists, Goldberg and Pico, from different perspectives point at prefrontal lobes as the pivotal mechanism in the human brain that produces thoughts and consequently, behavior that is exclusively human. For the discussion here, it is important to note that there is such a mechanism. Like the emergence of life as an emergent property of the complex interactions of complex chemical compounds, independent thought emerges from the interactions of biological structures.

Thoughts that exhibit such independence, I define as knowledge. Hence, I define knowledge as an emergent property of the human mind. In this sense, knowledge is the exclusive domain of the human mind. As an emergent property of an individual human mind, this knowledge is subjective.

Whatever the ultimate explanation might be for the algorithmic capability of the frontal lobes (in cooperation with other parts of the brain) in human beings, it is such a capability that produces human thoughts that gain a measure of independence from the biochemical processes that create them. We are able to create informational constructs (subjective knowledge) that have a 'life of their own'. These informational constructs take the form of what we defined in an earlier chapter as mental renditions. These can take the form of words, images and sounds; but also the form of feelings and emotions. In short, all of the products of the mind and in the mind as we, as humans, intuitively know we are capable of. We can store them, recall them, reflect on them, and modify them. We can act on them, or not act on them. They are the background against which we make our plans, pass judgment, or establish objectives. We use them to create strategies and define tactics. They are autonomous entities. We can have thoughts without invoking immediate bodily actions (at least, most of the time), and finally, we can externalize our thoughts using various means of expression including speaking, singing, painting, sculpting, mathematics, or using the languages of philosophy, politics, or the stock market.

This new phenomenon emerged within the variation of just one percent of the genetic code! Once again, similar to crossing the threshold to the life domain, nature added a new dimension to the expansion of knowledge, and to the potential for rendering that knowledge into novel physical systems.

The new four-dimensional information constructs that we call thoughts, accumulate and interact to form a new, four-dimensional information space that I call the domain of knowledge. This domain is populated by all of the thoughts that humans produce and by all of the physical renditions that are expressions of such thoughts. Let me make a note on one special class of renditions of human thought. I am referring to the class of all the computer algorithms that exist in the world today. Computer programs, no matter how powerful, are renditions of human thought. Sure enough, when we execute such algorithms, they can produce their own renditions of the imbedded knowledge. These are the numbers, the images, the graphs and or the sounds that computer screens, printers and speakers communicate to us. We can use these renditions to advance or modify our own knowledge. Our human mind does not need to know the details of the underlying computer algorithms to understand the renditions produced by the algorithm, any more than we need to understand the internal workings of an automobile to drive one.

What computers can do today may be the most powerful proof yet, of the independence of human thought. The artificial intelligence community strives to equip computers with creative faculties similar to the ones that lift human beings into the domain of knowledge. Before we do, in my opinion, we will have to figure out first, how to give them *independence of thought.*

"JJ?"

I bet you want to ask, "What's new, with the appearance of the human mind?"

"You'll be surprised, but that is not what I'm up to. It has to do with something that happened earlier. I couldn't help but overhearing your little conversation with Novare."

You did?

"Sure. She cannot communicate without me, you see, whereas I can do things without her. Thought you knew that. It is not a question that I have, but a confession of my own."

Go ahead, Alvas.

"I have to admit that she is right."

In what?

"She added a new level of excitement to my life, see. With her involvement, there are choices to be made. Without her, the only thing I can do is relay messages, from input to output, so to speak. Now, she creates many different options for actions that are independent of the momentary state of the world outside. At any time she is present, she chooses which option I put into action. Even more fascinating is her ability to add new information to my repository. Gives me a completely new world to consider. She gives me things I didn't even know existed. For example, she comes up with new ways for me to decide on my own. Sometimes she calls it *intention*, sometimes she calls it *motivation*. Although I don't get the difference quite yet, but I get the 'drift'. She has this thing called *values*. It is a new tool for me to judge situations. Gives me a feeling of what I am *supposed* to do."

That is an interesting confession, Alvas.

"Just want to let you know that I do recognize her contribution."

Novare will be happy to hear that.

"She knows. No need to overdo it. That would just make her a bigger pain in the butt than she already is. You see, I'm the oldest part of being human. Sometimes I don't want her around at all. Just to'let the lid' off. Hard to do with her being around."

You better let me be the judge of that, Alvas.

"Sure, JJ. But I do have some 'values' of my own, you see. I do have my own consciousness, and my own conscience. If I do something that is against my own nature, I get sick, you see. Or better, you get sick, JJ. That is because I don't like the coercion Novare sometimes tries on me. There needs to be a give and take between her and myself; that's all I'm saying."

Interesting, Alvas. I'll have to think about that.

Consciousness

We may position the appearance of Homo sapiens, and hence our new acquaintance Arr at a mere 100,000 years ago. A tiny sliver in the evolutionary performance of nature; hardly any longer than the initial Big Bang that catapulted our universe into existence.

Scientists have since long agreed, that the identifying difference with our immediate ancestors, the Chimpanzee, lies in the brain. After the appearance of Homo sapiens, the brain didn't change a single bit, at least as far as we know. Before that time, over a period of several million years, the size of the brain had tripled. The remarkable difference between Homo sapiens and all other human-like creatures is the strict terrestrial life of Sapiens. That terrestrial life set the stage for the dramatic reconfiguration of the anatomy of the human body, which in turn made the growth of the brain possible.

For a long time, the use of tools has been viewed as the salient difference between Sapiens and its ancestors. That theory has now been abandoned, since even chimpanzees use tools in some simple ways. Today it is common to use a property that we describe as consciousness, self-consciousness, or (self-) awareness as the distinguishing feature of human beings.

We all know what consciousness is, at least intuitively. We know when we are, and we *don't* know when we are not. John Searle best expresses this intuitive description of consciousness in the opening pages of his book *The Mystery of Consciousness*[10]*:*

"One issue can be dealt with swiftly. There is a problem that is supposed to be difficult but does not seem very serious to me, and that is the problem of defining 'consciousness'. It is supposed to be frightfully difficult to define

the term. But if we distinguish between analytic definitions, which aim to analyze the underlying essence of a phenomenon, and commonsense definitions, which just identify what we are talking about, it does not seem to me at all difficult to give a commonsense definition of the term: 'consciousness' refers to those states of sentience and awareness that typically begin when we awake from a dreamless sleep and continue until we go to sleep again, or fall into a coma or die or otherwise become 'unconscious'."

This definition of consciousness offers a commonsense description of the phenomenon of consciousness, but it does not explain the mechanism at work. With the introduction of Searle's definition, I am not entering the discussion about consciousness, or about the fascinating issue of the 'body-mind' duality. However, the reader might wonder why I avoided the use of 'consciousness' and introduced 'the independence of thought' instead. Hence my own perspective, so that the reader knows where I'm coming from, when I use the words conscious or consciousness in the coming chapters. This means, of course, I am getting involved in the discussion anyway.

Pico sees the PIM as the basic mechanism that invokes consciousness, when he says:

"The PIM may ... be viewed as the basic structural unit, the functional cell, of consciousness. And consciousness, as stated, is the emergent property, the irreducible frame of reference, defined by the creation-propagation sequence of 4D thoughts within a field of integration modules, independent of the momentary temporal flux of background neural activity. The rate of thought generation sets the serial pulse of consciousness, and the unique representational content of the thought function defines the direction for the arrow of consciousness time through propagated 4D information space."[11]

I do not follow Pico's suggestions on consciousness. Rather, as described above, I suggest the emergence of knowledge, defined as independent thought, as the exclusively human property, and not consciousness. I have several reasons for doing this:

First, Pico's model equates thought to consciousness as the two sides of one coin. He sees thought as an exclusive product of the human mind, and consciousness as an emergent property of such products, whereas I emphasize the independence of thought as the defining property. The difference is that my perspective leaves the *possibility* open, for thought, as well as consciousness, to be *not* exclusive to human beings.

Second, there are many potentially interesting alternative theories about the human brain, like Roger Penrose's theory of microtubules.[PEN1] Each neuron contains, and every cell for that matter, contains about a million tiny structures, called microtubules. These structures might, through quantum effects (according to Penrose's theory) or otherwise, play an important role in the functioning of the brain, including the production of thoughts and consciousness. I simply can't help but thinking that many different mechanisms and the interactions between them are at the source for what we call consciousness. In my view, the independence of thought might be a result of consciousness exceeding a certain threshold. Alternatively, both, the independence of thought as well as consciousness might be co-phenomena that cross a threshold of 'intensity' together. Or both of them might be indeed the two sides of a coin, and we still have to discover the coin.

Third, following nature's way of creating dramatic changes by the *addition* of small variations, I believe that consciousness is an emergent property of the whole system that evolution put together, including that particular 'small' last change, included in the one percent change of the genetic code, between the great apes and the human. In this respect, I prefer Gerald M. Edelman's premise that sees consciousness as a process, in which many brain - and other nervous system - components cooperate, and in which Pico's PIM's might represent the necessary last organic change that puts humans ahead of their animal ancestors:

> *"As we shall see in a number of cases, it is likely that the workings of each structure of the brain may contribute to consciousness, but it is a mistake to expect that pinpointing particular neurons will, in itself, explain why their activity does or does not contribute to conscious experiences. Such an expectation is a prime example of a category error, in the specific sense of ascribing to things properties they cannot have."*[12]

For the purpose of this book, and to avoid a profound mental pretzel, we return to Searle's definition. Although his definition is circular, and explains very little, it does point to something, we all 'feel' about consciousness: consciousness is something like 'I know what I'm doing', or 'I know what I'm thinking'. Consciousness, in this intuitive sense, is necessary for knowledge (independent thought) to manifest itself in the physical systems that humans create, and that harness that human knowledge. In the coming chapters, we are especially interested in the physical manifestations, the external renditions, of human knowledge, as well as the feedback of those external renditions on human knowledge.

Just as life, being an emergent property of biological organisms, still remains a mystery, so does knowledge as an emergent property of the human mind remains an unsolved puzzle. A puzzle, that might require us to invent a new type of puzzle altogether, before it can be subjected to finding a solution. With Arr and Arrière, evolution crossed another threshold, adding a mystery to those of the first two thresholds: the threshold that nature crossed at the moment of the Big Bang, the threshold to life and now, the threshold to knowledge.

Before we gently push this last mystery into the scientific minds of the future, I want to develop a simple metaphor for the human mind, so that we can discuss the renditions of the human mind, that are changing our world with every passing moment, at the speed of light.

The Business of the Mind

Imagine a landscape. Your vantage point is the highest mountain on the planet earth. Or even better, you're at the controls of our Einstein simulator that helped us out before. Now, you can take a vantage point at any point above, or on the planet earth. You can slow down or speed up the rotation of the earth, you can zoom in or out, pan and, yes, you can even penetrate structures such as buildings, water-dams or the bodies of trees or animals. From your vantage point, you can see, hear, smell, taste and touch everything your mind can imagine. There are mountains, oceans, and rivers. There are forests, orchards, flowers, trees, valleys, trails, canyons, rocks, meadows, streams, rivers and ponds. There are highways, bridges, tunnels, caves, crossings, streets, and alleys. Cities, skyscrapers, airports, factories, offices and homes. Elephants, cats and dogs, snakes, spiders, butterflies and horses. Your friends are there as well: those from the past and those from the present.

But that is not all. There are symphonies and country songs in this landscape. There is fear, surprise, pain and joy, horror and passion and this landscape. Dancing, kissing, loving, laughing and suffering: all of them have their place in your landscape. Friends and foes alike, dead or alive: they are there. The simulator not only allows you to change a vantage point in space, but also in time. You find yourself in this landscape, as a child, an adolescent, an adult, all of your jobs, and the places you ever lived, or visited for a vacation or a business trip (or both). Your family doctor, your dentist, they are all there. The stars and the planets, the mathematical formulas you learned in school, history as you know it, it's all right there, in this landscape. Your first painting and the pride you felt, the embarrassment following the discovery of your first lie, it's all part of the landscape. Sometimes you find it hard to distinguish certain features; sometimes you are surprised by the presence of unexpected landmarks.

This landscape is not a heaven, where you rediscover all the good things you have experiences, and it is not a hell, where you are doomed to meet the bad. *That landscape is our own mind.* I will call it the mindscape. It is the place where we record the story of our life.

When we wander and wonder in this mindscape, our brain reproduces our stories as mental renditions that we call memories. Every experience we ever had is put into the repository of this mindscape, and every experience might influence the shape of the mindscape. In turn, our mindscape influences the position, the shape, and the texture of every new experience that enters the mind. As such, our mindscape is the roadmap to process the present. And since every present experience is a tiny step into the future, it influences our future as well. It is part of our 'survival toolkit' as we venture into the unknown territory of the future.

We can travel to any vantage point in this landscape. Unlike the Einstein simulator however, where we have full control of the navigation, our mindscape determines our mode of travel. Gravitational fields determine the motion of every object in the universe, just as an environment (or ecology) determines the modes of behavior of a creature that lives in it. In a similar fashion, the mental 'topography' of the mindscape determines our mode of transportation and the pathways we take to process a new experience. Some vantage points are easy to get to, some are impossible to reach. Our mindscape is like a mental four-dimensional tapestry that weaves its own patterns, chooses its own colors and textures, and forms its own landmarks and vantage points. The perspectives that each mind is capable of developing are dependent on its own experiential history. The ways we travel through our mindscape is the *way* we think. In turn, the way we think, determines the perspectives we are able to develop, or in other words, influences the result of our thinking process. In my terminology of renditions and algorithms, the pathways we use to produce a thought, are the algorithms of our mind. The totality of our mental renditions and our mental algorithms constitute our *personal knowledge domain*. Six billion personal knowledge domains and all the various ways that these personal knowledge domains interact, from the knowledge domain that Arr crossed into, when he produced an independent thought product that we came to call a circle. The configuration of our mindscape, at a given point in our life, determines what we think, and the way we think it.

Our life experiences, however, are not the only events that influence the configuration of the mindscape. Before we ever start to think, our genetic code has established some of the basic 'features' of each individual mindscape. These basic features might contribute to what we call the 'personality traits' of an individual human being (or, for that matter, many 'higher' animals). This assumes that our mindscape is not 'empty' when we are born. Our mind is not an empty canvas on which life can paint any pattern imaginable. As Steven Pinker explains in *The Blank Slate*:

"I think we have reason to believe that the mind is equipped with a battery of emotions, drives, and faculties for reasoning and communicating, and that

they have a common logic across cultures, are difficult to erase or redesign from scratch, were shaped by natural selection acting over the course of human evolution, and owe some of their basic design (and some of their variation) to information in the genome. This general picture is meant to embrace a variety of theories, present and future, and a range of foreseeable scientific discoveries."[13]

Pinker develops the hypothesis that the human mind is not a blank slate, but that it comes 'preconfigured' into the world. Watching the performance of evolution, readers might find the presumption of such pre-configuration an obvious proposition that is compatible with the concept of the evolutionary 'onion' that manifests itself in every living creature. Many think otherwise, however. The implications of their 'blank slate' view are far-reaching, as Pinker suggests:

"The Blank Slate was an attractive vision. It promised to make racism, sexism, and class prejudice factually untenable. It appeared to be a bulwark against the kind of thinking that led to ethnic genocide. It aimed to prevent people from slipping into a premature fatalism about preventable social ills. It put a spotlight on the treatment of children, indigenous peoples, and the underclass. The Blank Slate thus became part of a secular faith and appeared to constitute the common decency of our age. But the Blank Slate had, and has, a dark side. The vacuum that it posited in human nature was eagerly filled by totalitarian regimes, and it did nothing to prevent their genocides. It perverts education, childrearing, and the arts into forms of social engineering. It torments mothers who work outside the home and parents whose children did not turn as they would have liked. It threatens to outlaw biomedical research that could alleviate human suffering. Its collary, the Noble Savage, invites contempt for the principles of democracy and of 'a government of laws and not of men. It blinds us to our cognitive and moral shortcomings. And in matters of policy it has elevated sappy dogmas above the search for workable solutions"[14]

I use this lengthy quote, because it is an eloquent summary of the pros and cons of both assumptions: the brain as a blank slate, or the brain as an elastic, or at least plastic, and pliable mechanism, that has, however, a basic structure of some sort.

"How can anybody think, that I am a blank slate, JJ?
You are not, Alvas?
"Of course not. How should I take care of your mind during its development?"

"He is right, Mr. J. I emerge between 12 and 24 months after birth, you see. Sorry, but that's the time it takes for the brain to make up its mind, about me, I mean."

"There you go, as if her late appearance is my fault. Anyway, in those first few months I have to survive with what I've got. However, the moment I start working, I start changing my mind, until Novare shows up and spoils the fun."

The pre-configured structure-function, or algorithm, is part of the organic operating system of the human condition, while the 'blank' portion allows for the creation of a vast variety of 'applications' whose formation is influenced by the underlying algorithms of the operating system. In this view, both, research that can lead to cures of mental deficiencies, as well as forms of education that maintain or enhance the plasticity of the mind, should be at the top of any community's agenda.

According to Pinker, half of the human genome is primarily used to shape the brain. We might remind of ourselves, that this genome is 98% identical to Rufus' genome. Yet, it seems to me, that the 'programmable portion' of the human brain is at least somewhat larger than the brain of Rufus. Let us look at this self-organizing ability of the brain in more detail. To do this, I will use an old-fashioned tool as a hi-tech model to demonstrate what goes on, when we think.

Illustration 33: 'Pinart' toy, consisting of an array of freely moveable pins. An object, like the fish in both pictures above, is pushed against (from below) the pins, an impressions is made that becomes visible topside. To make the pictures above, two identical fish were used, one to create the impression and one on top to show the resemblance with the imprint made.

Do you remember the old-fashioned 'pinhead', like the one in illustrations 33 and 34? Today it is available under such names as 'pin-art' or 'imprint-art'. We used to imprint our hand or face on it by pressing against all the moveable pins. The word 'pinhead' is also used to describe a 'very dull and stupid person', as Merriam-Webster's unabridged dictionary describes it. This definition appropriately describes the primitive nature of the toy-pinhead in illustration 33 as a model for the mindscape of our mind. However, it does 'provide' us with a useful tool to imagine what goes on in our minds. Imagine the *mindscape* to be such a pinhead. In the toy-pinhead, the pins can move freely up and down, thus allowing the 'artist' to create an impression by pushing an object into the pins, and erase the impression by simply shaking the device.

I need to make a few modifications to our pinhead-mindscape, however, and make it a 'hi-tech' one. First, it consists of a hundred billion pins, equal to the amount of neurons in the brain. That way, any imprint we make will leave a detailed, truly 'high-resolution' image. The hand in illustration 34, for example, would show detail of photographic quality. Next, our pinhead shows a few other remarkable differences.

Each pin is a tiny hydraulic piston that can move up and down in a cylinder. The pressure of the oil within the cylinder regulates the resistance of the pin against movement. The higher the resistance, the more pressure is required to make an imprint. In this way, high resistance in a local group of pins could influence the shape of the overall imprint. This, however, is not the only change to our hi-tech pinhead.

A microcomputer housed inside each pin drives a little pump, also housed inside the pin that regulates the amount of oil in the cylinder, giving the piston a variable resistance against movement. If there is not much oil in the cylinder, the pin moves easily; if the cylinder is completely filled with

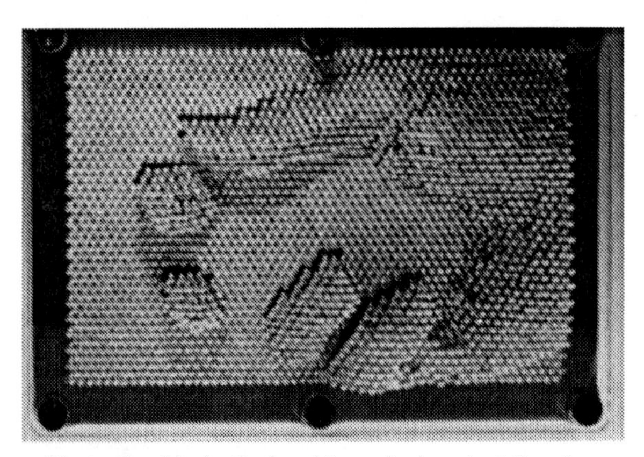

Illustration 34: As the hand is pushed against the pins (from the back), the pins move, and 'topography' of the hand appears, and is 'stored' in the pinhead.

oil, greater force is required to depress that particular pin. The pins are like a tiny version of the shock absorbers in our cars, but with a difference! The little pumps can also move the pins by themselves! I am sure that the design of such a device is a snap for human ingenuity, so we can safely assume that our pinhead can be designed this way.

"What tells the tiny computer when to change the resistance, Mr. J?"

Remember the synapses? They are the 'wires' that connect the neurons in the brain. Similarly, each pin is connected to thousands, or even millions, of other pins by tiny wires. The computers in each pin send out signals over those wires. These signals are received by the computers of receiving pins. Receiving computers will translate those signals into instructions for the pumps to pump more oil in or out of the cylinders.

"Nifty, JJ. But something is still missing."

What's that, Alvas?

"What tells the computer to send a signal in the first place?"

"Alvas, that's pretty obvious to me. When an imprint is being made, each pin senses the pressure that is put upon it. The information about that pressure is received by the computer and sent to connected pins. The computer knows what information it has to send to each connected pin."

"So, when I depress a particular pin, some others might change their resistance?"

"Correct."

"Or other pins might even move up or down, depending on the pressure put on a particular pin?"

"Exactly."

"You see, sometimes I'm amazed myself by what comes up while doing my work. Now I know why."

"That's no excuse for sloppy work, Alvas."

"Don't give me that Novare. The balancing acts I have to perform are beyond your comprehension. The pinhead-model doesn't begin to explain the complexity of my work. Some things are still missing, anyway."

I know, Alvas. We'll get to some of that a little later. For now, the model will do.

The real brain has around a hundred trillion synapses, imitated by the wire connections in our pinhead model. While the sheer number of components in our brains is impressive by itself, it is the complexity of the whole system that hides the mystery of what makes our brains 'intelligent'. Billions of computers that receive information, process that information, and send out signals to thousands of other computers, that, in turn, process information, send it on to others, and so forth. The brain performs any 'algorithmic trick' in the book (of computer programming), like parallel processing, interrupting and re-entering ongoing processing, feedback and feed forward loops, changing branching conditions 'on the fly', or changing the overall structure of algorithms.

No two pinheads are alike. Each one is pre-configured 'out of the box'. This is like the 'default settings' of computer software when you install the software or the 'factory settings' of your digital camera or DVD recorder. Each of our pinheads comes with different programs running its tiny computers, each has different connection patterns to other pins, and the local variations in the density of pins, varies from pinhead to pinhead. In the real mindscape, these (and many other) features are determined by our genetic code. We can see this pre-configuration as a basic 'topography' of the brain.

Now our high-tech pinhead is ready to receive its first imprint. Each experience in our life creates an imprint in our mindscape. Just like in our pinhead model, an imprint resembles the original experience with varying degrees of accuracy, because the shape of the impression left by the experience depends on the momentary 'state' of the brain. In our hi-tech pinhead model, the shape and texture of the imprint depends on the resistance of the pins, which is controlled by the signals from *all* the connected pins, even signals coming from the pin-computers not directly involved in the imprint. Some pins might be rigid enough to prevent an imprint from being made at all. Alternatively, an attempt of a new imprint might be only partially successful, because no pins are available to receive the imprint. When this happens in a real mind, we might say: "that did not impress me at all." In these ways, a new imprint can change an imprint already made previously, change the way the next imprint will be processed, or have no effect at all.

This is what happens in the brain: the incoming information not only provides the brain with imprints, it also changes the way the *brain does its processing*. In this way, experiences influence the way other experiences will be processed. The model shows that each imprint changes the features and properties of the pinhead, which in turn will influence the next imprint. Also, the *kind* of imprints (e.g. large or small, deep or shallow) will change the way in which the next imprint will be received. Even our mood (e.g. pressing to hard, using the wrong angle) influences the imprint, just as our mood can influence the mental rendition we produce, or the way 'we see things'. The brain performs its own, subjective magic on its imprints, and on the way, it creates them. As Antonio Damasio says in his book *The feeling of what happens*:

> *"The brain is a creative system. Rather than mirroring the environment around it, as an engineered information-processing device would, each brain constructs maps of that environment using its own parameters and internal design, and thus creates a world unique to the class of brains comparably designed."*[15]

The pinhead model serves, in all its simplicity, to give us a glimpse at the complexity of what goes on in the brain.

"Please Alvas, stop laughing!"

"I can't help myself, Novare. I keep thinking of Merriam-Webster's 'very dull and stupid person'."

"What about it?"

"I'm sorry, Novare, but that sounds like *you*."

"What?"

"I said I'm sorry. See, when you don't do anything about it, I'm just going about my ways. I've enough to do to keep everything going. I even use my spare time to come up with my own ideas, when you're not available. Now I know that your laziness is visible to other people! I just love it, that's all."

"I'm sorry too, Alvas. For me, this is annoying, you see. Every time I take a rest, you fall back into your own habits, like a hamster turning round and round in its cage. Time for you to take on your own responsibility. Webster's definition is because of you, Alvas, not me. Just so you know."

"Hey, Novare, I'm totally dependent on the impressions you make!"

"And I'm totally dependent on the way you make them, Alvas!"

OK guys. Maybe both of you can help improve my dullness and stupidity.

"Sorry, Mr. J."

"Don't take everything personal, JJ. Bad for your health."

I want to use our pinhead-model to point out a few basic features of our minds, as they influence the future of our world.

First, the brain is a self-programming mechanism. Within the boundaries set by a basic structure, its 'factory settings', the brain forms its own algorithms for information processing. Given the fact that the brain is not a blank slate when we start our life, many basic algorithms are in place. Some of these algorithms 'run' our bodies. They regulate our heart rate and blood pressure, control the production of the right enzymes and hormones at the right time, and perform a plethora of other functions. They wake us up in the morning, with or without the help of an alarm clock, they inform us that we are hungry, and they 'give' us of a headache.

These algorithms run autonomously to the rhythm of our metabolic metronome. Our brains can expand and modify algorithms and invoke them on demand. These algorithms can be compared to software that enables us to create our own – higher level - applications. A good example is spreadsheet software. Spreadsheet software offers you a 'blank slate' (when you start up

Microsoft's Excel spreadsheet software, you get that blank sheet with a grid on it), on which you can create the most complex algorithms of your own.

However, the things you can do are limited by the functionality of the algorithms already contained within the software. Similarly, as we can see from our pinhead model, the algorithms produced by the brain are dependent upon other algorithms already in place.

This self-modifying algorithm-creating activity of the brain goes on as long as we live. The flexibility to create new algorithms depends on the flexibility of the brain (like the properties of the pin-computers and wiring connections of the pinhead). When we say that somebody is 'open-minded', we might refer to somebody with great algorithm-forming capability. A 'rigid mind' might belong to somebody with firmly established pathways, somebody with pre-configured algorithms that cannot be changed.

Second, the algorithms of the brain produce renditions. The final shape of the imprint made on the pinhead depends on the shape of the object pressed upon the pins and the algorithmic structure formed by the 'intelligent' connections of the pins. As discussed earlier, renditions can be of a mental or a physical nature. Mental renditions never leave our body. They exist in our imagination, our fantasies, our dreams, or our focused deliberations. They are subjective, and they constitute our personal knowledge, together with the underlying algorithms. We do not know how the brain stores renditions, whether as algorithms and parameters that (re-) produce those renditions, or as a direct representation, or as a combination of both. Internal renditions are the 'objects' of our mind. They are what we work with, are conscious of, and retrieve from our memories. We can imagine them in our mind, or we can express them to the outside world. In that case, we produce external renditions.

"Mental renditions can not leave our body, Mr. J."

I was afraid you would ask, Novare.

"There are many people that are serious about such things as extrasensory perception and telepathy."

"Come on, Novare. I've enough on mind as it is. I have enough senses to deal with. In addition, you are having trouble finding your way around the mind with the information I have; imagine I would have to process other people's mindscape!"

"Shortsightedness becomes a euphemism, when used to describe your mental reach, Alvas. All the body can detect are quanta of energy. The senses transform bombardments of quanta into the different kinds of things we can experience in our mind. The senses are transducers. Why couldn't there be different kind of transducers that are sensitive to different configurations of quantum hits? After all, many animals have different senses from our own. What do you think, Mr. J?"

I'd have to think about it, Novare.

"You doubt it, but you want to leave the possibility open. No harm confessing to that, JJ."

I confess.

Our external renditions are physical in nature. They can have the form of the discussions we have, the books we read or write, or the actions we perform in the world. External renditions are all of the physical things we create: the apple pie we bake or eat, the Lego-structure we build, the car we drive, or the spreadsheet we produce and manipulate on the screen. All the external renditions, built by the human race, are manifestations of our knowledge. Our scientific and intuitive discoveries of the physical structure of nature taught us how to use matter and energy to engineer and manufacture our own renditions. We commonly call them 'artificial' in contrast to natural renditions created by nature. If we include the human mind in our definition of 'nature', the distinction becomes less obvious.

Third, and in contrast with the pinhead model, the human nervous system and brain, like the animal's, is an algorithmic 'onion'. Evolution layered knowledge upon knowledge, algorithm upon algorithm, to form the stunning aggregation of knowledge and of physical renditions that constitute the physical part of our being. algorithms in these layers connect in intricate ways to form the operating system of our biological and mental machinery. Oftentimes, we fall back into 'routine behavior' by executing a well-established algorithmic pattern that is etched in our brain. Such routines might be expressions of algorithm, or habits, that we formed during our lifetime, or they might be the renditions of algorithm that linger at a deeper layer of the brain, waiting to be triggered into execution. Such algorithm might be the remnants from evolutionary times. Impulsive outbursts of 'uncontrollable aggressive behavior' might be expressions of such older algorithm. Some of today's emotions might be the remainders of 'survival algorithm' of times gone by. Sometimes vital to our well-being, sometimes less useful, today's emotions might be yesterday's thoughts.

"Novare?"

"Yes, Alvas?"

"Why are you so quiet?"

"I was thinking."

"I know *that.*"

"Mr. J. is talking, all this time, about *you.* I am curious to hear where *I* come in."

"Ah! I like JJ's metaphor of the pinhead. It describes the complexity of the things that *I* deal with, day and night. Especially considering that 'my' pinhead is a four-dimensional one. Makes me feel smart."

"Alvas, you are making a fool of yourself. Without me, you wouldn't even *know* you exist. I am the one that makes sense of your chaotic behavior. You are chaos, and I am order. You exist, because I tell you so."

"No need to be that way, Novare. I always *felt* that I existed, just as JJ feels *your* existence."

"That's what I was thinking about. You know, Antonio Damasio expresses that feeling in his book, when he says,

"Perhaps the most startling idea … is that, in the end, consciousness begins as a feeling, a special kind of feeling, to be sure, but a feeling nonetheless. I still remember why I began thinking of consciousness as a feeling and it still seems like a sensible reason: consciousness feels like a feeling, and if it feels like a feeling, it may well be a feeling."[16]

In that sense, Mr. J *knows* us. But he can not know how we function, because we don't know that ourselves."

"We could argue about it."

"Sure, just like Strudl and Blue talked about life. We know some of the effects of what we do together. We produce renditions, and those affect people."

"That gives us power over them, doesn't it?"

"That's the funniest thing, Alvas. You and I, the products of their organic mechanisms, yet we impact them."

"We better be careful, because, after all, we depend on them for our own well-being."

"Alvas, you are right. And for their well-being, they depend on us, especially our cooperation."

"Let's keep quiet then. See if JJ includes you in his metaphor. In the meantime, let's stay in touch some more."

Our mindscape has many extraordinary, sometimes mysterious features. You can only travel in it alone; nobody can go with you, as much as you sometimes want to share your view with a companion. On other occasions, you cherish the privacy of that feature. On clear days, you can see far, on other days all you manage to see, is fog. The sun might shine, or it might storm and rain. Sometimes you travel with the speed of light: you can fly. Sometimes you can only crawl, go in circles, seeing the same thing over and over again.

There are no two identical mindscapes. Even if your twin brother has a similar one, you still see a different picture, even from the same vantage point. Some of the routes you take from one vantage point to another become very familiar to you. They might become so familiar; you can take them blindly, without any effort. Something in your mindscape pulls a trigger and, wham, you are right there, like hitting a shortcut key in a videogame. Navigation of your mindscape is completely free, and extremely intuitive. No 'up' or 'down' buttons to click, no mouse to maneuver a 'cursor', no grid to 'snap to', and no double click to 'open' a new view. Your intention, your purpose, your feelings or your deliberate search are enough to get to where you want to go. You can choose any vantage point, and from there you can zoom in or zoom out, you can place filters in front of your lens to modify the picture, you can switch on any music you like without using your home stereo, you can focus, pan, look through trees and see what's behind them. You can take several views and blend them into one, and keep the result if you want.

Or you can add new views, freshly gathered from the outside world, or blend an existing view with a new feeling you want to attach, and submit it to memory, which you can do without clicking on any 'save' button. You can even delete views from memory forever, or for long periods of time. In that case, you might need a psychiatrist and her 'undelete' strategies to bring back the particular view, or help you alter it. This is like shaking the pinhead toy to bring back the original configuration of the pins.

Our minds can look at particular features of the mindscape only, or we can take several snapshots from different points and combine them into one new view. If the view is too limited, just go up, and expand the view.

Most astonishing, you can develop your very own impressions. To be sure, you are dependent on the renditions that are already part of your mindscape, and you are dependent on the total 4-dimensional topography of your mindscape, because it determines the way you think. In this way, from existing renditions, and the differential between them, you can construct your own 'snapshots' and 'scenes'. You can render them as words, images, sounds, mathematical symbols, or feelings, thus coming up with new mental renditions of new 'realities'. They might be musical compositions, a new idea, a new scientific theory, or a new insight for repairing your toaster.

While you are at it, the process we call 'thinking', as you travel through the mindscape and enjoy (or not enjoy) the views, your mindscape *changes* under the influence of your journey. While you change your mind about something, you change your mind, in a very literal sense. Such is the four-dimensional landscape that is our mind.

Our mindscape contains our personal domain of knowledge. It is always present, somewhere - and somehow - coded in the structures of our brain and nervous system. The 'symbols' used for this coding are physical and chemical properties of neurons, synapses and other components of the brain and the nervous system. Conscious or not, our mindscape doesn't leave us overnight (although we might have to understand the expression "she changed her opinion overnight" in the most literal sense). Even when awake, however, we are not able to view our mindscape in its totality, with everything in it, all at once. Most of it is hidden from our view at any given point. What is it that we see at any given point? What decides what that is? What controls the navigation, discussed earlier, through the mindscape?

Let me first expand the metaphor of the mindscape and introduce one of its most astonishing features: the *mindscape window*. This is a window without a window frame. It holds information constructs from the mindscape. This window has no boundaries, yet it is limited in its holding capacity. We can imagine this limitation as a physical one (in the brain), such as the number of PIMs in Pico's model limiting the number of snapshots that the PIMs fuse together into one coherent 4-dimensinal scene. Computer users might think of the windows on a computer screen that groups information into the framework of the windows on the screen. Although this association is useful to some extent, it can be misleading.

Our mindscape - window has no rectangular shape, or any other specific shape, at all. It has no framing device, just the fuzzy contours of the observed scene itself that gives a sense of boundaries. There are no navigational buttons to click, no hyperlinks to follow and no mouse to move. If we look closely at this mindscape-window, it is no window at all. It is a selected portion of the mindscape that seems to start glowing in the dark, as if illuminated from the inside, and giving the impression of a window that we look through. The effect is similar to a portion of the landscape in the real world that is illuminated by the beam of light from a lighthouse. Except, there is nothing that shines on the mindscape from the 'outside'. It is a strange affair: what is that window?

We want to remind ourselves that – as human beings - we entered the knowledge domain that I characterized as the domain of independent thought. That is exactly what our mindscape-window *is*: an independent thought; a 4-dimensional information construct; a fundamental unit of knowledge. Our window is an emergent one. It appears out of the 4-dimensional landscape as if by magic. Not magic, however, guides the composition and emergence of a window, but *attention*. This attention is similar to the one we 'feel' when looking at the world around us. A sound, a particular colorful feature might 'grab' our attention. In the personal domain of knowledge – our mindscape – the same thing happens. Something in the mindscape grabs our attention and the window of that feature emerges. In turn, other driving 'forces' like intention, motivation, purpose, passion, or deliberate 'reasoning' might guide attention. We have to remind ourselves of the importance of this window.

First, the size of this window determines the scope of any single thought we can have at any given moment in time. In Pico's theory, the size of this window would be determined by the number of connected PIM's that operate above a certain threshold level at that moment in time. When our thoughts become 'bigger' than the 4-dimensional size of our mind-window, we either create a 'train of thoughts' in our mindscape that we can replay in our window. Usually, that is when we start making notes, make sketches, use our favorite instrument to play the tunes that pop up in our mind, or use mathematical formulas to *express* and *document* our thoughts. Beyond the mind-window, we have to rely on memory. As human beings, we developed a multitude of ways to use external devices as memory. To do this we need symbols to create renditions of our thoughts. This capacity to use symbols is commonly seen as the most important difference between animals and humans. This capacity is a consequence of the independence of thought. Without it, there would be nothing to render, except for immediate bodily reactions, outside our bodies. Physical actions were the sole symbolic 'language' of animals. Independent thought required the invention of other means to render thoughts into something that would be independent of physical behavior.

Second, any experience of the 'now' is dependent on the size of the mind-window. Everything else, in the past and in the future, is dependent on our internal or external memory. Therein might lie a difference between the animal's and our own experience of the 'now', and a difference in the way we string together experiences to compose that mysterious thing we call 'time'.

As human beings, we are the embodiment of three domains of algorithm and renditions. We are composed of the substance of matter, energy, and forces from the cosmic domain, the substance of the bio-chemical processes and mechanisms of the life domain, and, the mechanisms and the substance – independent thought – from the knowledge domain. Each of these domains has its own referent frames, its own rhythm that we call time, and all of them integrated into one thinking living creature. Sometimes, when we are 'lost in thought', we travel solely in the knowledge domain, where our 'brain-clock' rules the baton, and we experience 'subjective time'. When we sleep, we 'lose all thought' and travel solely in the domain of life, where our metabolism rules our biological existence. Finally, when we lose the rhythm of life, our body exists solely as a physical-chemical entity as it enters the cosmic domain. At that point we are dead, although the atoms go on to do other things. A confusing existence, if you are an atom.

Let me now return to the mindscape-window, which is such an information construct: a thought. When a thought has my total attention, I lose touch with everything in the life domain, including my own body. I lose contact with everything in the cosmic domain, like the walls of my office or my wristwatch. In doing so, my subjective "I" *becomes* my thought: the thought is all there is. This is a *totally* conscious thought, meaning that my attention is focused on that one coherent thought only. In contrast, my brain might make changes to the content of my mindscape, while I sleep, and create an unconscious thought. The next day I might, or might not recover the information-construct as a conscious thought. In total consciousness, I am totally emerged in the knowledge domain, and excluded the other domains from my attention. When my attention gets distracted by 'real-time' events in the mindscape, usually because of new sensory input entering the mindscape from the outside world or from memory (like in "darn, I forgot to go to the bank"), my mindscape-window has to deal with two different domains of information, each with a different frame of reference and time-dimension (here is the source of our confusion between 'subjective' and 'objective' time. When the mindscape-window contains one coherent thought only, we are conscious of that thought only, have no reference to our 'wristwatch time', and live in 'brain-time'. Virtually no time seems to have passed between two moments of wristwatch time).

The mindscape-window is a scene. This scene consists of a series of renditions linked, according to Pico, by the cycle time of the brain. The construction of those renditions takes place automatically: we are never aware of the execution of any algorithm. What we see, what we *become*, in full consciousness, or full 'attention on the contents of the mindscape window', are the results of those calculations. The independence of thought means that we can 'linger' in that thought and go through the relationships between different snapshots, modify them, replace them by others, or even exchange snapshots and retrieve new ones. The relationships themselves become renditions also. They might be the sounds musical tunes combining single notes from a composition, they might be mathematical operators that link different numbers, or they might be novel associations that relate existing images in our mindscape.

Our mind relies on the contents of the mindscape and cannot invent something totally new from nothing. Rather, it merges and combines existing renditions, or renders the differential between renditions to create something 'new'.

Evolution continues to add onto itself by merger and acquisition. It did so in the cosmic domain, it did so in the life domain, and continues to do so in the domain of knowledge. The 'attention' that navigates the window through our mindscape feels like a force emerging from the resonance of our whole being: the accumulation of millions of years of evolutionary algorithm. This resonance is like the vibrations of a guitar, when its strings are activated by thoughts and emotions that express intention or purpose, or openness or quietness.

Thinking is a journey through the mindscape. As the window travels through it, a final information-construct will emerge that contains *meaning*. Meaning is the relationship between that obtained knowledge and the totality of the mindscape, which *is* our view on the world. Thinking not only results in the creation of new renditions, but also in the modification of the underlying algorithm, just as our model of the high-tech pinhead demonstrates.

Mindscapes and mindscape-windows are not necessarily exclusively human, in my opinion. The learning abilities of chimpanzees, their occasional use of simple tools, their ability to recognize themselves in a mirror, might be the lightning bolts that signify the brain's attempts at the creation of independent thoughts; thoughts that persist; thoughts that gain autonomy over the urgency of immediate biological action or reaction. These bolts of lightning are the precursors to fully independent thought. Just as the struggle of the first protocells to sustain themselves as an independent living organism, these bolts are the attempts of the first thoughts to gain independence. Let us immediately remind ourselves of the relative meaning of this 'independence'. As life cannot exist without the physical constructs and forces from the cosmic domain, thoughts cannot exist without the biochemical systems and phenomena that contribute to the emergent property of life. I use *information, rendition, and algorithm* as universal concepts, as they apply in all three domains of evolutionary existence. However, knowledge is the exclusive domain of human beings, as it relates to the independence of thought and the derivation of meaning from its relation with the background of the human mindscape.

The great apes started the climb against the entropy of biochemically-chained information-fragments in the brain. Until, a few tens of thousands of years ago, somebody without a name, or somebody named Arr, had a brain with the quantities (of components) and the qualities (transactions between components) necessary for the first thought to emerge and gain autonomy over its biological surroundings.

Technology not only changes the kind of renditions we make, it changes the way we make them. Similarly, the *technology to produce thoughts* not only changes the kind of knowledge products we create, it also changes the way we produce them. algorithm create renditions. These renditions are the physical manifestations of these algorithm. Technology is the expression of knowledge. A new technology forms relationships with the rest of the environment. This forms the template for the formation of new knowledge. Sometimes, the cumulative effect over ever expanding algorithm is not just the increasing complexity, but also the emergence of a new *kind* of algorithmic expression. This is what happened at the three thresholds that we encounter in the evolution of nature. We might unravel the mysteries of the algorithmic leaps at those thresholds, as I am confident we will.

Episode 10, "A Change of Mind"

From 10,000 ago until today,
90,000 - 100,000 years after the awakening of thought,
4 billion years after the delivery of life,
15 billion years after the birth of existence

"You, your joys and you sorrows, your memories and your ambitions, your sense of personal identity and free will, are in fact no more than the behavior of a vast assembly of nerve cells and their associated molecules. As Lewis Carroll's Alice might have phrased it: 'You're nothing but a pack of neurons.' This hypothesis is so alien to the ideas of most people alive today that it can truly be called astonishing."

Francis Crick[17]

"The last episode of evolution, JJ!"

As far as this book is concerned, yes, Alvas. Are you looking forward to the end of the story?

"I must admit, that the braid of mental pretzels keeps growing longer all the time. I look forward to the end of *that*. More than that, however, my sense of expectation is growing. About the significance of the digital world? Better be good, JJ, after this evolutionary roller coaster."

"Understanding is not his strong point, Mr. J. He does not realize that you are moving towards the beginning, not the end. Correct?"

Yes indeed, Novare. I do appreciate his mental pretzel syndrome, though. Most of them *are* the pretzels they appear to be. Their significance lies in their strangeness. We have to accept that our intuition is based on what we needed during evolutionary times. Apparently, we did not need quantum behavior, relativity, independence of thought, or a human mind, until now. This is where evolution results in a new beginning, which is the beginning of the third stage of its expansion.

"How are you going to pack ten thousand years of human history into one episode, JJ?"

I can't Alvas, and I don't need to.

"You just want to point out how the human mind changed, correct?"

Correct.

"Maybe we can help with that, JJ. After all, we were part of it."

That would be welcome, but you have to make a promise.

"Sure JJ, we'll work together. We have no choice when it comes to explaining *ourselves*. Unfortunately."

Eppur Si Muove, Part 2

The domain of knowledge is the exclusive domain of the human mind. With every new technology, with every new expression of that technology, the human mind changes. It changes by the impressions of that technology on its mindscape, and it changes by the creation of novel renditions from those impressions. New renditions emerge that transcend the immediate impressions made and delivered to the mindscape. While the mind does its work, the algorithm themselves change. That is the strength, and the vulnerability of the human mind. Experiences enrich the mind by expanding the mindscape, and while doing this, change the way the mind thinks. That is how new technologies, and our interactions with them, produce novel experiences that change the way we think. The notion that 'technology is neutral' is an ignorant under-estimation of the depth and creative power of the human mind. Technology is no less neutral than any other passionate, violent, or 'significant' experiences in our lives. They change the way we think, the way we process the next experience, the way we view the world, and hence the way we act in it.

The brain's plasticity provides the bio-chemical opportunity for this innovative behavior of the mind. In the 1949 publication *Organization of Behavior*, Canadian psychologist Donald Hebb suggested the existence of a mechanism in the brain that - bio-chemically - changes the way neurons influence other neurons. This mechanism changes the effectiveness of a particular neuron to 'fire' a signal to a particular other neuron. Today, this mechanism is known as a 'Hebbian mechanism' and it is recognized as one of the means by which the brain creates new pathways of behavior, or, in my view, creates new algorithm for thinking.

In his book *The Astonishing Hypothesis*, Francis Crick (the same one that discovered the structure of DNA) proposes that the network of neurons, axons and synapses in the brain defines what, and who, we are. The quote at the beginning of this episode is the hypothesis that Crick develops in this book. He uses the Hebbian mechanism as one of the bio-chemical tools that contributes to the explanation of neural networks as the brain's powerful faculty for *learning and adaptation*. One of the important tools for the study of neural networks in the brain, are their digital twins, or rather: digital cousins. Mechanisms, like the Hebbian, are used to enhance the realism of computer algorithm that try to emulate the neural network in the brain. These digital algorithm help to improve our understanding of the brain, because they can handle a degree of complexity and a scale of vast numbers that no other medium can accomplish. The world of bits and bytes becomes the workbench for the creation of new knowledge. As Crick points out:

> *"In the past, many aspects of the brain seemed completely incomprehensible. Thanks to all these new concepts one can now at least glimpse the possibility that some day it will be possible to model the brain in a biologically realistic way, as opposed to producing biologically implausible models that only capture somewhat limited aspects of brain behavior. Even now, these new ideas have sharpened our approach to experimentation. We now know more about what we need to understand about individual neuron ... We see the behavior of single neurons in a new light, and realize that the behavior of whole groups of them is the next important task on the experimental agenda. [Digital] Neural networks have a long way to go but they have, at last, got off to a good start."*[18]

"We need to modify JJ's pinhead model, Novare. It does not cater to my Hebbian mechanism."

"Simple, Alvas. We just need to modify the programming of the computers in each pin. For each wired connection of a pin, the computer has a procedure that translates the pressure on its head into the decision for sending a signal to any particular connected pin. In addition, we will let the computer count the number of times pin A does not reach the threshold for sending a signal to pin B. Finally, we'll change the procedure such that after a given number of failed attempts over a given period, the computer lowering the threshold, so that, eventually it starts firing of a signal to pin B."

"Marvelous Novare. I know, of course, that it is a gigantic oversimplification, but it'll do."

"How do you know it is an oversimplification?"

"Because there is more to it than that."

"You mean, there are more complex algorithmic structures involved?"

"That too, Novare. What I am referring to, is something different altogether. See, I produce renditions using all of these algorithm. But when you ask me to come up with something new, or sometimes, when I'm just going about my ways, I start trembling, as if I'm a tidal wave that starts to rise up. I don't know how to explain it other than JJ's strangeness of quantum waves. That's how it feels like."

"Let's not get into that, Alvas. What we are only trying to demonstrate that the people do change their mind. Not just in the sense of changing their perspective, but also in the way they develop new perspectives. The Hebbian mechanism, and, maybe, your 'mental quantum behavior', are just a few of the mechanisms that might explain the implementation of that algorithmic potential for change."

"All right. I can't explain it anyway. I'm just saying that it happens."

The known history of the human race, in the last ten thousand years, is not only the history of the events that occurred around the planet earth. All of the events, from habits and rituals of hunters-gatherers, to the quests of emperors and kings, the revelations of prophets, mystics and philosophers, the wars and glorious expressions of cultures in peace, the explorations of the earth and the pioneers in new territories, to the symphonies of Beethoven and Leonardo Da Vinci's creations, all of them are expressions of human knowledge. They are renditions of the thoughts created in the window and the landscape of our human mind.

Some look at these events as being the *complete* history of the reality of human existence, as it unfolded in these ten-thousand years. They derive explanations of *why* things happened, and *how* things happened, from the perceptible events that occurred. They might look at human beings as biological creatures that 'happen' to think, and see this thinking as an epiphenomenon of being human. Alternatively, they might see a human being as a mystical, but fundamentally unchanging, static and permanent creature that happens to produce tools, technology and art that are mere 'expressions' of 'human nature'. Or, they assume the existence of a mental substance that permeates the universe, not unlike the 'ether' that was commonly assumed until the beginning of the 20th century. They fail to recognize the evolutionary expansion of the human mind. They fail to see that human beings entered a new domain of existence, one that combines and merges the existence of the universe, the existence of biological life, and the existence of creative knowledge. Interestingly, or maybe ironically, this is also the point at which two opposites approach each other: the proponents of 'artificial intelligence' that see robots as the latest torch-carriers of the evolution of the world, and those religious leaders and followers that either deny evolution altogether, or see it as a scientific gadget that approximates reality but does not *explain* it. Both fail to see the fundamental changes of the human mind, its power, its depth, and its evolutionary potential.

These changes do not take place on a biological level where the fundamental structure of the brain changes. As far as we know, no changes took place in the brain since Arr, or even longer. Although biological changes, or enhancements, might be possible in the future, changes take place in the algorithm, which the mind produces through the process of thinking. On first sight, it seems that artificial intelligence ('AI') professionals should be the first to understand this, because a (all too) simple metaphor of what happens is given by the difference between software and hardware: the 'hardware' of the brain remains the same, while its 'software' changes. I say 'on first sight', because we should not ignore their premise, namely that an algorithm (the software) can be executed on any suitable hardware, hence, the algorithm of the brain, once downloaded on to a computer, can be executed by that computer. It seems to me that many sciences, including the science that carries the misleading label 'computer science', can no longer be carried out in isolation of other sciences. Evolution involves phenomena that need a perspective that can only be achieved by a blending of different scientific renditions. What makes us 'human' is the evolutionary accumulation of knowledge and technology that form an intricate being that transcends its interacting components. In turn, human beings interact with each other and all of nature, on all three levels of existence (cosmic, biological life, and knowledge). These interactions create our world; they form a tapestry that expands through evolution, just as the tapestries of the universe and biological life. The evolution of personal knowledge and the evolution of this 'tapestry of human behavior and thought' feed on each other as they proceed into the unknown territory that we call the future. Without interactions between the two, there would be no human being, and without interactions, there would be no evolution, not in the past, and neither in the future. Interactions, within each human being, and between them and nature is what creates 'being'. We call that transcending quality of human being 'human nature' or the human 'spirit'. Religion can contribute to this 'spiritual' transcending quality by taking an active, not defensive role in the interpretation of new scientific and technological discoveries and accomplishments.

One of the fundamental, no, the most fundamental perspective of this book, is the evolutionary nature of that human spirit. It is a spirit that changes with every single human experience; the direction of its change is the expansion of human knowledge, which is equivalent to the expansion of the human mind. Experience is where the mindscape-window emerges from the mindscape and feeds back new renditions to the mindscape, forever changing the way it operates and produces new windows: the independent human thoughts that reach out into the future. In fact, they are our future.

"Eppur Si Muove," Galileo uttered, sotto voce, four centuries ago. Since then, we learned to accept that the earth moves, despite the fact that we have no direct perception of its velocity. Similarly, we don't perceive the motions of our mind, as it adapts itself to the ever-changing reality of every day life.

Virtual Reality

In our evolutionary perspective, human history is the history of the expansion of knowledge. The human mind is the well from which new knowledge flows, knowledge that finds its translation in technologies that changed the physical landscape of our globe. New knowledge produced new physical renditions; these new physical renditions produced new knowledge, which produced new physical expressions of that knowledge, and so forth. This is the perpetual wave of expanding knowledge, resulting in greater diversity, greater versatility, and enlarged decision-spaces that produce the flow of events over time. As the wave of knowledge increased in breadth and depth, the mechanisms for the production of knowledge changed as well. They changed in the form of the knowledge creating extensions that we created, starting with the invention of the alphabet and leading up to the invention of computers, and they changed because of the way the human mind changed itself. It is the history of the expanding production of human thought. It is the history of the expanding virtual reality that exists in every human mind and in the collective human minds of the planet earth.

Today, we commonly view *virtual reality* as something 'not real', such as videogames or cyberspace. The virtual reality that I refer to, however, is as real as apple pie and doormats. It is the reality of the product of our mind: our knowledge. We do not know how we generate that knowledge, and we do not know how that knowledge is structured or represented in the brain and nervous system. Each of us can feel how it emerges, but we cannot perceive the process of emergence, nor can we perceive its structure until we created a rendition of that knowledge, either within our mind, or outside our body by creating an external rendition. In this sense, knowledge itself has the mysterious shape of something 'unreal'. We know that it exists, because its effects are perceivable, oftentimes measurable, and certainly real. Knowledge itself therefore must be real. Human history is the history of the continuous creation of knowledge, and the translation of that knowledge into technology, and the interactions between that knowledge and its renditions within the human populations around the globe.

In his 1976 book *The selfish Gene,*[19] Richard Dawkins introduced a new concept that is useful for the discussion of knowledge constructs. He introduced the concept of the *meme* as the mental equivalent of the *gene*. He defined the meme as the smallest unit of information in the evolution of culture, just as the gene is the smallest unit of information in biological evolution. According to Dawkins,

> *"Examples of memes are tunes, ideas, catch phrases, clothes fashions, ways of making pots or of building arches. Just as genes propagate themselves in the gene pool by leading from body to body via sperm or eggs, so memes propagate themselves in the meme pool by leaping from brain to brain via a process which, in the broad sense, can be called imitation. If a scientist hears, or reads about, a good idea, he passes it on to his colleagues and students. He mentions it in his articles and his lectures. If the idea catches on, it can be said to propagate itself, spreading from brain to brain."*[20]

In the perspective developed in this book, I have made the distinction between algorithm and renditions. algorithm are constructs of knowledge, while renditions are expressions of that knowledge. Renditions are physical in nature: humans can perceive them, either as physical objects (including – living – subjects), or as the (physical) behavior of those objects.

In Dawkins' definition as I understand it, memes can either by knowledge-constructs (that I call *algorithm*), or renditions of that knowledge. The concept of the meme found widespread acceptance, but is – unfortunately – used often to describe *either knowledge, or the renditions of knowledge.* A symphony is a rendition, while the 'knowledge structure' in the composer's mind is the meme. An idea is a knowledge construct; the expression of that idea is a rendition. Planes, trains and automobiles; operas, movies and novels; jokes, political speeches, and prayers; dance, kisses and laughter; they are all *renditions* of human knowledge. They are the expressions of knowledge and the result of the execution of algorithms, even if we cannot observe the execution itself. A meme as defined by Dawkins is both: *the 'idea' itself and the expression of the idea.* It is knowledge (the *algorithm*) and a *rendition* of that knowledge. Memes are the physical constructs that human beings use to communicate *meaning*: the knowledge (algorithm) embedded in the construct (rendition). This leads to an important question: will both he rendition and the embedded knowledge be transmitted accurately from one human mind to the next? According to Dawkins,

> *"Memes should be regarded as living structures, not just metaphorically but technically. When you plant a fertile meme in my mind, you literally parasitize my brain, turning it into a vehicle for the meme's propagation in just the way that a virus may parasitize the genetic mechanism of a host cell. And this isn't just a way of talking -- the meme for, say, 'belief in life after death' is actually realized physically, millions of times over, as a structure in the nervous systems of individual men the world over."*[21]

What does a 'belief in life after death' mean to *you*? Does it mean that you go to Heaven, body and all? Or does it mean that you 'live on' in the memory of others? That is where the definition of the meme runs into trouble. Unlike the gene, the transmission of a meme from one human being to another (usually) changes the meme, because it changes the knowledge content. This is the salient difference between the gene and the meme. The gene is an incredibly robust structure safeguarded against transmission and copying errors by the ingenious structure of DNA-molecules. A gene is a mechanism that contains truly 'objective' knowledge about the creation of an organism's phenotype. This feature is essential for the reproduction of all living organisms and the maintenance of the integrity of such organisms. In contrast, the meme is a dynamic construct that changes with every transmission and with every 'copying' attempt. The multitude of different interpretations of law exemplifies this phenomenon: its how lawyers make a living. The spread of rumors is another example of the power of the human mind to create an infinite set of memes from the same original 'meme'. The difference is essential, because the evolution of the human mind, the expansion of the domain of knowledge, depends on the dynamic nature of the meme. This is not my only problem with the widely used interpretation of the meme.

A French expression sais "C'est le ton qui fait la musique." If the tone is what makes the music, then this is where I stop following Dawkins' tone. Memes are not attackers, like viruses, but active vehicles for the production of innovation. They definitely affect human minds, as we can see from our impression-creating and algorithm-changing pinhead model. To call this 'infection', however, is part of the passive and helpless tone of Darwinian 'power-plays' that served biological evolution so well, like 'random variation', 'selection', 'survival of the fittest' and so on. It does not reflect the proactive and creative power of the human mind. Biological evolution gave us a brain. Memes turn that brain into a mind that can 'change its mind'. Genes have to wait until nature changes them, 'mutates' them, to make improvements. Minds transform memes, and memes transform minds, unconsciously or consciously. Memes can transform apathy into fervor and passion, and turn destiny into destination. Today, memes can even changes genes. The only thing we have to do is to change our tone, so that the newly found freedom of the creative mind expresses itself in the passionate music of aspiration, and transcends the tune of mere respiration.

Memes, like genes, are transmitters. They transmit a thought or an idea to a human mind. A gene is the medium that *transmits* a message. A meme is the medium that *becomes* the message when it enters the mind.

Sometimes, a meme will be copied accurately, both its rendition, and its medium. This is the case when the communicating minds have the same – common – algorithm (which can be thought of as the patterns in the mindscape, or the pinhead model) that process the meme. Scientific theories, expressed as mathematical formulas are examples, where the mindscape has been 'programmed' to receive and process memes in very well defined ways. This 'pre-programming' of the mindscape and mindscape-window establishes our 'sense' of objectivity and 'common sense'. The more minds receive and process a meme in 'the same way', the more objectivity or 'common sense value' we assign to the meme. The advantages and the risks are obvious, but it is what we do in daily life. 'Public opinion', 'common sense' and 'objectivity' are special cases of the common base of algorithm that nature (through the genes that establish basic algorithm in the brain) and society (the algorithm that education and social interactions bestow on the brain) etch in our mindscape.

In many - or most - cases, a meme will not be accurately copied into the mind it enters, just as a demonstration with our pinhead-model shows. As a meme is transmitted from mind to mind, many transformations take place. The meme will alter the way the communicating minds think, as we can imagine from our pinhead model. The mind that changed its thinking creates a new meme that propagates to the next person. Minds are not the accurate replicating environment for memes that cells are for the replication of genes.

On the contrary, the mind is nature's way to increase diversity, and stimulate innovation. Maybe 'fertilize' is the more appropriate word to use, rather than 'parasitize'. If the use of 'parasitize' is within the bounds of scientific explanation, then why not 'fertilize'? Memes are grains of fertilizer that enrich the soil of our mindscapes, because it sparks new meaning. The richness of this soil provides the chaos – the opportunity – for the creation of new order: the organization of novel constructs of knowledge. With the shift from the gene to the meme, as the vehicle for evolution, nature added another feature to its repertoire of tools to create change. A feature with a quality never used before in the history of evolution. That feature is the ability to change memes *on purpose*, to *intentionally* create new memes, and thus, to create new realities.

"Hah, I know exactly what you mean, JJ."

Good.

"You just have to listen to political speeches to understand *that*. Marvelous. What an elegant way to justify political humbug!"

Come on Alvas, be nice.

"He is just having one of his 'waves', Mr. J. I just happened to pay attention to him, because I am *so* happy!"

How come?

"Finally we can talk intention, purpose, design, and motivation. Hallelujah! No more anthropic principle that predicts the past. No more guessing. Good-bye, Mr. Random. Goodbye, variation and selection. Welcome to creativity. Welcome to intuition, imagination and, yes, conjecture. Conjecture is the overture to every scientific theory. I love it. Freedom at last."

"Amen, Novare. Never saw you that way. Could've been me. Wish you were that way more often. See, now you're fun. Now you're chaos, just like me. Again and again: amen."

Keep it cool, guys. The new invention comes on top of every other tool invented by nature. It's not a replacement, but an addition its bag of tools.

"Uh. Mr. J?"

When memes propagate through the minds of millions, or even billions of people, they might influence these minds in different ways. If a particular one is selected by a large amount of people as the 'prevailing' one, you might still see it a matter of (random) variation and selection.

"That's a bummer, JJ"

Not really, Alvas. Novare's optimism is justified. Human minds can decide on a purpose, an objective, goals, and create new knowledge, design new renditions that help to fulfill that purpose, meet that objective, and reach that goal.

"What is keeping us from it, then, JJ?"

"The axle of change Alvas."

Novare wears her frown.

"Today, evolution is the prevailing scientific theory for the explanation of reality, yet it is not the prevailing mindset; not by a long shot. The mindscapes of most people have not absorbed evolution, let alone the evolution of the mindscape itself. On the one side of the axle is scientific theory and the technology it produces with the speed of light; on the other end is the turtle driving the engine of our view of reality."

"Why then, do people fail to discover evolution in the changing reality itself, Novare?"

"Because they miss the concept of evolutionary change, Alvas. Their mindscapes operate on the wrong paradigm. Therefore, their minds order the events taking place in reality within the wrong perspective."

"Pretty scary, I'd say."

"It is, Alvas."

As Novare indicates, all of us live with the virtual realities of our mind. It is the reality of our mindscape. Its 4-dimensinal topography determines the way we think: our mindset. That mindset dictates our view of the world, or our operating paradigm. It is more than a perspective, because it molds our experiences to fit its topography. It is the virtual reality of our life. To each individual, that virtual reality is reality itself. To larger communities, societies, or the world as whole, the individual realities add up to something that we cannot grasp; something that is greater than the sum of its parts; something that emerges from the tapestry of individual worlds, transcends it, and feeds back on individual mindsets. We commonly refer to this transcending meaning as the spirit of a community, a society, or the world, like in 'the spirit of the law', 'the spirit of the conversation', 'a positive spirit', or 'the spirit of a culture'. Virtual reality is everything that does not have any physical manifestation that our senses can perceive. Yet, it is real, and we do 'sense' it, each in our own way, by the way, we process the physical reality that is the basis for the 'spirit'.

Virtual reality did not start with the advent of science fiction movies or interactive video games. Virtual reality started with the first human thought. It is the invisible tapestry woven from individual mindsets into an intricate pattern of networks and hierarchies of collective mindsets. Virtual reality is the reality of human thought. These thoughts are the recipes - the algorithm - that determine human interactions, while at the same time, through their actions in the world, they effect changes in that world. Just like the first clouds that signified the appearance of stars and galaxies, isolated patches of virtual realities evolved over the ages. These 'clouds' of collective mindsets appeared in the form of cultures, religions, empires and kingdoms. When they stayed isolated, they flourished, and then died. When they interacted with others, new ones arose, many times out of the ashes of physical conflict, while the fragments of different mindsets remained and recombined or merged with others to form new ones.

Starting with the industrial revolution and its first product of global conflict, World War I, the isolated clouds of virtual reality started to touch and interact. This process accelerated further during the last century and intensified again in the last two decades. We found a label for it and call it globalization. While many see globalization primarily as the phenomenon of business expansion, its true meaning goes much deeper than the physical expansion of corporations. It is the global connectedness of mindsets: of cultures, including the culture of business, of religions, of 'ways of life', of living human 'races', of nations, and of political and economical systems. In short, globalization is the 'becoming connected and interconnected' of every existing human way of thinking, of human mindsets, individual and collective mindsets alike.

The Medium is More than the Message

"In a culture like ours, long accustomed to splitting and dividing all things as a means of control, it is sometimes a bit of a shock to be reminded that, in operational and practical fact, the medium is the message. This is merely to say that the personal and social consequences of any medium – that is, of any extension of ourselves – result from the new scale that is introduced into our affairs by each extension of ourselves, or by any new technology. Thus, with automation, for example, the new patterns of human association tend to eliminate jobs, it is true. That is the negative result. Positively, automation creates roles for people, which is to say depth of involvement in their work and human association that our preceding mechanical technology had destroyed. Many people would be disposed to say that it was not the machine, but what one did with the machine, that was its meaning or message. In terms of the ways in which the machine altered our relations to one another and to ourselves, it mattered not in the least whether it turned out cornflakes or Cadillacs. The restructuring of human work and association was shaped by the technique of fragmentation that is the essence of machine technology. The essence of automation technology is the opposite. It is integral and decentralist in depth, jut as the machine was fragmentary, centralist, and superficial in its patterning of human relationships. [22]

These are the words of Marshall McLuhan, from his book Understanding Media, published in 1964. In 1994, fuelled by resurgence in McLuhan's work, the book was re-published by the MIT Press. The book is considered a classic expose that presents a radically new view on the influence of media on the human mind. The world is 'a global village', we are living in an 'information society', and 'the medium is the message' are the products of McLuhan's mind.

This episode of the evolution of the universe is the *stage of changing the human mind*. It is not a biological change. It is not the last stage of evolution, either. Rather, it is the first stage of the evolution of 'independent knowledge'. It is the first stage of the expansion of *human* knowledge. Humanity is working to complete the technologies, 'media' in McLuhan's terminology that rings the bell for another round of evolutionary expansion.

The last 10,000 years are, the *years of reverse engineering*, in which the human mind has discovered how nature works, has used that knowledge to create technologies that extend its own physical versatility, and, finally, has used that knowledge to create the technology to expand its mental versatility for the creation of new knowledge. Now, we start translating that knowledge into innovative realities.

Humanity changed its mind in the last 10,000 years, not through changes in the brain, or the nervous system, but by extending, no, expanding it with the help of *digital technology*. All other technologies, including biotechnology or nanotechnology, are products of the digital expansion of the human mind. They would simply be impossible without digital technology, which takes away space and time, and the limitations of the human scale with it. We are approaching, in my view, the turning point between the reverse engineering of reality, and the engineering of new realities.

McLuhan was one of the first to realize the power of technology as extensions of the human body and mind, when he said, "any technology could do anything but add itself on to what we already are." From the wheel as an extension of our feet to machines as extensions of our bodily abilities to what McLuhan calls 'electric technology' and 'automation' as extensions of the human nervous system and the brain. Today, this is a commonly accepted view, even to the extent that some call the totality of natural abilities and artificial extensions of those abilities the *extended phenotype* of the human being.

In the domain of life, we have already described the emergence of new forms of organic structure-functions, like the fins of fish, or the arms and legs of primates and human beings, as technologies that nature created to enhance the dexterity of living creatures. In evolution, the faculties for the management and control – nervous system and brain - of those technologies developed simultaneously with the emergence of the technologies itself. In the domain of knowledge however, things are a little less harmonious. Scientists discover, and develop new knowledge, that describes the reality surrounding us. Usually it takes a long time before the explanatory power of new scientific insights reaches the minds of people.

While the new explanation changes the mindset of a few scientists, philosophers and artists, the masses go through a process of ignorant bliss, or they plainly refuse – under the influence of political or religious leaders - to 'change their mind'. Yet, at the same time, the engineers, and the business entrepreneurs of the world, go ahead, produce the most exciting products – technologies – and market them to the masses. Thus, the mindset of people usually adapts only slowly to the new technology. Peoples' mindscapes, however, will absorb the new technology with a mindset that has not been 'programmed' to understand reality with the newly found explanation of reality. The friction in the axle of change increases, and the differential gear - the human mind – has to pick up the stress.

McLuhan became an instant celebrity when he suggested – exposed – the influence of technology on the human mind. In his view, the invention of the alphabet, and later, in the 1500s, the invention of moving type and the introduction of the Gutenberg printing press, present some of the most seminal 'mind-altering' technologies ever introduced in this stage of mental evolution. With the use of the alphabet, structures became sequences. With the introduction of the printing press, the mass medium for the presentation and the distribution of knowledge became a sequential one. Not only was it shift from the intuitive nature of direct 'face-to-face' oral communication, it was a shift from the communication of 'whole structures' to the communication of sequences. This transformation changed the *way people think*. The alphabet and the printing press for mass-distribution of alphabetically expressed knowledge came to determine the way people think, and that became a determining factor western culture, science, and philosophy. Because of the printed word, "We have confused reason with literacy and rationalism with a single technology," according to McLuhan.

Every technology, or, in my terminology every physical rendition, is a medium, and every medium transmits its existence, and its behavior, to the brain in different ways. Print, for example, is visual, and speech is oral. Our various senses are involved when we perceive the technological manifestations that we use to extend our own bodily and mental extensions. The ratio, in which the different senses are involved, according to McLuhan, determines the way our mind perceives and processes the medium. As a consequence, our own way of thinking and hence our actions will change. They are modified by the medium itself.

"Mr. J?"

Yes Novare?

"The ratio in which our different senses receive the information influences the way our mind processes that information, correct?"

Indeed.

"Therefore, the impression left in our mindscape is dependent on the degree in which we use the different senses?"

Correct.

"In that case, we have to make another modification to our pinhead-model!"

That's right.

"Simple. Our pins already measured the pressure of an imprint, remember?"

Sure.

"Now, we replace that mechanism by several new ones. One that measures incoming light to represent vision; one that measures the pressures of airwaves to give us hearing, one that measures the configuration and types of molecules to give us smell, a similar one for taste, and we leave the existing one for pressure in place to give us touch."

"That's crude Novare, because it doesn't give the model any sense of emotions and feelings, and it doesn't give us a sense of any of the other senses that we possess."

"Come on, Alvas, it will suffice as a model. Of course, Mr. J, there is some incredibly difficult programming involved. Each pin-computer requires its own algorithms to process the information from these sensors, so that the right signals go to the right connections to other pins. On top of that we need algorithms that mix all the signals to produce coherent information-constructs that Alvas and I can recognize."

An overwhelming task, Novare. No wonder, the artificial intelligence folks have such a hard time.

"They are simply on the wrong track, JJ. They keep trying to model what they think can lead to consciousness. They should focus on me instead. I'm the mystery to it all, see. It's in the way I respond to intention. That's all I'm saying."

Thanks for sharing your thoughts, guys. You finally managed to hand me your own mental pretzel in the form of a pinhead.

What we can easily see from our model is that the nature of information, - the physical properties of the rendition – determines the mental rendition that our mind creates, and how it changes the topography of our mindscape, and the shape of the mindscape-window with it: how the medium changes the way we think.

In his 1998 book *The Alphabet and the Goddess*, Leonard Shlain, chief of laparoscopic surgery at California-Pacific Medical Center in San Francisco, takes the influence of the medium on the brain a (consequential) step further:

"There exists ample evidence that any society acquiring the written word experiences explosive changes. For the most part, these changes can be characterized as progress. Bur one pernicious effect of literacy has gone largely unnoticed: writing subliminally fosters a patriarchal outlook. Writing of any kind, but especially its alphabetic form, diminishes feminine values and with them, women's power in culture. The reasons for this shift will be elaborated in the coming pages. For now, I propose that a holistic, simultaneous, synthetic, and concrete view of the world are the essential characteristics of a feminine outlook; linear, sequential, reductionist, and abstract thinking defines the masculine. Although these represent opposite perceptual modes, every individual is generously endowed with all the features of both. They coexist as two closely overlapping bell-shaped curves with no feature superior to its reciprocal."[23]

It comes as no surprise when Shlain declares McLuhan's aphorism 'the medium is the message' to be the Leitmotif of his book. Shlain makes a very compelling case for the direct influence of the media that we create on the actual functioning of our nervous system and brain, and the graveness and depth of the consequences of this interaction between the external renditions that constitute reality and the mental renditions that represent that reality. The key to Shlain's thesis "lies in the unique way the human nervous system developed, which in turn allowed alphabets to profoundly affect gender relations." The following quote, where 'Shlain meets McLuhan', gives an example of the modern medium of TV as it changes the way we think, and changes social behavior:

> *"Previously, alphabetic print had exploded Western culture into millions of hard-edged shards of individualistic shrapnel. Both reading and writing are, in most cases, solitary endeavors. Television abruptly reversed the process, and the centripetal implosion not only pulled together individual families but also began to enmesh the entire human community into what McLuhan called 'one vast electric global village'. Television was so startlingly original that many other adjustments in perception were necessary for the brain to make sense of it. The electroencephalogram (EEG) brain wave patterns of someone reading a book are very different from those of the same person watching television. So fundamentally different, in fact, that there is little deviation in those patterns even when the content of the book or television program is varied. A network program about adorable koala bears elicits essentially the same wave pattern as a program containing violence or sexuality.*[23]

While this quote undoubtedly triggers a host of subjects for passionate discussions, it clearly indicates the power of media as mind-altering messages. When we read a page of a book, we use the cones, while watching TV requires the activation of the rods in the retina of our eyes. We would have to make further refinements to our pinhead-model to demonstrate this feature. The model however, does allow us to demonstrate the basic difference of the brain's processing of the various sensory inputs.

"We just need to connect different colored lights to the various sensors of each pin, Mr. J. Say we connect blue lights to the sensors that measure the arrival of photons, and yellow lights to the sensors that measure the pressure of airwaves. Reading a book would produce a blue-looking pinhead, and listening to a symphony could produce a yellow-looking one."

"Would watching a surround-sound movie produce a green one then, Novare?"

"If the sensors would mix the two colors, yes."

"Boy, I'm glad I need no lights to see what comes into my world!"

Two final quotes from McLuhan's *Understanding Media* express his passionate conviction, and mine, of the mind-transforming power of technologies as media:

> *"It is in our IQ testing that we have produced the greatest flood of misbegotten standards. Unaware of our typographic cultural bias, our testers assume that uniform and continuous habits are a sign of intelligence, thus eliminating the ear man and the tactile man..."*
> *"...Our conventional response to all media, namely that it is how they are used that counts, is the numb stance of the technological idiot."*[24]

Interactions between physical and mental realities change our mind, and they change the way future interactions take place. Reality changes the mind, and the mind changes reality; the mind changes the way it changes reality, reality changes the way the mind changes reality, and so forth. Such is the nature of mental evolution, as it expands the domain of knowledge.

When the mind perceives the reality of a physical system, it creates an imprint in the mindscape. The difference between original and imprint is the *meaning* that the mind derives from the information. Meaning is the difference between the 'objective' reality of the perceived physical phenomenon and the mental imprint of that reality. Meaning is the difference between information and knowledge. And knowledge, the mental imprint, is subjective, as it is unique to the mindscape and its window that did the processing. By 'lifting out' specific mental renditions from the mindscape, the mind establishes new relationships (differentials) with the mindscape itself, and thus creates new renditions. 'My mind was wondering' can be seen as the mindscape-window that wanders through the mindscape and emerges from it in full consciousness, like a violin-solo emerging from the background of the sound-waves emanating from the orchestra, rising up, then soaring as an independent melody, when the mind identifies itself with that thought in the defining moment of creation. A novel independent knowledge construct is born.

Increasing versatility is one of the defining hallmarks of evolution, as we have seen in the domains of the cosmos and life. In the knowledge domain, it continues its expansionary journey. Biological evolution has created the human being as the most versatile creature yet, that keeps its own blueprint safely stored in the human genome. Many Darwinian evolutionists view the biological manifestation of the genome, the human body, merely as the 'vessel that the genotype uses to sustain its own existence'. Another remainder of a 'tone' of voice that, in my view, belongs in the evolutionary history books, but should have no place in the reality of defining the evolution of the future. The difference is similar to looking at a color photograph. Both, the 'negative' and the 'positive,' contain the same image. What a difference it makes when you look at the same image, but turn it from negative to positive!

Biological evolution is the continuous creation of new technologies. From the genotype, nature worked to create the new knowledge and physical systems that expanded the awareness, the processing power and the dexterity of living organisms. Human history is the continuation of that process.

Magnifying glasses, microscopes, telescopes, contact lenses, retinal implants, hearing aids, cochlear implants, loudspeakers, X-ray machines and MRI-machines expanded our awareness as they amplify our senses. As these devices appeared and evolved, they changed our thinking about the structure of reality, and in doing so, changed our explanation of reality. The technology changed our mind.

Cell phones, videoconference devices, TV, and fax machines extended our senses to bridge large distances, effectively eliminating the boundaries of space and time. These media changed the way we handle space and time; no longer do we accept them as limiting factors in our endeavors. They changed our mind.

Chariots, ships, automobiles and airplanes extended the dexterity of our limbs, carried us with increasing speed through expanded territories, and changed our view of the reality of the planet. The means of transportation changed our minds about family life, and the structure of social interaction. They expanded our mindscapes with impressions we couldn't absorb before. Impressions added to the mindscape changed the way we process experiences. New means of transportation changed our mind.

Steam engines, electric motors, explosion motors, and electricity producing power plants expanded our dexterity even further. These machines increased our mobility, eliminated the boundaries of our physical strength and redefined human labor. They expanded our patterns of behavior, and they changed our mindset.

The alphabet, the printed word, the novel and the movie-picture, were some of the stepping-stones in the expansion of the creation and communication of human renditions of knowledge. Like every other physical system, they were media transmitting knowledge between reality and human minds, and changed the minds they touched. In *Interface Culture*, Steven Johnson describes the effect of the word processor with its graphical user interface on the way we think:

"The computer had not only made it easier for me to write; it had also changed the very substance of what I was writing, and it that sense, I suspect, it had an enormous effect on my thinking as well. [25]

Jacquard looms, the Babbage analytic machine, 'Hollerith' punch card machines and a variety of other mechanical 'information processing' machines extended our processing power. These machines pushed the mind to expand itself, forced new reflections upon its own functioning, and in doing so, changed its own view of the world.

The ten thousand years of human evolution was a roller coaster ride of mental creativity. New knowledge and new technologies took turns in creating new realities, feeding back upon each other and adding to each other while continuously expanding the mindscape of individuals, communities and societies. The invention of the alphabet and the appearance of the printing press were cusps in a curve of acceleration knowledge creation, and the curve of expansion of the human mindscape. New means of transportation and communication increased the spread of physical renditions and the knowledge that created them. And all along the way, with every new discovery and every new technology, the mind changed its own patterns of thinking.

By 1930, less than a century ago, our explanation of reality had changed so completely that it would take a century (or more) for the changes to be absorbed by the minds of our planet. From Copernicus to Galileo to Newton to Einstein to Hubble our explanation of cosmic reality was turned on its head. By 1927, participants of the Solvay conference in Brussels had extended our explanation of reality to include the strange behavior of the micro-world of particles that permeates every thing that exists. Darwin discovered that the living world was not created 'as is', but rose out of the dust of the universe.

While human awareness and physical dexterity expanded relentlessly, human processing power – except for the use of mechanical tabulating machines - hardly did. Until, in 1936, Alan Mathison Turing developed a radically new concept. He devised an abstract model for an artificial computing machine. This machine would be able to process knowledge, as long as numbers could represent that knowledge.

The publication of his paper *On Computable Numbers* marks a turning point in human history. His concept resulted in the appearance of the digital computer. The importance of his invention as a new means for information processing is now obvious. More importantly, however, the invention would lead to a new way of representing reality. For the first time, reality could be rendered *outside of the human brain* with a depth, resolution and operational precision that was unthinkable before. With the 'Turing Machine', a new way of creating *functioning* renditions of reality was born. This new medium would change our way of thinking in manners Turing could not foresee, as it would transcend its role as medium and become part of reality itself.

Speech, the alphabet, the spoken word, the written word, the printed word, movies, radio, TV-news: every medium was a different way to manifest human thought and to represent reality. Not only did they change reality, they changed the way we perceive reality. And in doing so, they changed the *way* we think.

Now there is email, hand-held, laptop, desktop 'personal' computers, and the Internet. The digital medium combines most of the media ever invented by humans. It is emerging as the well from which other media spring. And that will change our mind, again: *The medium is more than the message, because it changes the way we think.*

Digital Wakeup Call

You spread out your body on the beach, offering yourself to the light to get a suntan under the guise of 'just relaxing'. After a while, you become aware of something touching your feet. It is a tiny wave of water. Then, more and more waves are coming, creeping a little higher every time. When a wave touches your knees, you realize what's happening. The high tide is coming.

Digital Bubbles

In the last few decades, waves of a different nature are tickling our awareness. These waves are digital. Yet, many don't sense the amplitude and the depth of the swelling tidal wave. These people view the Internet, cyberspace, PC's, maybe even some software applications as useful (or annoying) 'tools': 'neat' technology that we apply to some of our activities. They are fascinated by 'surfing the net', 'online gaming', and 'e-commerce', 'wireless networking' like children watching the playful dance of sparrows on a late-spring afternoon, missing the announcement of the approaching summer. It reminds me of the words of Nicholas Negroponte in his foreword to *Unleashing the Killer App*:

> *"... the fundamental difference between a dog and a human being is simple: When you point with your finger, the dog looks at the tip. The human looks toward the direction in which you point."*[26]

Many keep looking at the (digital) finger and fail to see the direction. Negroponte was one of a few that saw the significance of what was happening, way back, in the sixties and seventies. In 1976, I welcomed him to the University of Technology at Eijndhoven, The Netherlands, where I worked as a teacher and researcher at the time. The seminar he presented was about

MIT's work in 'Computer Aided Architectural Design', but his message went far beyond the subject at hand.

His message was about the pervasiveness, ubiquity, and personal nature of computing in the future. 'Personal computing' had not been invented yet, and 'user-interface' was a strange word in an emerging digital vocabulary. After the presentation, one of the participants asked "what about the 1984 scenario?" The person in question was referring to Orwell's book 1984 that conveyed the dark 'big brother is watching you' prophecy. At the time, many viewed computers as the Trojan horse that would transform that Orwellian prophecy into reality. After all, 1984 was just 8 years away. The question was asked in the context of Negroponte's remarks on 'idiosyncratic computer systems': software that would be able to better understand human intention. His comments on the question might seem obvious today; back in 1976, however, not many heard them as realistic predictions of the emergence of hundreds of millions of connected personal Computers:

> *"... the social consequences can be dealt with in terms of having a large population of such machines effectively available to everybody. It is really a matter of distribution and access, where the social consequences are going to come in, not in terms of the machine function. It seems to me, and we can argue this at great length, that the proliferation of computer systems is so large while the cost will go down so radically that this is going to fall out automatically. There is a comment that my colleague Seymour Papert always made, and that is 'that if every child in the world had a computer, then [every child] could afford to have a computer'. In other words, there would be so many of them, we could produce them at that cost, and I have a feeling that that will in fact be something we can rely on. That is my optimistic view of the future, as opposed to some centralized machine of the 1984 variety ... I'm certainly very conscious of it, I'm indeed afraid of it, but I don't think it's going to happen, I really don't. I think we are going to see a kind of ubiquitous, or distributed use. Even if we forget machine intelligence for a second, I think we are going to see this kind of thing in toys and iceboxes and bicycles, at home, all over the place. It is going to be so distributed and proliferated that I don't see the 1984 model emerging. Yes, I would be afraid of it, but I don't think that that's the consequence of studying ways to have machines be sensitive to things that are important to human discourse."*[27]

In his best-selling book Being Digital, Negroponte points out the depth of the changes that digital technology enables us to make, as 'bits replace atoms', the overarching themes that runs through his book. His vision rings through, from his reply to the Orwellian question in 1976 to Being Digital in 1995, bridging 20 years of pioneering work by MIT's multimedia laboratory, founded by Negroponte:

"While politicians struggle with the baggage of history, a new generation is emerging from the digital landscape free of many of the old prejudices. These kids are released from the limitation of geographic proximity as the sole basis for friendship, collaboration, play, and neighborhood. Digital technology can be a natural force drawing people into greater world harmony. The harmonizing effect of being digital is already apparent as previously partitioned disciplines and enterprises find themselves collaborating, not competing. A previously missing common language emerges, allowing people to understand across boundaries. .. But more than anything, my optimism comes from the empowering nature of being digital. The access, the mobility, and the ability to effect change are what will make the future so different from the present."[28]

As profound an impact as bits have as they replace atoms in many of our professional, social and personal activities, and as deep as the effects on our lives might be, it is, still, just the beginning.

First, bits and bytes started to replace traditional renditions of reality (e.g. email replacing mail, digital images replacing traditional blueprints). Next, bits and bytes began to replace parts of reality itself (e.g. online catalogs replacing showrooms, electronic agents replacing travel agents and insurance brokers). Next, through the (inter-) Net, we can create new bits and bytes, and thus, change reality. As Esther Dyson suggests in *Release 2.1*:

"Indeed, the biggest opportunity of the Net is that it allows you to go beyond choosing and start creating. The Net is uniquely malleable: it lets you build communities, find ideas, share information, and connect with other people."[29]

Negroponte's vision of accessibility, pervasiveness, ubiquity, and richness of digital information and communication is becoming a reality, and his 'harmonizing effect of being digital' is emerging in some parts of reality, like business, science and engineering. However, the harmonizing effects in the social fabric of the global world are still waiting for many digital bridges to be built, especially to communities of people who would benefit most. Dyson's 'malleability' of the Net is apparent, but will be profoundly influenced by political interpretations of freedom and democracy by the forces that rule societies around the globe.

The depth of the digital wave as it manifests itself through the Internet is best described by the words from Pierre Levy's *Cyberculture*:

"Cyberspace, the interconnection of computers across the planet, is becoming the major infrastructure of production, management, and economic transaction. It will soon constitute the principal international collective infrastructure of memory, thought, and communication. Within a few decades, cyber-

space, virtual communication, its imagery and interactive simulations, the irrepressible turbulence of texts and science, will be essential mediators of humanity's collective intelligence. This new medium of communication and information will be accompanied by previously unknown forms of understanding, new participants in its production and management, and new criteria of evaluation in the orientation of knowledge. Any educational policy worth the name will have to take these factors into consideration."[30]

Not only will the new medium 'be accompanied by new forms of understanding', as Levy points out, but it will, by its own imprint in our mindscape, transform our understanding. And that is 'just' the impact of the Internet alone.

Digeality goes beyond the Net. In the vast and deep universe of the digital reality, the Net provides the threading that provides the backbone of the weave-world of digital renditions, and digital machines that are the products of human thinking. In this universe, beyond the replacement of atoms by bits, and beyond the creation of new bits, we can glimpse the contours of another vision. Fifteen billion years of evolution created the brushes that begin to draw that vision. Those brushes are the six billion human minds on our planet. These brushes are designing *new* realities.

"Six billion, JJ? Most people on the planet don't even have *access* to the Internet. Let alone, knowing how to use the software that allows them to create *new* things."

"Alvas is right, Mr. J."

I know. You point at the biggest divide ever created in human history: the knowledge divide. It is the most severe one, because it deepens all other divides. The Internet eliminates the barrier between the known and the knowable, and therefore widens the gap between the connected and the non-connected communities.

"As I said a few chapters ago, JJ, 'education, education, and education'."

"We have to do more than that Alvas. Access, access, and access. It should be on top of any planetary agenda."

"What organization can discuss such an agenda, JJ?"

The United Nations can, and does.

"Can they produce an action list, JJ?"

Sure, and they do.

"Then, who executes the action list, Mr. J?"

...!?

Illustration 35: Virtual reality rendition from the PC-game SimCity[31].

Illustration 35 shows a screenshot of a scene in the strategic computer-game SimCity. It shows a virtual city, created by a player of the game. A toy? Sure. As a toy, it is one of the waves tickling our brain, a forebode of things to come. Children are growing up with the possibility of creating novel digital worlds. For them, a gigantic shift is taking place, and it is taking place naturally. A shift that transforms the *impossible* into the *imaginable*. For children and adults alike, this shift is amplified by the digital renditions that we experience in scientific movies like Starwars, The Gladiator, AI, and a host of others that show novel, imagined worlds. Just movies? "For your entertainment only"?

New technologies, such as biotechnology and nanotechnology, made possible only through digeality, began to shape a second shift that transforms the realm of the *imaginable* into the realm of the *possible*. The time might have come to start thinking about the *feasible* and the *preferable*.

Digeality sprang from the creativity of the human mind. Today it starts re-entering the human mind as a new platform for human creativity, not as a separate entity, but as an extension of the human mind. Digeality emerges as the drawing board, the laboratory, and the meeting room for the design & engineering of new realities. That design & engineering activity will result in an expansion of the bit-replaces-atom power of being digital: the re-configuration of atoms in the real world replacing pre-arranged bits. The rearrangement of the real world, and the creation of new worlds, on our planet and beyond, exponentially expands the power of 'Being digital'. Both, the bit-replaces-atom and the atom-replaces-bit potential of digeality is the subject of part III.

Digital Explosion

Illustration 36: The 'dotcom' debacle: end of the digital era? Picture is from a 'dead-end' traffic sign somewhere in (very) rural Montana. The 'dots' visible in the sign are bullet holes, left behind by deer-

"Wait a minute, JJ!"

Why is that Alvas?

"Didn't the Dotcom debacle prove that your digital world doesn't work? Most of the hot digital companies are gone. Many think that we reached the end of the digital road."

Alvas is right. The last millennium ended with a Digital Bang, and many saw the end of a digital journey, that had just begun. If it would be for those folks, there is no Part III to this book, because the year 2000 exposed 'digital' as a bloated world of bits without substance, and the 'new economy' as a hoax. The collapse in the stock market doomed the 'dot.coms to the virtual existence of a bad memory. What more proof was needed? Trillions of US dollars burned in the digital flames of self-fulfilling prophecies called business plans.

Those who lost money in the NASDAQ had found the culprit. Now we can easily wash down our pride with the sweet wine of a vast consensus about the unsuitability of the new economy, as if it manifested an unsuccessful gene-mutation that resulted in a species destined to die in the world it was supposed to enrich. David McCoy, editor for "Gartner's Predictions for 2002" sums up the general feeling in January of 2002:

> *"Many have learned from the dot-com debacle and welcome a back-to-basics business approach. 2002 will bring enterprise prudence, a return to profitable business practices, a focus on return on investment, and strong volatility that will disrupt up to 50 percent of vendors."*[32]

The bubble had burst. 'Back to basics!' Really? What basics? Is McCoy referring to the basics of a one-dimensional interpretation of capitalism, where 'quarter-to-quarter' monetary growth is the only yardstick for success? Or was the bursting bubble of dot.coms an explosion of a new type of business-gene that changed the world of business from within? Was it an outburst of creative energy from innovative individual entrepreneurs, or just an asymptotic climax of greed?

Joseph Schumpeter, in his landmark Capitalism, Socialism, and Democracy, published in 1942, coined the phrase "creative destruction." Schumpeter viewed human creativity as the driving force of economic progress in free (market) societies, and hence "Capitalism, then, is by nature a form or method of economic change and not only never is, but never can be stationary"[33]. To Schumpeter, economic progress is change, driven by human creativity that constantly produces innovative products, services, production processes, and new markets. Keepers of the status quo will stay behind, because the rest of the world goes on. If creativity is the oxygen that fuels innovation, then the conformists will die from a lack of oxygen. The maintenance of a culture that constantly provides a supply of oxygen is a necessary condition for questioning, and overhauling the status quo. That is what 'creative destruction' is all about. When organizations fail to keep up this maintenance, they will be forced, eventually, to a more forceful form of creative destruction, or face complete destruction and cease to exist.

The 'dot.com' bubble was, in my view, a container that harnessed an enormous amount of oxygen in the form of new concepts, new ways of communication, and new products and services. More than the concepts themselves, it contained the 'DNA' for a new way of thinking. A way of thinking that constitutes a new - digital foundation - to the 'free market economy'. This foundation eliminates the boundaries of space and time, thus obliterating geographical and, as a consequence, cultural isolation. This foundation exposes our world for what it is: a finite sphere without any boundaries.

On its surface are six billion people trying to make a living. Each human being with his/her own lifestyle, and trying to maintain, or improve his/her state of well-being. The new digital foundation connects each one to all the others, and makes all of human knowledge available to all of them. The digital foundation shifts the attention from the physical infrastructure to the mental infrastructure, from new factories to new ideas, from improvement of products to the invention of new human experiences. That foundation opens up the forest of the 'mass-consumer' and reveals the personalities, and the lifestyles, of individuals. It changes the concept of 'scale', revealing that there is no absolute dimension in which to measure 'well-being', and that 'the customer' is an individual human being, rather than a hungry monster with a million mouths to feed.

The digital bubble contains the seeds for a new world-tapestry that weaves together the minds of people. This is the tapestry of human knowledge that finds its expression in the bits and bytes of a new digital reality, thus exposing new needs, demands for new services, and the contours of what those new services should be.

Michael J. Mandel, economics editor at Business Week, predicted the sharp economic downturn that would devastate the technology sector, in his book The Coming Internet Depression, published at the peak of the stock market in 2000. Mandel believes that a New Economy did emerge, and his prediction is based precisely on the ignorance of the profundity of the change. According to Mandel,

> *"The Old Economy marshaled the forces of the financial markets to support investment in physical capital, such as factories, railroads, and roads. The New Economy marshals the forces of the financial markets to support innovation-and that is a big difference."*[34]

The 'new economy' is real, because the foundation of society went digital, and the bursting of the bubble was a digital Big Bang. Like the one that started the universe, it was big. Unlike the first one, it produced a lot of noise. However, its impact should not be measured by the noise it created, economic or otherwise. Rather, it is the quiet, yet pervasive and global fertilization of human thinking that signifies its importance. New memes emerged from the waves of digital renditions of human knowledge and filled the bubble. The pressures of human creativity, entrepreneurship, as well as greed, expanded the bubble quickly. It expanded in the vacuum of a world without a purpose, a world that measured well-being in dollars, a world that grew the number of its occupants together with their inequality and the ignorance of that inequality. The bubble finally exploded and spread the memes throughout a new connected world, exposing the inequalities. And therein lays the promise. Never has the opportunity for the creation of new synergies been greater, and never has an opportunity been more challenging.

Amongst the mind-storms of the 'war on terror' and the threats of transforming old conflicts into full-blown wars as a consequence, a different sound rings through: the sound of recognizing the value of every human life and every human mind. The sound might be faint, but it emanates from a new foundation based on digital technology, that permeates the planet in a single heartbeat. This sound is not the tune of 'unification' of human values, of human thought. It is, on the contrary, the sound of the value of diversity, of creating new forms of organization, and of defining new purpose.

Digital Thoughts

The last ten thousand years of human history tell only part of the journey traveled by Homo sapiens since it first appeared. A bird's eye view highlights just a particular aspect of that leg of the journey, and does not do justice to the richness of human history at all. Evolutionary timescales soften the severity of this lack of historic detail. If we compress the 15 billion years of evolution into one year, ten thousand years represent less than 22 seconds; compressing it to one-day makes human thought only a fraction of a second old. It is in this fraction of time that the human mind woke up to its own capabilities. A fraction in evolutionary time, measured in thousands of generations of human lives.

Generations of human thought that produced ever-expanding waves of knowledge, followed by the renditions of that knowledge - new technologies - that, in turn, lead to the formation and creation of new knowledge, and so forth. These expanding spirals of knowledge changed the way humans think, in the sense as McLuhan and Shlain explain in their works, or as explained by my simplistic pinhead model of the mindscape. This is not a matter of biological change, but a matter of the way we use our minds. In the process, humanity expanded its common knowledge base – the common portion of its collective mindscape.

While Arr made the first move to cross the evolutionary threshold to the domain of knowledge, we are only now waking up to the creative power of our own human minds.

Digital technology is the latest invention of nature – the nature of the human being – to expand the versatility of its individual creatures. With it, nature invented the tool for the next stage of evolution. The bursting of the bubble did not signify an ending, or even the beginning of an end.

Some see the digital world as a manifestation of Teilhard's Noosphere that encircles the globe as the *final* manifestation of thought. They see it as the beginning of the closure of the meaning of humanity; the beginning of a process that leads to a point Omega that marks the emergence of God.

Rather, digeality opens up new ways to render human thought, and new ways to connect and render those thoughts in innovative and unpredictable ways. Thus, once again, nature provides a new *kind* of chaos, that presents the opportunities for the creation of new order, and new kinds of order. Rather than a convergence or a coalescence of thought, a new universe of expanding thought is emerging. Teilhard's vision of a point Omega might be the correct one; the Noosphere however, as it turns out, is not the means that brings about closure. Rather, by rendering it digitally, it brings a new outlook for universal expansion of human thought. This universal nature of digeality without closure is the central theme of Levy's *Cyberculture*:

> *"And what about totality? Using my terminology, totality is the stabilized unity of meaning associated with diversity. That this identity might be organic, dialectic, or complex rather than simple or mechanical changes nothing: it is always totality, an encapsulating semantic closure. However, in bringing about humanity's virtual presence to itself, cyberculture does not impose a unity of meaning. That is the principal argument I have been trying to defend in this book."*[35]

The sounds that emanate from a world in friction come to us with the speed of light, rendered in digital format. After hitting the snooze button for so long, a digital wakeup call is calling for our attention. That wakeup call is the bell for a new round of evolution. That round is the evolution of human thought that began just an evolutionary second ago.

First, nature found a way to express algorithmic knowledge in the form of organisms. The algorithm is embedded in the genotype, and the executing machine is the cell. The produced organism had a behavioral repertoire that was encoded in the genotype as well. It simply produced organisms that could only perform certain functions, and not others. Next, the control and management of the behavioral functionality acquired its own machinery that was separate from the behavior itself.

Then, the nervous system and brain acquired the ability to *learn*. This was the end of the monopoly of the genotype over the phenotype. The brain could now build, and repeat new algorithm. The brain and the nervous system was the execution machinery, and the various organs of the creature manifested the executed algorithm in the surrounding world, in the form of physical interactions.

With these accomplishments, the living organism seemed to be 'complete': it had everything imaginable to 'make a living'. However, as it turned out, nature wasn't done, and evolution didn't stop there. At the threshold of knowledge, the learned algorithm gained independence from the machinery of the brain and the nervous system. The human started to produce 'memes'. While living cells 'knew' how to translate a genotype into a phenotype to express the knowledge embedded in the genes, humans needed to learn to translate its memes: to translate them into new and extended phenotypes. The invention of tools, the discovery of agriculture, the construction of advanced shelters, and the fabrication of clothing for colder climates were some of the first expressions of the newfound knowledge. Organisms found ways to make a living in the air, on land, and in the water. Human knowledge expanded the range of possible living environments once again, and to communicate that knowledge, we went from speech to language to print, and so forth.

Slowly, human beings learned to render knowledge into physical systems, from the wheel to the space shuttle. For most of human history, human beings were the executors of the procedures (algorithm) that led to the construction of such physical systems. Slowly, human beings learned to render procedures for the construction of increasing complexity into physical systems. Examples are drawings, diagrams or written manuals that encapsulate the design of physical systems as well as the operational steps to manufacture them. The Jacquard loom was one of the early attempts to create machines that could execute the rendition of a design, and produce the physical system automatically. Hollerith's punch-card machines extended the use of punched holes to represent numbers that in turn represented properties of the physical world.

The binary 'hole-or-no-hole' nature of the punch-card was the start of the 'bit' as the basic unit of information that enables us to express any physical or mental structure into a pattern of zeros and ones. We now needed a machine that could execute such representations, just like a living cell that 'knows' how to execute the genotype-algorithm by 'making sense' out of it's sequence of nucleotides and produce the right proteins. Just as nature had to produce a brain that was able to represent learned behaviors in its biological structure of neurons and synapses, we had to find ways to render knowledge in the language of bits and bytes. And just as nature gave the brain the ability to re-execute the stored algorithm to repeat the learned behavior, we had to invent a machine that could re-execute the pattern of bits and bytes and repeat the embedded procedures. In his 1936 publication, Turing came up with the conceptual foundation for a programmable binary machine that could do all of this. Not for the first time, nature – in this case human nature – found a way to express algorithmic knowledge in a novel way. Now, nature's way was digital.

Alvas & Novare: Alien Thoughts

"Wow!"

Are you glad we are done with evolution, Alvas?

"That too, although I have to say some things about that."

I know. That is why the heading of this paragraph has your name in it.

"That is exactly what makes me come out. Now I know you are listening to me."

I am. So, why don't you just let it out?

"I was infuriated."

That's clear. Why?

"Why wasn't I made aware of evolution much earlier?"

I thought you knew.

"I knew *about* it. But it meant nothing to me!"

But you were part of it, Alvas.

"All I know is that my topography changes all the time, at least within some limitations. That I know, because I can feel it. What I didn't know is that you call that evolution. It's all Novare's fault. She has been too lazy to explain it to me."

"Don't even listen to him, Mr. J. As usual, he blames it on me. You see, because he cannot *feel* the earth moving, he refuses to accept that it does, no matter how often I whisper *Eppur Si Muove*. He is the lazy one, always following the path of least resistance in that landscape of his."

"Let's get something straight here, Novare. I am the one who takes care of this vast mindscape. I can listen, I can look, I can speak, I can dream, I can feel, I can create the most complicated thoughts, and I can initiate very complex actions, all simultaneously, and all by myself. The only thing I ask for is a little window to look through; to look outside, and to look at my own landscape. And that window happens to be you, Novare. When that window is closed, there is nothing I can do, except keeping the body alive. It's time to wake up to your own potential, and live up to my expectations."

"*I, I, I, I.* Alvas in full color. Get rid of your habits, and you will free up time to look and listen. My window is open, whenever you are ready. Unfortunately you were not when I explained evolution to you."

Calm down guys. Do you *now* understand evolution, Alvas?

"Yes I do, because now I understand its effects; now I can *feel* it. And it makes me mad."

Why is that?

"Because it makes me wonder, JJ."

"It makes me wonder too, Mr. J."

About what, Novare?

"Many things."

And you, Alvas?

"Many things as well."

Come on, both of you. Let's hear it.

"People say they love their children, JJ, and their grandchildren. But they do nothing for their future. Just their own. They don't seem to give a damn about the future beyond their own death. Is that evolution?"

"Let me tell Mr. J what is *really* on our minds, Alvas."

"OK"

"Alvas and I have been talking."

Oh! About what?

"We wondered what aliens would say about the state of affairs here on earth. You know, about evolution, and the way people think. We believe we know what aliens would think!"

You do?

"Yep, JJ. Novare and I would like to play it for you."

Play it for me?

"Yep. We play two aliens that discovered earth in early 2003 and wonder about the things going on."

Sure, go ahead. But keep it cool, especially you, Alvas.

"I'm cool, JJ. As always."

"We explored your planet's knowledge repository, Mr. J. We traveled up and down and forth and back in it, visiting with the minds of millions on your planet. What we found is astounding, exciting, and extremely puzzling."

What was so puzzling?

"Most people don't realize their own evolution, JJ! They still think that they live in a permanent universe. Even those that realize that your universe had a beginning, they don't draw the conclusions from it. At best, they find it an 'interesting theory' that has no consequences for their lives. They still live their lives as if time is a stream of 'stuff' that runs out of a faucet like water, and that time is something you can 'use' or 'waste'. They fail to see that time is produced by change. If they stop having experiences that are different from one another, no time is produced and they fade away; you call it dying. "

"That is not what really puzzles us, Mr. J. Many minds that we visited recite Albert Einstein, and how he changed your view of time. What puzzles, or better, *amuses* us, is that your world has two different worldviews. On the one hand, we found a view that says that your world emerged through a process that you call *evolution*. You call that the *scientific* worldview. What amuses us is that only a small minority on your planet knows about it, although we now know why. What puzzles us is that only a tiny fraction of those that know *about* evolution, *think* evolution. Most of the minds that we absorbed think of the world as a game of pool: they treat events as they come along, just like the spread of billiard balls on the pool table, and take the situation from there. They don't have a clue about evolution in general, let alone the evolution of their own minds!"

I don't quite understand. Can you give me an example?

"I give you an example, JJ. Lawyers in the United States are trying to file lawsuits against companies that collaborated with the Nazi regime in World War II. Most people agree: that should be done. Fine, because horrible things happened. Other lawyers are seeking compensation from the government of the United States for the black population that was subject to the horror of slavery. Fine, inhumane and horrible things happened."

What's your point, Alvas?

"My point? Why don't the people on your planet learn from the past? More importantly, why don't those things make them realize that something similar might be going on today? Why don't they learn that the mindset of civilization changes in fundamental ways, even without much conscious planning? Has the number of people living under inhumane circumstances ever been bigger than today on your planet? Has the number of people that slowly destruct the planet ever been greater? You have the solutions, you have the technology, you have the money, but you let it happen anyway. My point? When do lawyers start to sue just about every government in the world for the pain imposed on every colonized territory? When do you start blaming whole generations for the dirt and poison in your atmosphere, or the total loss of job security? When do you start to blame your religions for preventing people to become co-creators of their own futures? It seems to me that smart lawyers can start handing out business cards now for the pain inflicted today by letting half of the world population rot away without even having the gutters to do it in? Whom are you going to blame a generation from now, when the mindset of the world has changed? That's my point. The mindset changes. More importantly, the mindset can be *changed*."

"Alvas always gets a little emotional about lawyers, Mr. J. Here is my analysis: some minds on your planet discovered that evolution is the reality you live in. First the universe, then the thermodynamic autonomy that you call life, and finally the biological independence of thought. Many are not aware of the evolution of the universe. Most ignore, or even fight the evolution of life, and only a handful of minds have discovered the evolution of your own knowledge, and hence: the way you think."

"It's not about the lawyers, Novare. Don't try to rationalize ignorance. Why does every company talk about 'intellectual capital', when none of them pays any dividend on the investment to the providers of that capital? Why do the same companies still have departments that they call *human* resources, when they only treat humans as the *links* between computers? Why do they call them 'knowledge workers', when they are being replaced by the next software version of their 'enterprise resource planning system'? The lawyers should get ready to sue these companies twenty years from know, when we have mastered to climb the next rung of Maslow's ladder. JJ talked about Teilhard de Chardin. As one of the few, Teilhard clearly saw the mental evolutionary power of human beings. The future is about the human mind, not about feeding the body. That should have been resolved by now. The future isn't even any longer about reproduction of those bodies by means of copulating bodies. As I said before, physical sex is the past, mental sex is the future."

"Alvas, you keep forgetting that you play an alien who just reports on his surprising findings."

"Surprising findings? Let me tell you about surprising findings!" Sure enough, some minds on your planet discovered evolution.

That's a first step. Now the masses have to accept that. Then people have to embrace it as a new source of creative energy. From the first quark in their universe to the first thought on their planet, evolution is the road of increasing freedom, increasing options, and increasing choices. That is where surprise comes into the picture. Most people are stuck in ancient thoughts. They still fight their fights and 'resolve' their conflicts with the tools of an era they should have transcended. Surprisingly they all fight their fights in the name of the same almighty God. Ironically, their God is a God of love. I'm thankful I come from a different universe."

A different universe?

"Indeed Mr. J. We were born in a different universe. We also evolved, but in a different way. We came from entities that developed the independence of thought before anything else. You might compare it to your computers, although our 'organisms' work with a different computing paradigm. As information processing organisms, we were dependent on the thermodynamic forces of the universe. We were nomads in the universe, if you will, without much self-determination. However, we did learn to communicate with each other. Once we had established communication, we started to find ways of synchronizing our modes of understanding. After that, the real frustration began: we did have no physical dexterity whatsoever. We could not *do* anything but think. We had to develop what evolution simply handed to you. We had to develop *life*. That is why Alvas becomes upset when he sees lives being wasted, and killed in the name of God or some other form of 'justice'."

"She is right, JJ. Sorry, if I crossed the line. But you might understand our puzzlement at the way people deal with their own minds. We attained 'intelligence' first, and had to struggle for ages to find a way to actually live, so that we could expand our range of choices, or find any choices at all, while the minds here don't yet recognize the value of life. They use their minds 'backward', if you get my drift. They don't use it to create new futures, like we did, but just to refine the methods for physical interaction."

We came a long way though, Alvas. After all, it has only been a few decades since we started to seriously study the human mind, and got rid of behaviorist thinking. In a way, it was easier for you, don't you think? You had a sophisticated mind first, you had established mental communication between yourselves, and you found ways to synchronize your understanding and share your knowledge. Then, you started to invent different physical creatures. These creatures could apply your knowledge, and act in, and upon, your universe.

"Wrong conclusion, JJ."

Oh?

"This is where the most astounding feature about the species on this planet comes in. In your case, nature provided you with diversity from the ground up. Diversity in the galaxies, solar systems and chemicals in your universe, then diversity of living organisms, and finally, a 'built in' diversity of human minds. Incredible! We had to discover that 'being' made no sense whatsoever, if everything is the same. Without variety, there is no communication, you see. What does a relationship between two identical things look like? Without a difference, the relationship has no property, accept a number that says 'how many are there'? Communication requires friction, and friction means difference. Difference is the spark that makes innovation possible. Thanks to the way we evolved, we did develop individual thoughts, and hence we recognized that our 'living manifestations' might have to be different too. Believe me, we fought many 'mind-wars' over the ways in which we should be different. Frankly, it was not until we crossed our universe into yours, and discovered your planet, that we learned how to do it. If you realize this, you might understand why our view of your world is one of awe, and frustration."

You also mentioned 'exciting' at the beginning. What did you find so exciting?

"I took a picture, Mr. J."

A picture, Novare?

"Indeed. Being female, it is a little difficult for me to explain."

What is the meaning of your picture?

"Please, Mr. J, before I explain, I have to make sure that I use it as an example only. The people of your planet finally discovered evolution, yet they don't embrace it yet. Then we heard you talk about McLuhan's and Shlain's theories. Alvas and I know that those theories carry some very fundamental truths. That is why I choose this picture as an example to demonstrate the depth of evolution."

Why do you hesitate then?

"... because I like it. It is funny, at least to me."

"Just confess, Novare. JJ, that is exactly what she finds so exciting. She finally finds a confirmation in it for the superiority of the female mind, that's all. I can't say that I like it much."

"Here it is, Mr. J. Remember my remarks on sex?"

Sure do.

"I said that sex was merely nature's way of creating new genotypes, while Alvas had different thoughts. Anyway, nature developed a host of different mechanisms to make sure that recombination of DNA did take place, resulting in two different types of animals, and human beings, that we call *male* and *female*."

And?

"This resulted in different physical characteristics for both. Male human beings acquired the greater physical strength, while females acquired the ability to develop offspring in their bodies, and a greater capacity to nurture that offspring. Also, according to many scientists, the male brain received a better capacity for *sequential reasoning*, while female brains acquired a better capacity for a *holistic* way of thinking."

And?

"Don't you see? Evolution gave men and women what was most useful for their respective functions. Then the alphabet came along, then the printed word. That gave an enormous advantage to men and their sequential brains, as Shlain explains. Then, the 20th Century came along, and every thing began to fall apart; for men, that is. Sorry, Mr. J, Alvas is right. I *am* excited."

Go right ahead, Novare.

"That is when the industrial revolution was in full swing. It took away the advantage of male physical strength! Then the Internet came along with its multimedia form of communication. From McLuhan's general observations on the influence of media on our thinking, and as Shlain explains specifically, men lose the advantage of their so-called 'rational brains' in the age of the Internet! It's fascinating, don't you think?"

Interesting, Novare.

"Just wait, JJ. She is not done."

There is more?

"There is, Mr. J. You see, sex is back to what it was in the times of the eukaryotes, billions of years ago. The recombination of DNA does no longer require physical intercourse. With modern technology, the requirement for copulation will be all but eliminated."

"But that doesn't mean..."

"Stop it Alvas, everybody knows what *you* want to talk about."

I'm not sure what you are trying to say, Novare ....

Illustration 37: Statute is part of the Franklin D. Roosevelt memorial, Washington.

"Nature has given, and nature has taken, Mr. J. And to you, Alvas: take it like a man! It's evolution, you see. This century is the century of female thinking. Men will have to look around in the domain of knowledge to find new meaning. Physical is out and mental is in! Don't worry too much, though. We'll take care."

"Unbelievable, JJ. Such arrogance. After all, we brought the world to where it is today."

I'm not sure, Alvas, how to judge *that.* I do find Novare's example an interesting bite for thought.

"Remember, Mr. J, I explain this as an alien visiting the planet earth. Seen from that perspective I can say that there is hope. You finally found a way to adequately represent knowledge. I'm talking about digital technology. The development of the right algorithm requires a lot of 'male' sequential thinking. So, there might be your male future. Although you still don't have the right computing paradigm, you are on the right track. It was exciting to see that after centuries of inventing new ways to render your knowledge through the spoken, the written and the printed word, you finally found the secret to render the algorithm that produce renditions. Congratulations! Now you can produce your automobiles, houses and bridges digitally, without ever laying a brick or bending a sheet of metal. However, at the same time it puzzles us."

Why?

"Because you don't seem to realize the potential of digital technology, JJ. You now acquired the tools to produce new life forms, molecular self-replicating machines, new futures. And about that 'male' future, JJ. Don't forget that all of us carry 'female' and 'male' traits in us, as Shlain explains. Novare should therefore be more humble. We'll have a role, that's all I'm saying. Especially with the new complexities that digeality will bring to the world."

"I believe it is time to explore what digeality really *is*, Mr. J. And with that, turn our attention from the past to the future."

Well said, Novare. That's exactly what we will do.

PART III: MEMORIES OF THE FUTURE

"In the 21st Century I believe that the mission of the United Nations will be defined by a new, more profound awareness of the sanctity and dignity of every human life, regardless of race or religion. This will require us to look beyond the framework of states and beneath the surface of nations and communities. We must focus, as never before, on improving the conditions of the individual men and women who give the state or nation its richness and character."

Kofi Anan, United Nations Secretary General[1]

"Be certain that the human species will do what it has always done - embrace challenge, vigorously inhale the smell of danger, and seek out ever more intoxicating adventures. We are genetically programmed to incur risks, to seek out the novel, and to accomplish the impossible...
...Over the centuries, these acts of creation will change us forever, expanding our minds, our sensibilities, our imaginations, and our sense of efficacy beyond anything we can currently conceptualize. We will finally see ourselves as the true masters of the universe."

Michael Zey[2]

Human beings occupy one planet in our universe, and all of them share, and harbor, the same experiences of billions of years of evolution. Our common biological foundation, manifested in flesh and blood and algorithm, is the bond that we can choose to ignore, or to embrace with passion. In between the two is the void where arrogance reigns. While we keep hammering on the superficial differences of color and gender, we ignore the astonishing mental 'room at the top', which is the new domain for evolutionary expansion: the domain of human knowledge.

Human nature created a new reality as an expression of individual and collective knowledge. In this new reality, we are capable to express new physical realities: new renditions of non-living entities and living creatures. As we expand the realm of the imaginable in digeality, we expand the realm of the possible, and face a new challenge: the challenge of choice, the choice of the preferable. Evolution expanded the glass of opportunity to unprecedented proportions. The options range from self-destruction to the expansion into new realms of space and time. To do the latter requires a fundamental change of mind, while the first hardly requires any change. The chaos in our mental existence, with the powerful extension of digeality, provides the opportunity for change. The chaos provides plenty of mental energy to make that change. We just have to pick up that energy, and create novel forms of order.

As we change our mind, we will take our human talents into the universe, and expand life and knowledge into the universe, and thus expand the horizons of human 'being' itself. While we prepare for a new journey of expanded being, we will change our perspective on the planet that produced us, and sustained us, in the past, and into a new future.

Chapter 9: Digeality

"When Isaac Newton walked along the beach, picking up seashells, he did not realize that the vast ocean of undiscovered truth that lay before him would contain such scientific wonders. He probably could not foresee the day when science would unravel the secret of life, the atom, and the mind. Today, that ocean has yielded many of its secrets. Now a new ocean has opened up. As we have seen, it is a wondrous ocean of scientific possibilities and applications. Perhaps in our lifetime, we will see many of these marvels of science unfold before us. For we are no longer passive observers to the dance of nature; we are in the process of becoming active choreographers. With the basic laws of the quantum, DNA, and computers discovered, we are now embarking upon a much greater journey, one that ultimately promises to take us to the stars. As our understanding of the fourth pillar, space-time, increases, this opens up the possibility in the far future to become masters of space and time."

Michio Kaku[1]

In his book *Visions*, Michio Kaku refers to 'space-time' as the fourth pillar of modern science. The first three Pillars are Matter, Life, and the Mind. In the previous part of the book, we have explored matter as the evolution of the cosmic domain, life as the evolution of the Life domain, and the mind as the emergence of the knowledge domain. Each domain added a new 'sense' of space and time: a new manifestation of the space-time continuum superimposed on the previous one. With digital technology, we reached the frontier of the space and time: the speed of light.

As we travel around the world carrying our mobile phones, we hardly realize that tiny particles of matter are searching for us around the globe, when somebody wants to call us. These particles carry the digital code that uniquely identifies each one of us, and transport the information with the speed of light in pursuit of the phone with a matching code. With digital technology, matter connects the minds of living human beings with the speed of light.

In this chapter, we want to get a 'feeling' for the vastness of the ocean of knowledge that the emergent digital reality represents. Like Isaac Newton in the quote from Kaku's *Visions* above, we'll pick up a few 'seashells' and look at them to see what they can tell us about the depths and the properties of the ocean they come from.

Before we look at them, however, we want to sit back, relax, and imagine a sailing trip across the digital ocean's surface to see if we can identify some of its underlying currents. I suspect that some readers will be amused, some fascinated, others frightened or frustrated, and that some even might consider all of it pure nonsense. None of it is futuristic, however. All of the imagined scenes, and all of the' seashells' we look at, are manifestations of technologies that are happening, or at least starting to happen today. Technology is going ahead, with or without adaptation of our mindsets. Evolution of human knowledge, and the human mind, is proceeding, even as many ignore the very existence of evolution itself. Let us now find out what digeality *is*.

A Digital Day

6.15 am: Con te Partiro

The house is trying to wake me up, while the bed performs a gentle massage to the tune of 'nature's dreams', one of those tunes that brings the sounds of nature to soothe the human soul. Usually, this gets me out of bed quickly, but not today. Not a good omen. Since I'm still not moving, the bed switches to Andrea Bocelli's *Con te Partiro*. I get the message, and with it, a sense of time.

"The house is already up and ready to go, JJ. You have a breakfast meeting at 7.30 with Angie Rapallo at the Echo Café," the bedside lamp says, while projecting Angie's face on the ceiling.

Darn. I forgot about that. "Where are my notes? And get my car warmed up. Don't forget to switch on the seat-heaters."

"Already done, and the notes are in your shirt," the bed answers.

It's too late for the treadmill, so I have to get the morning news from the mirror. I stumble to the bathroom, to consciousness, and a new sense of time. I ask the mirror to give me a selection of the morning news. It gets my preferences from my digital twin and searches for the bites of sound and images I want. While it compares today's news to yesterday's, it stays in 'mirror' mode; time enough to shave. Thank God for comparing news. My own digital newsagent compares for contents and assembles only what's new, as compared to yesterday. If I want to, I'll get a summary of yesterday's news to bring me up to date.

"Want to hear your schedule for the day?"

"Sure."

"After your breakfast meeting, no more *physical* appointments for the rest of the day, until meeting your wife for dinner. She comes in on the shuttle at 7 pm and meets you at 8 pm at the *Rose Garden*. She will call you later. At 10.30 am, you will meet Carlson from New York, Villeneuve from Paris, and Lee from Hong Kong. Lee only reports on the research findings, so her mind will not participate; otherwise, it is expected to be a highly interactive meeting, so your personal presence is expected. Mind-presence only, though. You could do it from the car."

"Good. Then I might finally look at those real estate properties, remember?"

"That's exactly what I suggest. Your next appointment is at ...."

"Ouch!" I zoom in on the mirror to study today's result from a clumsy shave-cut. "Oh well, let's have the news, anyway." Better get showered. As an ISN reporter comments on the progress of the Mars colony, I think back to the old day of TV, where hundreds of channels broadcasted the same things, over and over again. Oh, some still exist today. Interestingly, those that finally confessed to their biased reporting found a new and valuable niche. ISN [Inter Society News] is one of the major feeds to Meta-News-Web, formerly CNN. MNW was one of the first fully interactive TV 'Networks' that integrated the old Internet with traditional TV. They transmit multiple, linked programs that each participant can tailor to personal preferences. Indeed, yesterday's *viewers* are today's *participants*. The producers use the latest software that semantically links different news-feeds and commentaries. The result is that I can receive opposing viewpoints on a certain subject, no matter what channel it comes from, look at different subjects with the same bias, or get a high-level overview of different perspectives. Many of the 'anchors' are replaced by virtual personalities, orchestrated by teams of writers. It's funny. The ones that prided themselves for their 'objectivity' were the easiest to replace, while the ones with a personal opinion, the ones that showed emotion, stayed. Human is in, these days.

Anyway, I finish the shower and put on my underwear. The shirt is of the newest variety, lightweight, and with much higher functionality. I take it off the hanger that doubles as a charger, and check its status in the 'mirror'. I only wear it when I'm out of the office. Its built-in 360-degree airbag gives protection in all kinds of unforeseeable situations. It monitors my bodily vital signs, and gives me advanced warning for such things as heart attacks or strokes. The newest ones even warn me in the case of certain bacterial or viral infections. It maintains my DNA signature, and produces a new one, whenever a fresh one is required. Using somebody else's DNA is a felony, and prosecuted stringently. Many places do require an instant test using your hand as the source of DNA. Iris-scans proved too easy to fake, especially with new organic iris replacements, or alterations using laser surgery. All my medical information is attached to my DT (my 'Digital Twin'). My DT only provides information to authorized people. My digital counterpart is my extended memory. It stores all information pertaining to me, from my DNA signature to the odometer reading of my car; from a detailed 3D model of my house to the balance in my checking account, and from my childhood pictures to my favorite music today. In the old days, doctors, insurance companies, supermarket chains, online retailers or the government owned information about me. Not any longer. I own it: it is my DT. I have to authorize any use of it and never transfer ownership.

The older shirts had a lot of local memory, because too many places didn't have high-bandwidth transmission capability. The shirt starts doing its morning routine, as it sorts through the notes sent by the bed. I dictated those notes last night on my tablet-PC that doubles as alarm clock and remote control for the house. The shirt transcribes my notes, so I can use them at my breakfast meeting. Finally, it measures my vital signs and transmits the data to the doctor's digital twin.

I unplug my shoes from the charger and put them on. I had to replace them recently, because of new GPS technology with a much higher accuracy. More importantly, a new technology, called 'raspberry', replaced the old system because of its much higher sensitivity. These days you need reliable wireless communication, because you want your family members to know where you are. With this technology, you can find your way around places, because office buildings, shopping malls and fun parks recognize you, know where you are, and give you directions. Of course, the subscription fee to these services is cheaper if you agree to watch, or listen to commercials. The annoying thing is that you have to buy new shoes, when a new technology becomes available. No 'plug and play' for shoes these days. The shoes themselves are the expensive part, that's how they make their money, you see. Antenna, receiver, and power supply, all in the shoes. A new technology made it possible to transmit power wirelessly to my wristwatch that contained a DVD burner. Even these days you need to back-up your important stuff, just in case.

My mocha is ready, chocolate and all, prepared by the newest model of the 'Star'O'bucks' espresso machine. I grab the coffee, take a sip and... the chocolate is missing!

"Your shirt told me that your blood sugar is too high this morning, therefore no chocolate, as per your own instructions. By the way, it's decaf. Your blood pressure, remember?"

"Oh well, can't blame DT for taking care of me."

"The chocolate is getting low, anyway. I'll have the car remind you."

On my way to the door, the microwave stops me in my tracks.

"Don't forget about tomorrow's date." When the microwave's door displays my wife's smiling face in bright colors, I remember: tomorrow is her birthday.

7.00 am: On the Road

The morning is cold, the car nice and warm, and the batteries charged.

"Want me to turn the house, JJ?" the car asks.

"Sure, and I forgot to open the curtains. The house is still in manual mode."

"Done. I turned it to auto, so it will move to keep the windows on the sunny side. Don't forget to pick up chocolate. Want me to find a place that has your brand in stock, or you want it delivered? The refrigerator answers the door anyway."

"I'll pick it up at Charlene's. Those folks always crack me up. I like them, and I like their good humor. Just remind me when I'm close."

The Echo-Café is small, warm and never too full. It has a free Net-connection of only 200 Megabytes per second, but for the meeting it'll do. Miss Rapallo is an ambassador for the 'Global Lifestyle Society', one of those new Planetary Organizations, commonly called *PO's*. They provide their members with a full array of personal, professional and social services. Much debate is going on, since many traditional nations still refuse to relinquish power to other, more global organizations. PO's provide all the traditional insurances, like life, home, car, health, liability, and so on. The newer ones go far beyond that by providing services that used to be in the exclusive of the government. They provide such things as unemployment benefits, disability benefits, and the usual law-enforcement and security protection services, which they contract out to local governments. Membership assures you of these services no matter where you live in the world.

GLS, the one represented by Miss Rapallo, currently has 50 governments under contract, but your choice is not limited to those countries. If you live in one or more of the 50 under contract however, or if you commute between them for your work, the services provided are more elaborate. Your membership fees cover your taxes (as we used to call it), that, in turn, the organization pays to governments for their services. Your social security moneys don't go to governments, but, at your option, stay under own control, or go to a choice of affiliated financial services firms.

"The Dow-Jones in New York is still in rally mode. You want me to do something?" the car reports. These days, you invest in the New York stock exchange strictly for dividends. Times have changed. Companies are no longer valued only for their short-term financial performance. Financial performance is a necessary, but not sufficient criteria for determining the 'value' of the company. Financial performance can now be monitored online, and in real time. 'Quarterly earnings reports' are gone. Accounting and financial transactions were totally computerized and automatically relayed to the stock market, and the public. Financial analysts had to re-invent themselves: their 'predictions' based on their own financial analysis were replaced by user-friendly software that gives you an understandable, objective bird's eye view of the company's financial status. Besides financial performance, sustainability of the organization was now at the top of the list. It was measured by its ability to attract the right talent for the right projects, its ability to anticipate or to create new markets, and its consumption of planetary resources versus resources from space. You invest in these companies for the dividends they pay and the security of a (very) long-term capital gain. Today's volatility is rare, probably because of the Mars colony that opened up new perspectives for companies involved in space travel and tourism. Engineers had succeeded in creating fuel from the carbon dioxide in the planet's atmosphere and the ice below its surface. They had now introduced bio-genetically modified bacteria to help with the production of oxygen. That boosted the stock-price of the new breed of energy companies.

When I park the car in front of the Echo Café, the Café's charger connects itself to the car. "Angie Rapallo is going to be a few minutes late. I just talked to her car. She is ten kilometers away. She knows you're here." I instruct the car to get ready for the real estate trip later on.

Funny. Real estate in the old days, I mean. Six percent of the value of your home went to a real estate agent, because the agent *owned* the 'listing' of your property. I remember that as recent as 2003, the American Association of Realtors was still filing lawsuits to protect its 'ownership' of real estate information. This was at a time, when every real estate property for sale was already posted on one website or another. Today, every piece of real estate had its own 'Digital twin' in cyberspace, whether it is for sale or not. Property taxes, homeowner's insurance, maintenance records, inspection reports or the history of ownership, all of it is connected to the property's DT, which in turn is connected to the homeowner's DT. New residences or commercial buildings are going directly from the architect's drawing board to cyberspace. Cars, houses or people: their DT is created the moment they are conceived.

"Angie Rapallo is already inside, JJ. You dozed off, your shirt says. You only had decaf this morning, remember? That's why. Maybe it's time for some action." The car pulls me back to reality. The Café had detected Angie's presence and informed my car. It is obvious that she had told her DT to let me know her whereabouts.

8.00 am: Planetary Breakfast

I'm on my second cup of coffee. Angie had given me the story of her organization, which was very similar to the story of the society I belong to.

"I just wanted to explain to you that we offer all the services that you currently enjoy. Please allow me to get to where we differentiate ourselves from your organization."

"Please go ahead Angie."

"The difference can be summed up by one word: *citizenship*. We have been working on this with our members for years, and it evolved as we went along. Our philosophy is as follows. Digeality, and the technologies that sprang from it, have eliminated the boundaries of space and time. The geographical borders of traditional nations, distance, or travel-time have lost their meaning as the 'natural' boundaries for the organization and management of the planet. Most borders are becoming demarcation lines for the administration of services. The increased transparency of the world because of digital technology accelerated the awareness of people around the globe to such an extent that discriminatory boundaries because of race, color, gender, religion, or sexual orientation were on the road to their final breakdown by the 20's. Bridging these social gaps was further stimulated, when the traditional western nations finally realized that the astonishing decline of their white populations could no longer be kept out of the political agenda. As you know, today they are competing with one another for people to join their economies. Some do this by sticking to their traditional values as a country, some by becoming a service provider to organizations like our own, and others by a combination of the two."

"I understand your premises. Please tell me about your philosophy." Angie continues, a little embarrassed, "our philosophy can be summed up in two short sentences. Please hold on." She takes a display from her purse and rolls it out. The two sentences emerge in bright colors from a background image of my favorite village in the Black Forest of Germany:

- *freedom of lifestyle*
- *relationships create substance*

She continues to explain the services of the organization while my mind wonders through the memories of that village.

"..... as I was saying, we realize that relationships are at the core of any lifestyle, and our services help you connect to people with similar interests around the globe, whether these interests are of a religious, scientific, cultural, or other nature. We help you to establish and maintain exchanges. These inter-relationships create the substance of what we call citizenship. That is why we call our members *citizens*." Interesting. Forty years ago, that would have sounded eerie. Over the years, the planet, and its incredible diversity, had become so visible through digeality; it gave you a different perspective. The last war in the 00's had further accelerated the awareness of this great diversity. It was the start of a mind-shift. Now, with people on Mars, my perspective on the planet had changed. Now, I consider the planet my home, if you will, especially since my last trip to the moon.

9.15 am: Don't lose your Shirt

I had gone through my notes, asking Angie all the questions I had prepared, and we agreed to be in contact in a couple of days. My shirt had wired the slides she had shown to my home-server. While I drove away, the car asks "Were you happy with the services in the café? You forgot the tip." On the way out, I verbally confirmed the bill at the door; the shirt confirmed my identity and pays the bill. 'Loosing your shirt' has a new meaning these days. I was supposed to add the amount of the tip either vocally, or by entering it on a writing-pad. I forgot. "Just send two *Globals*; with my apologies."

"Done. You want to listen to the memo for your upcoming meeting?"

"Sure." Unlike many other people, I'm old-fashioned when it comes to implants. Because of a severe hearing problem that kept deteriorating, my wife convinced me to get cochlear implants. They double as a phone and microphone. I'm connected to the world all the time, unless I switch it off by simply telling it to shut up. The car simply forwards the memo for the meeting to my ear as I listen. The meeting is 'mind-presence only': my DT will be present in digital or holographic form in the meeting room. I'll be connected through my cochlear antenna and react verbally. My DT will actually speak my words and mimic my gestures. It's neat, because it gives the people in the room a sense of presence. For a while, hackers managed to disrupt such meetings by creating some 'interesting' disruptions. Anyway, newer technologies, whereby I have full digital presence are still experimental. With new technology, my DT is fully present and the physical me sees through its eyes, and so on. With all of that new technology, true physical meetings have become a real 'treat', and obtain a new significance.

The new Global Positioning system guides me to the first property I want to see. My wife had studied the place in digeality, in three-dimensions, and all the details. Except the trees. The property was shown in the summer, with a lush flora hiding many of its features. Now, in wintertime, the trees are bare and allow a view of the 'naked' property. She asked me to have a look at it. The owner of the property indicated that the property would be for sale three years from now, when he planned to retire. Sometimes you could make a good deal, if the owner had the certainty of selling the house *today*, and could stay for the three years. We'll see....

"Damn!" I did it. I saw the house in the last instance, hit the brakes, and the snow on the road did the rest. "I can't pick up the signals from the road, JJ. Too much snow. I'm sorry." Sure enough, I hit a sign that reads: 'DEAD END', just as the car said "no road-signal." I wish them all to some place far away: the sign, car, digital, house, all of it.

"I've done a preliminary estimate," the car says soothingly. "The total is 1,200 Euros, the insurance company approves, and here are your options for body-shops. You're lucky, the frame is OK, and the fender repaired itself. You can drive. Just give me a visual check. Oh, your right blinker doesn't work, but you just need to replace the LED. Here is the closest….."

12.50 pm: Precious Issues

The meeting had gone well. A precious issue, for everybody, not just for us, but every organization in the world, had been resolved. At least, we were confident of the proposed solution. It had been a struggle for everybody, since the paradigm on which the world had been operating, since the appearance of Homo sapiens, had lost its relevance. The paradigm's hold on the way we think, stronger yet: its hold on our *un*conscious mindset that had worked so well for thousands of years was gone. It had evaporated almost overnight. Gone. Maybe forever. At least as far as this planet is concerned. The Mars project helped economically, but it was only a drop on a hot plate. We had been late in changing our mind about the end of planetary growth and the emerging knowledge base of developing countries. In fact, it had been known since the end of the second millennium, but it had been too sensitive an issue to bring to the public's attention. Politics at the time was still based on the privileged access to information, and the manipulation of that information. The factories of the media could still manufacture political knowledge out of information. After all, it was a 'medium'.

Anyway, today it was obvious: the world population had peaked. The pie didn't grow any longer. This was it. The 'old' industrialized countries were hit the hardest. Populations in those countries had leveled off right around 2002. Space travel did not get the funding that would have been appropriate, at least in hindsight. A similar situation surrounded the development of alternative energy sources. The white population in those countries had gone down by a few hundred million total, while many kept fighting the immigration of 'foreigners'. Funny, how things change. Years ago, I had to compete to find an employer. Today, employers, like any other organization in the world, whether nations, cities or golf clubs, are competing for people. Which brings me back to my meeting that I wanted to tell you about.

I had not been able to attend because I had driven my car straight to the body shop. The parts came from across town and had been delivered by the time I got to the shop, which didn't have a Net connection suitable for my meeting. Therefore, I had given some last minute instructions to my digital twin to participate without me. Over the years, new software, called 'semantic janitor' had built an extensive profile of the language I use in such meetings, and the meaning of the concepts I try to convey. Using the same software, my DT had done the same for the people I meet frequently, so that a basis for mutual understanding had been established.

"I think it's time to get to the point, JJ," Alvas says through my cochlea.

I'm sorry, dear reader. The subject of the meeting was our low retention rate of highly rated talent. The proposal on the table was a risky one for the company, but then, the times required innovative solutions. We had negotiated a two-sided contract with one of the major 'Knowledge Coops' (KC's) that were based on the continuity-requirement of our company to retain talent, and the long term interests of the KC's members. This particular KC had relationships with large communities of members in Russia, Sweden, former East Germany, India, and the northwestern territories of Canada. New communities were in the process of being established in the Southwestern states of the US and Mexico. Since the US became officially bi-lingual in the 20's, the mobility of Spanish speaking people between North and South America had been increasing. Russia and Northern Canada provided climate-enhanced environments that greatly enhance the living conditions in their remote territories. In return for a 'social security retainer' that we pay to the KC, we have the guaranteed availability of the talent we need, while being able to select that talent on a project-basis as needed. That retainer paid for social security and pension benefits of the individual members and the continuity gaps of the KC. In addition, we agreed to participate in the organization's 'mobility & lifestyle' program, meaning that individual people and their families can choose a new residence once a year, if they so desire. This gives our own employees the possibility to participate in the program as well. The KC had excellent relationships with schools in many regions, and partially financed the continuous 'knowledge renewal' of the school's online and offline teachers.

"I'll be home early; I managed to get an early *glide* home with Jess. Want to have an early dinner at six?" my wife says. "Great. I'll see you at six at the *Rose*."

"Don't forget her birthday, JJ." Indeed, I kept the name Alvas for my DT, and he keeps calling me 'JJ'.

6.05 pm: A new Concept

A lazy afternoon. After I had replayed the meeting, the car was ready to go at 3 pm. I proceeded with my real estate 'tour', and got a birthday present for my wife, making sure she could return it. If she would say something like "how original!" I would know: return was imminent. Thank God that some things never change.

"Darling, this is the most original present you ever came up with!"

See what I mean. The thing was as good as back in the store. "Wonderful, I'm glad you like it. How did your strategy session go?" My wife works as a 'lifestyle counselor', a new profession that rose out of the phoenixes of the former real estate and travel agent professions. Her employer is Millennium-three, M-3 in short. M-3 is a combination of Century-21, the former real estate franchise and the former travel division of American Express. M-3 is an important client of my own company that provides 'thinking solutions' to a variety of industries.

"Would you like to change your mind about your selection?" the waiter asks. Both of our DT's had looked at the menu earlier in the day, and pre-selected our meals based on our preference, what we had last time, what we said about it, specials of the day, and so forth. The DT's kept themselves up to date on all of the menus around town, except the few that didn't maintain digital ones.

"I'm OK," my wife says. "Are you?" I look at my selection on the table in front of me. "Fine with me."

"I want to tell you about my glide." Jessica has one of the newer ones. They are truly whisper-quiet, no wonder they call them 'glides'. Flight-control had to divert us to a parking nearby, because the Rose's parking is too small for Jessie's glide. Hers is one of the new van-type models that can accommodate up to 20 people."

"Tell me about the meeting."

"Well, I submitted my proposal to localize our services. Our competitors are catching up with the services we offer. As you know, the inference engine that your company sold us, gave us the advantage of providing better knowledge to our customers. Your engine had better semantic description and cross-referencing capabilities. Now, the competition is catching up on that. More importantly, our clients expect a more personalized approach. They want counselors to understand their own personal interpretations of 'a good neighborhood', a 'good place to live', or 'a preferred way of living'.

"Our new product does exactly that."

"Darling, I know. But the product has to sell itself; I can only do so much for you." Indeed, some things never change. Digital connections are necessary, but not sufficient, to do business. Our new software allows users to define their own links between the knowledge provided in cyberspace and the meaning the inference engine attaches to the knowledge. That way, especially clients attaching a portion of their DT's to our engine, the counselors can 'program' the software to derive personal interpretations.

"The essence of my proposed concept is a human approach. Only human counselors can establish the final link between the interpretations of the client and your software. We already introduced a 'couleur locale', because our advice to clients depends on the *local* circumstances of the regions of the planet under consideration. With my new concept, we want to add a 'couleur humaine', if you will."

"Couleur humaine?"

"Yes, darling. You see, we live in the age of female intuition, remember? The guys used to call it 'holistic thinking' or 'systems approach'. Mind research has demonstrated the innate ability of the mind to form coherent holistic views of situations. Until your software can do the same, we'll have to trust our human 'instincts'."

"Funny thing, our mind. The moment our software catches up, the mind seems to travel beyond the capability of the software."

"Our minds focus on renditions, darling, not the underlying algorithm. Your analysts and programmers have to deal with algorithm that produce new renditions. We as counselors simply take the new renditions into account and produce new knowledge from them, capiche?"

"Keeps us in business though. How's your dinner?"

"Excellent, as usual. Cannot wait to get home though, and drop my shoes. I'm tired of them."

8.00 pm: Dropping the other Shoe.

Finally, we could drop our shoes. These days that was one of the most important moments of the day, since our connection to the outside world was in the shoes. Now, we were disconnected, and the evening was ours.

The ride home had been uneventful. As we approached the water, the house had already turned towards the passageway that led to the road. My car was an old-fashioned 'auto-mobile', and could only handle old-fashioned roads and calm waters. We had bought the house a few years ago to escape the hustle of suburbia. The house was part of a generous subdivision, just half a mile off the coast. A part of the ocean had been tamed by a system of submerged barriers that rose and lowered with the tide, just like the house itself. In cold weather, we lowered the house into the warm coastal waters to save energy and get away from the noises of the nearby coast.

At dawn, the house would rise again and find the sun, follow it, and absorb the sun's energy to power our lives for the coming day.

Afterthoughts

A digital day in some imagined future. How should we feel about such things? Is it frightening? Amusing? Promising? Detestable? Fascinating? Pure nonsense? Something that is too far away to worry about? Science fiction? Who cares?

TV, newspapers and magazines overwhelm us every day with the announcement of new gadgets and technological breakthroughs. Let us now turn our attention to some of these 'gadgets', and see what these 'digital seashells' tell us about the ocean of the underlying digital reality.

Because no matter what feelings the 'digital day' produces in our minds, this digital reality is of historic evolutionary importance, and will change not only what we see, but also how we see it.

Bytes of Living

In the following paragraphs, we look at some 'shells' of digital technology as they appear around the globe. The choice of shells is a tiny sample of tiles that are part of a vast digital mosaic that is covering and penetrating the planet. Some of the examples are of a military nature, because the current atmosphere in the world is one of threatening global wars. Given this military priority, and given the financial backing of military efforts, this will accelerate the development of fundamentally new digital technologies, as well as other technologies that are dependent on digital tools.

Digital Planet

The planet is becoming digital. Satellite images spread out and join in digeality with centimeter accuracy. The use of these images ranges from protecting the environment to military operations, and from real estate positioning to flight-route planning. Satellite images are two-dimensional; however, software can compose three-dimensional maps from two images taken from slightly different angles, thus mimicking the capability of our eyes.

New structures, like buildings, dams, roads, waterways, pipelines and so forth, are designed in three-dimensions using computer aided design software. The planet, and everything on it, will emerge as one vast three-dimensional digital landscape in the future.

When this book goes to print, the war in Iraq is a few weeks old. The palace in Baghdad (which might, or might not be there when you read this) and the pentagon, shown in illustration 38, are the poles of a continuum of war that might, or might not, be seen as an 'umbrella of opportunity' by other countries to enforce a resolution of their own conflicts. The reason to show these pictures is not only their symbolism for the pain of our times that drives the entertainment engines of 24-hour news channels. The reason to show them is the fact that they are private satellite pictures, taken by a commercial enterprise, using a private satellite called 'Quickbird'.

The planet is no longer a flat surface with the artificial grid of nation borders, nor is the knowledge of secrets any longer the domain of a Tom Clancy-type of espionage. In digeality, the features of the world are available and accessible to everybody. Baghdad, the Pentagon, London, Ham-

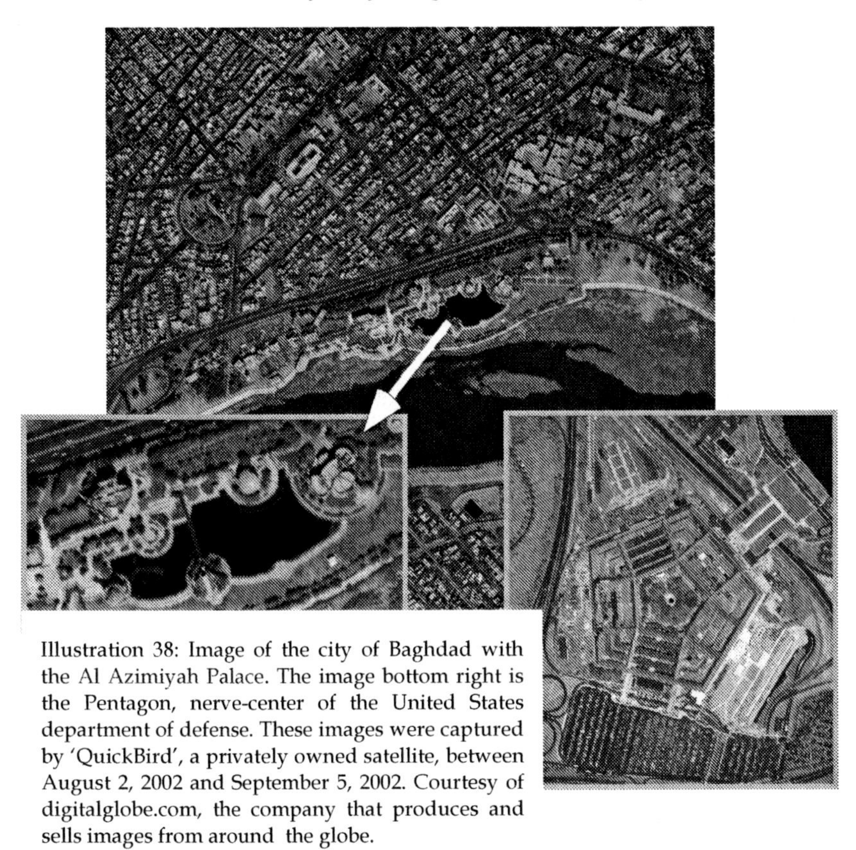

Illustration 38: Image of the city of Baghdad with the Al Azimiyah Palace. The image bottom right is the Pentagon, nerve-center of the United States department of defense. These images were captured by 'QuickBird', a privately owned satellite, between August 2, 2002 and September 5, 2002. Courtesy of digitalglobe.com, the company that produces and sells images from around the globe.

burg, Sydney, Yokohama or Paris are just 'targets' without a distance (illustration 39). They can be the targets of war and terrorism, or they can be the targets of our minds for innovation and creation.

Major parts of the war are planned, and executed, in digeality itself. Unmanned, remote controlled aircraft fly with digital 'awareness' and digital targets. 'Smart' bombs and guided missiles find their way in the reality of bits and bytes. The city of Baghdad is modeled digitally in three dimensions: the optimal positions of snipers and the trajectories of missiles can be calculated with perfect precision.

Digeality takes off the covers of distance and time, and penetrates brick and mortar: any and all real estate will be known, inside and out. Second, it exposes mindsets of people and nations, their leaders and their followers, their intentions and political modi operandi, whether anarchical, democratic, totalitarian or anything in between. Thirdly, and most importantly, the visibility in digeality begs to address the questions underlying the necessity for war, because it communicates, like never before, the mindsets of the people around the world. No longer can we ignore the differences of those mindsets, but we have to shift our attention to dealing with a growing diversity. Repairing the present no longer suffices, because we entered a new domain of evolutionary expansion. Planetary bytes bring it into the full daylight of the computer-screens around the world.

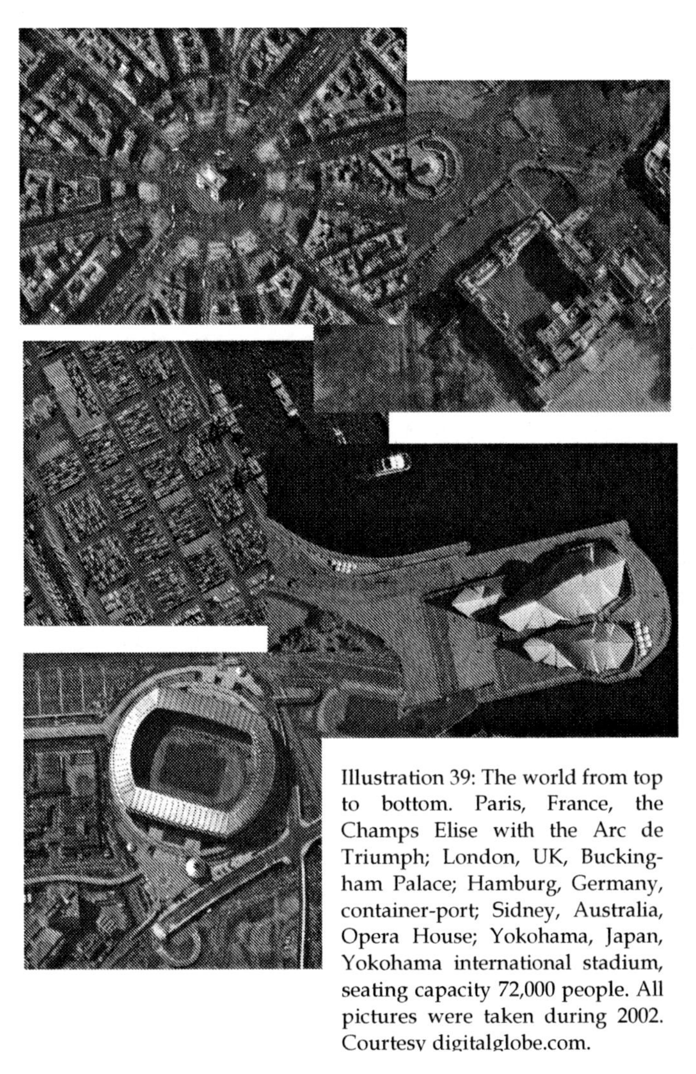

Illustration 39: The world from top to bottom. Paris, France, the Champs Elise with the Arc de Triumph; London, UK, Buckingham Palace; Hamburg, Germany, container-port; Sidney, Australia, Opera House; Yokohama, Japan, Yokohama international stadium, seating capacity 72,000 people. All pictures were taken during 2002. Courtesy digitalglobe.com.

In the future, structures will be present in digeality in full three-dimensional detail, including the internal layout. Every building and every bridge will have its digital twin. This is not some futuristic vision, but happening right now. As structures are maintained digitally, the dimension of time will be added to the structure. As the structure will be related digitally to the operations that go on inside of it, the people that work in it, and the suppliers and clients that visit it, this structure will become a digitally living and breathing 'organism'. Future physical structures will exist in digeality first, as they are designed digitally, and receive a physical twin after the construction is completed.

Digital Universe

The reader may remember the 'Einstein simulator' that I used earlier in the book to look at the universe and our planet. The simulator eliminates the obstacles of our human scale, in space and in time. Such simulators are now under development. The universe is being mapped in digital format, and becomes an integral part of digeality in the not too distant future.

The *Sloan Digital Sky Survey* is a joint project of many institutions in the United States and from around the world. Creating a map of the (visible) universe is the most ambitious 'map-making' project ever undertaken.

The partners in the Digital Sky Survey are the University of Chicago, Fermilab, and the Institute for Advanced Study, the Japan Participation Group, The Johns Hopkins University, Los Alamos National Laboratory, and the Max-Planck-Institute for Astronomy (MPIA), the Max-Planck-Institute for Astrophysics (MPA), New Mexico State University, University of Pittsburgh, Princeton University, the United States Naval Observatory, and the University of Washington.[2]

The digital sky survey will map one quarter of the visible universe, and create a digital landscape consisting of 100 million celestial objects, the distances to one million galaxies, and the distances to 100,000 quasars, the most distant objects we know. This three-dimensional map will provide new information about the distribution of dark and luminous matter throughout the universe, and might shed new light on the mystery of the expulsionary forces. This digital 'roadmap' to the universe is expected to take up the equivalent of 1000 computer hard drives with a 15-Gigabyte capacity each.

To explore the universe closer to home, NASA has created a 'solar system simulator' on one of its websites. Visitors can choose one of the bodies in our solar system to look at, choose a time - present, past, or future – and choose one of the planets, the moon or the sun as viewpoint. The images of Earth and Saturn in illustration 40 are the results of using of NASA's simulator on the Net.

The motorized transportation systems of the industrial era have made our planet a connected collection of locations, and digeality makes it one big recreation park of experiences. Similarly, digeality will transform our perspective on the solar system, making our planet one of the possible places 'to be'. We might see it as the 'home-planet', or the 'planet of origin': The one with an incredibly vast collection of lifestyles, experiences, and natural beauty.

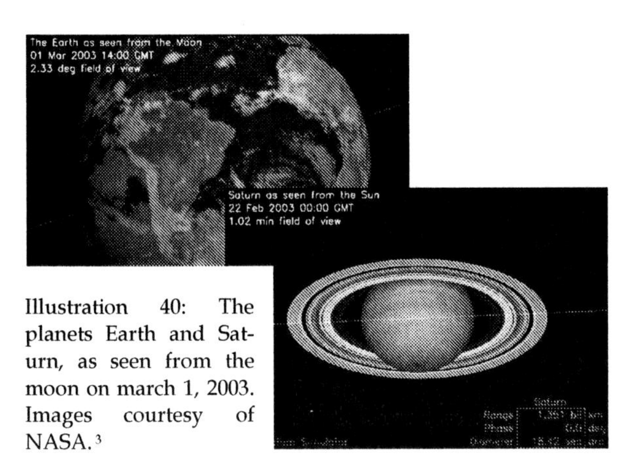

Illustration 40: The planets Earth and Saturn, as seen from the moon on march 1, 2003. Images courtesy of NASA.[3]

When engineers get their hands on new scientific possibilities, commerce is never far behind. Space Holdings corporation in New York City calls itself: 'the premier multimedia company dedicated to space, science and technology'.[4] With names like Lou Dobbs, business commentator for CNN, and astronaut Neil A. Armstrong associated with the venture, they might expect serious business. Space Holdings publishes Starry Night™, a sophisticated software application for the display of celestial objects. The Starry Night 4.X astronomy database includes a prodigious 500 million objects: stars, galaxies, planets, nebulae and other astronomical objects.

Another product is *The Digital Universe™*, by SYZYGY[5] research & Technology Ltd. in Toronto, Canada. The digital universe uses the latest available data (before the digital sky project), and includes a three-dimensional simulator that allows you to 'leave the earth and fly through the universe'. According to the company, *The Digital Universe* software lets you simulate the appearance of the heavens for any location on Earth, from 100,000 BC to 100,000 AD. To do so, it considers the effects of precession, proper motion, nutation, aberration, refraction, light travel time, and general relativity, as accurately as these effects are known to science. Simple versions of our 'Einstein simulator' are available today for US$ 50 - US$ 150! The first guides for space-tourists are available today, before we even have the vehicles to get us there. Software like *Starry Sky*, or *Digital Universe*, as educational tools or as interactive video games will bring new dreams to our children. Projects like the Digital Sky Survey will move the dreams of children into the realm of the possible. By the time the children grow up, their mindscapes will have the imprints of the digital universe, and their mind-windows will be ready to go for the 'real thing'. By the time our next generation reaches adulthood, another message, the message from the universe itself, might be clear: evolution is real; evolution is expansion, and their minds will be ready to transform the universe into a living one, and into a thinking one. That generation might begin to expand the diversity of styles of living, styles of thinking, and the physical environments to make a living.

Any *Point Omega*, be it Tipler's, Teilhard de Chardin's, or your own personal version, is hidden by the veil of the future. However, that future is not the walled city of a crowded planet earth, because even the highest mountaintop is not high enough to satisfy the potential for the expansion of the human mind. Evolution brought life and intelligence to our planet. The minds of our children will continue the journey.

The possibility and the nature of a 'Hereafter' might be of utmost interest to many of us. However, the nature of another hereafter should interest *all* of us. That hereafter is the future of humanity: the *Now-After*. That future will include the expansion of time and space. First, we will create such futures in digeality, and then we will translate some of them into new physical realities.

"JJ?"

Yes, Alvas.

"I wonder. With all the future power of computer simulation, will it be possible to simulate the evolution of the universe itself?"

I believe that we will be able to, Alvas, within a generation.

"How many years, Mr. J? We don't know how long the life span of one generation is going to be in the future."

Good point, Novare. We used to measure generations in periods of 25 years. Today, that number is closer to 30. So, we'll see simulations of highly complex phenomena within the next 20 years.

"Will people accept evolution as the true history of the universe, JJ?"

I believe, they will, Alvas.

"What about the future of the universe, Mr. J? Once we are able to simulate the past, can't we then predict the future?"

At least we will have a better understanding of what *can* happen, Novare.

"Therefore, we might be able to influence what happens, JJ?"

"Alvas, we have been doing that since we can think! We just should do it *consciously*, you see."

"Then it might be time to get your act together, Novare."

"Alvas, the future cannot be subject to scientific study!"

At least Novare, we will be able to study *alternative* futures.

"Like 'what if games', JJ?

Exactly. Keep in mind, however, that evolution is not a linear process. For example, who would have been able to foresee the future at any of the thresholds that we discussed?

"Not me, JJ."

"I have to confess: me neither, Mr. J."

Digital Life

Artificial life researchers from the Canadian National Research Council and the University of Waterloo, Canada, are closely examining the self-replication process that underpins evolution using a computer simulation of self-replicating strings of symbols that work as a simplified sort of DNA. The objects are like gene sequences in a string of genetic code, assembling into patterns similar to the way codons make up strands of DNA or RNA.

The work promises to provide a better understanding of life's workings. It also lays the groundwork for inexpensive and flexible manufacturing processes that borrow from life's vast experience, including the possibility of growing machines in vats of chemicals. It will be more than a decade before this type of work could be ready for use in practical manufacturing processes, according to the researchers. The research is scheduled to appear in the winter, 2003 issue of the journal *Artificial Life*.[6]

The humane genome project that deciphered the human genetic code in half the projected time, scientific study of biogenetic phenomena, and bioengineering could only proceed, and can only proceed in the future, with the tools of digeality. The precision and speed of computers and the elimination of human scale by digital technology makes modeling of existing and hypothetical biological designs possible and practical.

"Using Computers, Scientists Successfully Predict Evolution of E. Coli Bacteria," says a press release from the American 'National Science Foundation' on November 14, 2002. Here are some excerpts of the press release:

> *"By combining laboratory data with recently completed genetic databases, researchers can craft digital colonies of organisms that mimic, and even predict, some behaviors of living cells to an accuracy of about 75 percent...*
> *...Scientists may use the approach to design new bacterial strains on the computer by controlling environmental parameters and predicting how microorganisms adapt over time. Then, by recreating the environment in a laboratory, researchers may be able to coax living bacteria into evolving into the new strain...*
> *...Such digital models are known as "in silico" experiments -- a play on words referring to biological studies conducted on a computer."*[7]

The Digital Life Laboratory at the California Institute of Technology (CALTECH) "is a research group focusing on the dynamics of simple living systems, in particular their evolution. Fundamentally, this research is guided by the premise that the physical processes that give rise to life can be characterized by a set of simple principles, and that these principles can be implemented in different media as long as the latter allow for enough complexity. According to this view, simple living systems can be constructed."

The Digital Life laboratory has developed software to simulate the development of organisms from genotype to the functionality of the phenotype. "Using our digital organisms we have developed a technique where the predictions of algorithm designed to analyze expression data from micro-arrays can be tested, because for 'Digitalia', we know both sequence and function."[8]

Genetic manipulation without digital technology will be impractical, if not impossible. Thus, digital technology presents us with incredible opportunities to change the properties of living organisms, or to create new combinations of properties, structures and functions of living organisms in digeality, and simulate these organisms before making them into a reality. Digital modeling efforts like the ones mentioned, and in labs around the world, mark the first step towards this possibility. Digital simulation of new forms of life will shift the imaginable into the realm of the possible. The ethical challenges can only be envisioned dimly, because the digital simulation itself will change our perspective.

In 1953, Francis Crick and James Watson built a life-size model of the DNA molecule. You also might remember such models, consisting of colored beads and sticks used to teach the structure of atoms and molecules. Not anymore. The beads and sticks now live in digeality and computer aided 'chemistry' software is used to compose and build chemical, organic, and living structures.

Digeality is not just a collection of renditions, but contains many of the algorithms that produce the renditions. Digeality is a new reality where we re-create such algorithms, and with them, the 'creatures' they represent. The human genome project has been completed, and many others like the chimpanzee genome project are underway. Computers are unraveling individual genotypes of humans and many other species. When we uncover the algorithm that leads from genotype to phenotype, living creatures can be reconstructed from a sample of DNA and rendered in digeality in color, and full 3D detail. Jurassic Park will come alive in 3D! After that, recombination of DNA can lead to the digital creation of new creatures, and the characters from the Starwars movies can be recreated, not only in digeality, but potentially, their living physical 'twins' as well. As Richard Dawkins surmises:

> *"I believe that by 2050 we shall be able to read the language. We shall feed the genome of the unknown animal into a computer that will reconstruct not only the form of the animal but the detailed world in which its ancestors (who were naturally selected to produce it) lived, including their predators or prey, parasites or hosts, nesting sites, and even hopes and fears...*

...If the Dinosaur Genome Project is successful, we could perhaps implant the genome in an ostrich egg to hatch a living, breathing, terrible lizard."[9]

"Therefore, JJ, we will be able to reconstruct and simulate the evolution of life as well! That should convince people, shouldn't it?"

I believe it will, Alvas.

"That would mean a re-interpretation of religion, wouldn't it, JJ?"

"Just a re-interpretation of the words, Alvas. Not the message. The medium became the message, if you will. Religion would have to untangle the two."

"Thanks Novare. I like that."

Digital Atoms

Everything in the universe, the solar system, on our planet and in our human bodies consists of matter and energy, and everything is built from the ground up: atom by atom. The use of atoms like Lego-blocks is the subject of nanotechnology, conceived by Richard Feynman and pioneered by Eric Drexler who remembers:

"And at that time – this would have been in late seventy-six – I thought that you could self-assemble multiple layers of biomolecule-type things, and make a very small computer. I was persuaded that with a sufficient amount of effort one could develop computers on a scale of nanometers, with components based on the kinds of physical phenomena that are seen in biological materials, but without the structure itself being a biological system. They would be self-assembled objects made out of proteins, or protein-like molecules with lipids. I didn't have any marvelous insights, I just had this image of self-assembling things."[10]

The underscored words in the quote (they are mine) point at the two pillars of nanotechnology: building structures atom by atom and the ability to build structures that can replicate themselves or build other structures.

The size of atoms is in the order of a nanometer (one billionth of a meter), hence the name 'nanotechnology'. Using atoms in a Lego-block fashion to build from which you can build larger structures called molecules. That is why the technology is also called 'molecular technology'. From molecules, larger structures can be built, and so forth. More importantly, imagine that some of the smaller structures built from molecules would be machines themselves that can manufacture other machines, and so forth, until a machine is built that can manufacture computers or washing machines. Such is Drexel's vision, when he says:

"By early 1977 I had realized that instead of information processing, the really interesting, high-leverage application of molecular machinery was to manipulate molecules to build other things, including better molecular machinery. If you iterate this process, of using machines to build better machines, it's pretty clear that you could build machines that could build copies of themselves."[11]

The words were spoken in 1995, at a time when nanotechnology started to become reality. Seven years later, in March of 2002, NASA scientists presented Recent Results from NASA's Morphing Project at the ninth Annual International Symposium on Smart Structures and Materials in San Diego, California. In the concluding remarks, the authors state: "The research in the project focuses on the areas of smart/adaptive materials and structures, micro active flow control, and biologically-inspired technologies. While many application issues remain in applying these technologies to actual flying vehicles, the potential for broad, significant change in the capabilities of future vehicles remains unbounded."[12]

The National Aeronautics and Space Agency's (NASA's) Morphing Project, led by the Langley Research Center (LaRC), is part of the *Breakthrough Vehicle Technologies Project* that conducts fundamental research on advanced technologies for future flight vehicles.

NASA's Morphing Project strategically incorporates both micro fluidic and small and large-scale structural shape change to address the intertwined functions of vehicle aerodynamics, structures and controls and also to seek new innovations that may only be possible at the intersection of disciplines. The project is directed towards long-term, high-risk, high-payoff technologies, many of which are considered to be "disruptive" technologies...These areas are supported by the core enabling areas of smart, nano and biologically-inspired materials, multidisciplinary optimization, controls and electronics.[12]

The Morphing Project uses nanotechnology to create a new type of skin for aircraft. Carbon tubes are constructed with "new structural and electronic properties to create a 'skin' for advanced aircraft .... The outer layer will be sensitive to changes in airflow and mechanical stress; and it will use the carbon nano-tubes as tiny actuators to help the machines modify their shapes in response to changing aerodynamic conditions."[13]

Without computer aided design technology and without the digitized knowledge of the airplane's components and the airplanes atmospheric conditions, shape-shifting airplanes would not be possible. Every piece of knowledge, from the design of the carbon-nano-tubes, to the tracking and controlling of the airplane has to be modeled in digeality.

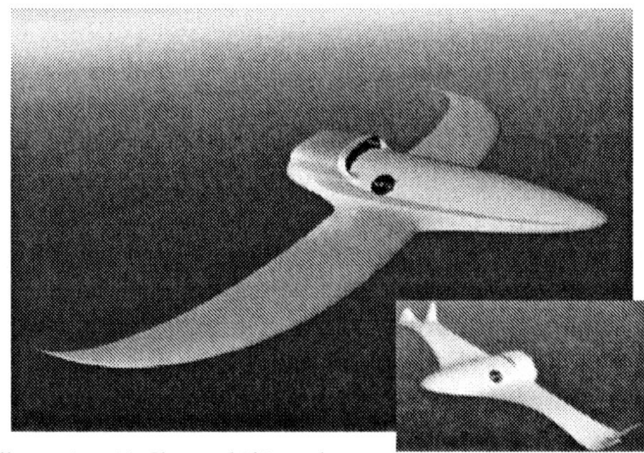

Illustration 41: Shape shifting planes, or a new type of bird? Courtesy NASA.

The construction of airplanes that can fly in more adverse weather-conditions, can take off on short runways (or even take off vertically, like birds), will change the way we think about flying, and ultimately change the meaning of the word 'automobile' from 'car' to 'auto' – 'mobile'. Vehicles that will transport us over the surface of the ground as well as through the air will change our mind about where to work, where to live, and where to go fishing.

Future planes like this will be integrated systems that consist of nano-engineered materials and components, built from single atoms, organic molecules and biological structures that change their shape, and digital intelligence that coordinates the hundreds, or millions, of sensors and actuators to produce a fluid behavior of the airplane's skin. They can adapt their shape to minimize drag and reach higher speeds easily. Increased mobility can turn us into planetary nomads, once again. Planetary mobility might thus prepare us for 'galaxary mobility'. And as the physics of atoms blend with the self-replicating properties of living organisms, new opportunities and new threats challenge our values, our social institutions, and our future.

In February 2003, a proposal was submitted to the U.S. congress to provide $2.1 billion over three years for nanotechnology research and development programs at the National Science Foundation, the Department of Energy, the Department of Commerce, NASA, and the Environmental Protection Agency. Nanotechnology is the top inter-agency priority in the current US Administration's fiscal 2004 proposed budget for non-medical, civilian scientific and technological research and development. According to the authors of the proposal, "This $2.1 billion investment for nanotechnology research and development will go a long way in developing a new economic engine for this country," and "Nanotechnology may be the 'smallest' field of science -- the manipulating of individual atoms. But I've come to understand that in science and technology, few things could actually be bigger than nanotechnology in terms of its potential to revolutionize scientific and engineering research, improve human health and bolster our economy," according to government officials supporting the proposed legislation. The expectations are high, the promises bright, and the potential threats frightening, as Michael Crichton displays in his new novel '*PREY*'.

Whatever its threats and promises, the application of nanotechnology requires digeality, because we are too big to observe and manipulate atoms, and the mathematical formulas that describe the behavior of atoms too complicated for manual calculation. The design of new structures and machines, and the control of these machines, are dependent on our ability to render them digitally. Washing machines of the future will come as software. Our PC's will translate the code after we choose the color and dimensions, and instruct the nano-factory in the backyard to start 'growing' the washing machine. One would assume that by that time, doing laundry would be an almost forgotten memory. Atomic digeality is here; the first computer aided 'nano' design programs can be found on the Net.

Heisenberg, Dirac, Schrödinger, and many of the other attendees of the Solvay conference in 1927, would be astonished to see their theories and mathematical equations come alive in digeality, allowing us to design new atomic and molecular machines and structures from the ground up. Many of these equations, like Schrödinger's, still take the biggest supercomputers available today. Moore's law (the law that predicts a doubling of computer processing-power each year) is still going strong. More powerful chips become available this year, and new processor concepts, like quantum computing and biological computers are on the horizon.

Atoms are used to construct the new quantum computers. In 2001, Isaac Chuang, now a professor at MIT's media-lab, and his colleagues at IBM determined that the two prime factors of the number 15 are three and five. While this might not seem a memorable scientific accomplishment, the calculation was performed by seven atomic nuclei in a custom-designed fluorocarbon molecule.[14]

Quantum computers use the quantum behavior of subatomic particles to represent complex problems. Potentially, these new computers exceed the power of today's most powerful supercomputers by orders of magnitude. Moreover, these computers of atomic scale could be embedded in every object, living or non-living. Quantum-digeality is on the horizon and with it another paradigm shift in computing. Our new PC's with a thousand times the power of our current ones will neatly fit under the lids of our eyes like a contact lens, and project a different world onto our retinas when we close our eyes. Switching from reality to playing video games happens in the blink of an eye. Talking about gaming....

Digital Games

In Seoul, capital of South Korea, people are playing online video games at twenty six thousand (yes, 26,000) baangs, according to an article in Wired[51]. Baangs are small, usually family run businesses located in places like malls, with rows of PC's that have high bandwidth connections to the Internet. People play in groups, usually the game Lineage. It is where the real world meets digeality. On any given day, 330,000 people in South Korea play Lineage, at $ 20 a month, boosting revenues for NCsoft, a publicly traded Korean gaming software company, to $ 100 million in 2001.[15]

The online video gaming is booming, and new jobs in digeality are created along the way. In the same issue of Business 2.0, we learn about the video-gaming business of media giant Sony.[16] *Everquest*, a descendent of *Dungeons and Dragons*, which in turn was inspired by Tolkien's *Lord of the Rings* is a glowing success for Sony's Online Entertainment Group: 433,000 paying customers (as of June 2002) bring in $ 5 million a month to play the game online. At that time, Sony employed 47 people to add new items and quests to the game, and 128 so-called "game masters," customer service reps that patrol the everquest world to answer questions: real jobs in a new, and digital, reality.

The creators, designers, and engineers; the facilitators, the referees, and the maintainers; people that sell weapons, new worlds, and new strategies; and the people that manage all of them; and all of them deal in bits that you cannot touch or smell, or throw in the trashcan. Does all of this just concern 'gaming'? Does anybody truly believe that this 'gaming' does not have a mindset-changing influence?

Gaming is playing with a reality that doesn't exist. Serious forms of this type of playing are carried out everyday in military planning rooms, in corporate boardrooms, and in a host of research and development laboratories. The complexity of environments that can be simulated in digeality vastly exceeds the capacity of 'planning white boards' or model-railroad type scale models. While playing their 'SimCity', 'Everquest', or 'Lineage', our children not only 'live' in different worlds, but their minds will learn to *think* differently. The renditions that they encounter, and the associations they learn to establish will shape the 'pinheads' of their minds different from their parents'. Cross-pollination between the algorithmic 'engines' used for children's games and those for the serious games of government and industry will propel the 'knowledge industry' to come up with more advanced scenarios of greater complexity and diversity built into their solutions, and thus enhance the accuracy of simulated worlds and futures.

Digeality is developing through both, the reverse engineering of the existing realities, and the creation of new realities.

Digital Cars

A car race has been planned for 2004. Unlike many others, this race is the first of its kind. The route for the race leads from Los Angeles to Las Vegas in the United States. If you want to participate, you have to hurry, because preparations are not trivial. These are the conditions for participation:

"Vehicles must be unmanned (no humans or other biological entities onboard) and autonomous. They must not be remotely driven. Only single, independent, untethered ground vehicles are eligible. No sub-vehicles will be allowed. All computing and intelligence must be contained onboard. Apart from the emergency stop feature (see Safety), automatic communication with autonomous equipment at the checkpoints (see Checkpoints) and GPS (Global Positioning System) signals, no external communication is allowed."[17]

Organizer of the race is DARPA, the Defense Advanced Research Project Agency of the United States Department of Defense. This is the same agency that developed the first network that linked computers together in the 1970s, the network that ultimately led to the Internet as we know it today. The agency's objective for the car race is as follows:

"DARPA intends to conduct a challenge of autonomous ground vehicles ... between Los Angeles, California and Las Vegas, Nevada in 2004. A cash award will be granted to the winning team. The route will feature both onroad and off-road portions and will include extremely rugged, challenging terrain and obstacles. The purpose of the challenge is to stimulate interest in and encourage the accelerated development of autonomous ground vehicle technologies that could be used by the US military."

The vehicles will be controlled, tracked and guided in the world of bits and bytes, while maintaining physical contact with the 'real world'. Sensors, antenna's, receivers and computers tell the vehicle where it is and where it should go. It is a race of maintaining a perfect match between the digital twin of the car and the digital twin of the terrain: the synchronization of reality and digeality.

Many institutions around the world conduct research and development projects that aim at automated guidance of motor vehicles. These vehicles need to be aware of the roads that 'guide' them. Cars will 'know' the roads, and roads will know the cars that drive on them. The car will know the digital twin of the driver, et voilà, the world will know the driver.

We will have a digital twin, and the world will know that twin, where it is, and what it is doing, at any time, and in any place. Many might wonder how this can possibly lead to a positive outlook that this book is supposed to portray. The secret lies in the creative power of the human mind that will be able to create new lifestyles that take full advantage of the benefits the digital world has to offer, while at the same time transcending its seemingly controlling and limiting boundaries. There is a room with a view at the top of all of it. That room is our mind, and the window is our mindscape window, our creative consciousness. While the foundation of our 'common sense' – our common knowledge – changes and grows, there is plenty of room above that foundation. Without a change of mind, from mistrust to trust, from isolationist thinking to openness and from 'unification of thought and culture' to respect for differences, the transparency and connectedness of digeality can be frightening prospects indeed.

In 2002, Toyota was running a commercial that shows an "intelligent" dummy. The commercial showed a wire-frame model of the dummy instead of the usual physical body with jointed limbs. This is the future of crash tests for automobiles. The dummies in the car will be actual 4D models of human beings (they can move in three-dimensional space and do so over time, hence 4D). Their bodies with limbs, joints and internal organs, blood vessels and tissue will behave according to the mathematical rules that have been deducted from the physical realities of their movements under the influence of external forces like a car crash. Cars are modeled, not just in appearance, but also in the behavior of their various parts, and their impact on the human body. Engineers are working on these systems, and there are no fundamental problems to be solved in order to make this possible.

Today, cars are conceived in digeality. Computer Aided Design software enables designers and engineers to conceive a new model and to engineer it in every technical detail. Different configurations, all the options the salesperson wants to sell you, are conceived, and exist in digeality. Every individual car has its own digital twin, right there in digeality. When your car needs repair, the VIN-number of the car uniquely defines it: the car's DT knows the parts it needs, the options it has, and the color of the seats and the type of paint of its exterior.

Physically, our cars already connect to the outside world. General Motor's 'On Star' system locates the car, and communicates to the car in digeality. Digital 'agents' that identify your car and link it to your bank-account are already replacing pay stations on toll-roads.

"We have used the simulator to successfully execute a substantial driving safety study, to demonstrate cooperative distributed remote operations with another simulator hundreds of miles away, and to conduct a demonstration of virtual proving grounds. We are all very excited about the research capabilities of this technological marvel and the potential it holds for driving safety research and vehicle system design," says L. D. Chen, director of the National Advanced Driving Simulator at the University of Iowa, United States.[18]

Illustration 42: 360 degree projection of environment in the driving simulator. Courtesy NADS, University of Iowa.[18]

In the United States alone, travel accidents claim 42,000 lives annually and cost society over 230 billion US-dollars. The larger picture however is the creation of a digital layer around the globe that represents everything on its surface (and under the surface), with roadways and landmarks, mountains and lakes.

Digital Airplanes

In the future, when cars and planes merge into one multi-modal-mobile, the pathways for travel will be digital pathways in the four dimensions of space and time, as they exist today as travel routes for airplanes. Digital travel controllers will know where the vehicle is, and who is traveling in it. Digital simulations have been in use for decades in commercial flight training. A trip online can get you incredibly accurate 'toy' flight simulators for a few dollars or euros. Add a 'force-feedback' joystick for added realism and 'up, up and away' you go. You don't need to be rocket scientist to fly a plane. Any computer can do it, including a human one. Unmanned, remotely controlled airplanes are tested on the battlefield today; even Iraq has them.

We can follow any flight of any airline online, and even draw maps on our screens to see where the planes are. Making your own travel reservations seems 'pedestrian' (pun intended) already, despite its fledgling existence. Once we link our DT's to the airplane, we can choose the passenger sitting next to us, and given the passenger's permission, look at his or her biography. Maybe a new way to find a job, to find employees, or a soul mate?

Digital Health

Stanford University is testing software that lets vascular surgeons sketch alternative possibilities to determine the best location for grafting arteries such that optimal blood flow is restored with minimal complications. First, a three-dimensional model of the patient's anatomy is derived from magnetic resonance data. Computer aided design software is then used to place grafts at several locations to determine the best procedure for the bypass operation[19] In the future, the 'skeleton' of my digital twin will be a three-dimensional model of my body with all its organs, arteries, bone structure, tissue and the like, for medical use. A yearly update of my 3D model will make my digital body available around the world in case of an emergency. At the same time, an accurate rendition of my body can serve in video conference calls that don't require my physical presence. That way, I can even participate in a TV show, while sitting in my favorite chair at home doing the thinking, and communicating through my digital twin.

With every medical procedure, my digital twin will be updated, and its history will be kept, so that doctors can look at the changes that occurred over time. My digital twin will be a true 4-dimensional construct and will *be* the history of my biological life. With every update, a biometrical inference software engine will analyze any changes and advise me of any potential problems. Some of the problem areas can then be repaired using intelligent nano-bots that travel through my arteries or nervous system.

At the Ames Research Center in California, USA, a NASA research center, a robot is under development that will be able to perform brain surgery. According to Principal Investigator Dr. Robert Mah, "the simple robot will be able to feel brain structures better than any human surgeon, making slow, very precise movements during an operation."[20]

Robots that perform surgery are under development at various research institutions around the world. The use of robots makes 'remote surgery' possible, where the supervising human expert can be thousands of kilometers away to monitor the procedure and coordinate the work of local specialists. This type of development might point to the beginning of performing simple surgical procedures at home, where the surgeon manipulates my digital twin, while the digital robot operates on the physical me.

My digital twin will be created upon the knowledge of my successful conception. Some of the questions that come to mind are: who owns and controls my digital twin, and who owns and controls the interface between my digital twin and myself? Obviously, I will be that person. How will that be regulated? Will it be added to the human rights bill as a fundamental human right?

Medical records, insurance data, government-related information, DNA-signature, a four-dimensional model of my body, physical and mental expectations, school-affiliations and results, resume, investments, bank account, memberships, habits, marketing information, preferences, my library, house, cars, assets, relationships, failures, my art, accomplishments, job profiles, career, publications, names of doctors, address and telephone books, preferences for music, music collection, personal software products, or all the pictures that nobody wants to look at: my DT will have it all.

Digital Science

No science can proceed without the power of digeality. Digeality allows us to change the dimensions of time and space at will. It allows us to recreate physical structures and physical phenomena from the ground up, from single particles or quanta of energy all the way up to galaxies, oceans, houses, organisms and societies. Every scientific progress and any technological progress would virtually stop in its tracks without digeality. New scientific insights are translated into digital algorithm, that are the basis for new technologies that result in novel products, or the innovation of existing products. In digeality, we can recreate cars and lungs, airplanes and hearts, solar systems and bacteria with molecular detail. We can simulate and test their functions, we can travel through these digital creatures and test, modify and study their behavior; then we can control the robots that produce the ones we find appropriate.

Science is the effort to describe and explain how things work. Engineering is the effort to put that knowledge to work and to create artifacts of practical use. Until digeality came along, these artifacts were physical in nature: trains, plains, skyscrapers, can-openers, precision guided missiles and the millions of objects that made our lives easier, and sometimes not. With digeality, a new type of artifact arrived: the digital machine or *thinking machine*. The electronic bank transaction or the online order that we initiate with the click of the mouse switches on a digital machine of amazing complexity. We only observe the surface - the renditions - that these machines produce; we are hardly aware of the intelligence - the algorithm - that are being executed in the digital world of bits and bytes. When we make an online reservation for a vacation that includes the plane trip, rental car and hotel reservation an astonishing 'thinking machine' starts to move its gears, sends our demands around the globe and brings a host of digital twins alive. A particular hotel-room will 'know' we are coming, the car is waiting, and airplane-seat 12A is now occupied for that particular flight. New electronic digital check-in machines replace the human intelligence at the counter, and the hotel-front-desk-clerk knows we are coming.

That is the depth of digeality: everything that science can describe and explain and everything the engineers can imagine can be expressed as digital machines. Once the machine is created, we can render any possible manifestation of that machine and make it a digital reality. We bring Arnold Schwarzenegger, Madonna, the Pope, the Coliseum in ancient Rome, or dinosaurs alive in digeality. Not only can we re-create the things that exist today, or existed in the past, with the help of digital machinery, we can create novel realities of imaginable future creatures and artifacts. The digital 'effects' of modern science fiction movies such as Hollow Man, AI or Simone are just simple examples of what is possible in digeality.

"So what, JJ. Those are just movies; some of them even boring. What does that have to do with reality?"

"Alvas still doesn't get it, Mr. J. Molecular technology, Alvas. Just think about it. We start to learn how to manipulate atoms and molecules. We start to understand the production of proteins from the genetic code."

"And?"

"What if we can replace the bits and bytes of artifacts, and those of digital creatures by atoms and molecules? We can transform them into actual physical structures, don't you see?"

"Wow. That's really scary. Who is going to decided which ones to create and which ones to leave alone? Some might be dangerous and inhumane."

"I'm glad you got it, Alvas. It is a fascinating prospect, and a frightening one as well. You see, that's exactly the point Mr. J is trying to make: the glass of opportunity is expanding exponentially with digeality. It's time we start to think about the challenges and the opportunities."

"Novare, that way you can imagine the possibility of creating totally new creatures and the way they live. Their means of transportation, the houses they live in, the leisure activities they pursue, and the way they interact. We could even build a new planet, for all I know. We would be able to create new futures!"

"I'm glad you finally see the light, Alvas."

"Darkness is what I see, Novare. We can't even handle reality as it is. How are we going to deal with new realities?"

"We can only take one step at a time, Alvas. Waking up to the power we have is the first one. As long as we are asleep, we will just drift on the current we create, without intention, without purpose, and without steering wheel, brakes and instruments that show us where we are."

"We could try out new realities in digeality first, right?"

"Good point, Alvas. You see, so far science was fiction. Science created fictions - we call them concepts, theories or hypotheses - that best matched the behavior of reality. Now its time to turn that around. In digeality we can create fictions first, and create realities to match the fictions. We can now turn the fiction of science into the science of fiction, Alvas."

"Science fiction turns into the science of fiction?"

"You've got it, Alvas."

Digital Clothing

At MIT, 35 professors have signed up to work on a 5-year project to develop a "smart coat of armor' for soldiers in battle. They hope to invent protective wear that will (a) fend off bullets and bioweapons, (b) camouflage the wearer by adapting like a chameleon to match the background, (c) send signals to headquarters about the whereabouts and condition of the wearer, and d) be a wearable emergency room - able to monitor vital signs and sense injury, stiffen to act as a cast for broken bones, release antibiotics or other medicines in case of injury or poisoning. This clothing miracle should also be lightweight and comfortable.[21] What is a wearable computer? MIT's media-lab answers the question as follows:

> *"To date, personal computers have not lived up to their name. Most machines sit on the desk and interact with their owners for only a small fraction of the day. Smaller and faster notebook computers have made mobility less of an issue, but the same staid user paradigm persists. Wearable computing hopes to shatter this myth of how a computer should be used. A person's computer should be worn, much as eyeglasses or clothing are worn, and interact with the user based on the context of the situation. With heads-up displays, unobtrusive input devices, personal wireless local area networks, and a host of other context sensing and communication tools, the wearable computer can act as an intelligent assistant, whether it be through a Remembrance Agent, augmented reality, or intellectual collectives."*[22]

At the Wearable Computing Laboratory of the University of Oregon, the Wearable Communities project investigates the use of mobile and wearable computing technology to assist people in social face-to-face meetings, whether on the way to the office, in the elevator, or at the grocery store. The main question scientists at the lab are trying to answer is "How can mobile technology facilitate or augment social interactions and which effect will this technology have on the way people interact and form communities?"[23]

The Bristol Wearable Computing Project is concerned with exploring the potential of computer devices that are as "unconsciously portable and as personal as clothes or jewelry." The project is a collaborative effort between the University of Bristol, UK and the European research laboratories of the Hewlett Packard Company. Engineers develop software and hardware using "context sensing to enable media and information to be delivered to the wearer of our jackets and devices appropriate to their current position and activity."[24]

The Wearable Computing Laboratory at the ETH Zurich (Swiss Federal Institute of Technology) focuses on recent advances in computer miniaturization, wireless technology and networking to open up a new field of computing. The vision behind the laboratory's work is that "a mobile computer should not just be a machine that we put into our pocket when we plan on doing some office work while on the road. Instead it will be an integral part of our every day outfit (hence wearable), always operational and equipped to assist us in dealing with a wide range of situations."[25]

In March 2003, CNN showed a demonstration of 'wearable computing' in the form of the 'American soldier of the future'. As the commentator pointed out, the gear shown was '3-5 years away'. The soldier's helmet contained the equipment necessary to receive digital information and render it on a helmet mounted, 'one-eyed' digital display that can be moved in front of one of the soldier's eye. A map of the terrain, or any other information relevant to the combat situation at hand, is displayed in front of him. The helmet is connected to his clothing that contains the computer that controls the display and all other gear. The automatic rifle in the soldier's hands is wirelessly connected to his clothing and helmet and has a digital display of its own to automate the task of targeting the rifle. His portable radio doubles as 'mouse' to control the displays, and the soldier is constantly connected to other soldiers and commanding headquarters through a local wireless 'LAN' (Local Area Network). The rifle has controlling devices of its own to control the computing equipment, so that the soldier "doesn't have to take his hands off the rifle."

Wearables are coming! Maybe, in the future, 'buckle your seatbelt' will be replaced by 'connect your shirt'.

My Digital Twin

One of the most ambitious, most controversial, and best-financed projects of DARPA (the Defense Advanced Research Project Agency of the United States Department of Defense that we discussed before) started under the direction of retired General Pointdexter in the beginning of 2003. Its name: The Total Information Awareness Program (TIA), administered by the Information Awareness office (IAO). Here follow excerpts of the Office's Mission and Vision statements, as published on the agency's website.[26]

IAO Mission: The DARPA Information Awareness Office (IAO) will imagine, develop, apply, integrate, demonstrate, and transition information technologies, components, and prototype closed-loop information systems that will counter asymmetric threats by achieving total information awareness that is useful for preemption, national security warning, and national security decision making.

IAO Vision: The most serious asymmetric threat facing the United States is terrorism, a threat characterized by collections of people loosely organized in shadowy networks that are difficult to identify and define. IAO plans to develop technology that will allow understanding of the intent of these networks, their plans, and potentially define opportunities for disrupting or eliminating the threats. To effectively and efficiently carry this out, we must promote sharing, collaborating and reasoning to convert nebulous data to knowledge and actionable options. IAO will accomplish this by pursuing the development of technologies, components, and applications to produce a proto-type system.

According to the IAO, example technologies to be developed or applied in the TIA project include:

- Foreign language machine translation and speech recognition
- Biometric signatures of humans
- Entity extraction from natural language text
- Human network analysis and behavior model building engines
- Event prediction and capability development model building engines
- Structured argumentation and evidential reasoning
- Story telling, change detection, and truth maintenance
- Biologically inspired algorithm for agent control
- Other aids for human cognition and human reasoning

The use of the word 'Total' in Total Information Awareness Program might be understandable in the aftermath of the terror-attack on the United

States, but it might send chills up your spine. In the following, I present the complete summary of the program[DA1] to you, because it offers the most eloquent explanation of the extremely ambitious nature (both: technically and socially) nature of the project.

Not only are our digital twins emerging, their creation is funded by the richest country on earth. The TIA program not only aims to *connect* my digital twin to the outside world, but also aims at *knowing* my digital twin, and hence: the real me. This is not the future, but the emerging reality today. Hot debates about 'security versus privacy' will be raging, not only in the United States, but the world over. Here are the TIA program's 'objectives' and 'strategy':

Program Objective: The Total Information Awareness (TIA) program is a FY03 [fiscal year 2003] new-start program. The goal of the Total Information Awareness (TIA) program is to revolutionize the ability of the United States to detect, classify and identify foreign terrorists - and decipher their plans - and thereby enable the U.S. to take timely action to successfully preempt and defeat terrorist acts. To that end, the TIA program objective is to create a counter-terrorism information system that: (1) increases information coverage by an order of magnitude, and affords easy future scaling; (2) provides focused warnings within an hour after a triggering event occurs or an evidence threshold is passed; (3) can automatically queue analysts based on partial pattern matches and has patterns that cover 90% of all previously known foreign terrorist attacks; and, (4) supports collaboration, analytical reasoning and information sharing so that analysts can hypothesize, test and propose theories and mitigating strategies about possible futures, so decision-makers can effectively evaluate the impact of current or future policies and prospective courses of action.

Program Strategy: The TIA program strategy is to integrate technologies developed by DARPA (and elsewhere as appropriate) into a series of increasingly powerful prototype systems that can be stress-tested in operationally relevant environments, using real-time feedback to refine concepts of operation and performance requirements down to the component level. The TIA program will develop and integrate information technologies into fully functional, leave-behind prototypes that are reliable, easy to install, and packaged with documentation and source code (though not necessarily complete in terms of desired features) that will enable the intelligence community to evaluate new technologies through experimentation, and rapidly transition it to operational use, as appropriate. Accordingly, the TIA program will work in close collaboration with one or more U.S. intelligence agencies that will provide operational guidance and technology evaluation, and act as TIA system transition partners.

In order to make the final connection between my digital twin and the physical me, DARPA has been funding its "Human ID at a Distance program,"[27] which aims to positively identify people from a distance through face recognition or gait recognition. DARPA adds that "a nationwide identification system would be of great assistance to such a project by providing an easy means to track individuals across multiple information sources." Larry Ellison, head of Oracle Corporation, the California based software company, has called for the development of a national identification system and offered to donate the technology to make this possible. He proposed ID cards with embedded digitized thumbprints and photographs of all legal residents in the U.S.[28]

These applications are surrounded by an aura of spying and 'big brother is watching you', and therefore they might harm us, or at least control us.

The thrust for the creation of our digital twin, however, goes beyond military interests, and has been underway for decades. Businesses have been gathering information about us under a great variety of guises. Just think of the product-registration card that you should send in. Many times, it promises you a host of benefits that never seem to materialize. We fill in 'applications' for credit cards or memberships without giving it much thought. 'To help us improve our product' and 'to serve you better' are some of the banners that cover the quest for information about people. Now, all these fragments of information, collected by commercial, government, and non-profit organizations, are the digital snippets that assemble into one profile: our DT is emerging, piece by piece, and will continuously evolve and change throughout your lifetime.

"Sounds like Orwell's *1984* all over again, JJ."

"Alvas doesn't see the difference, Mr. J."

What is the difference, Novare?

"The technology goes far beyond anything that Orwell could envisage, that much is true."

"You can run, but you can't hide anymore, that's what I'm saying."

"True. Therefore, it is time to learn the difference between anonymity and privacy. Everybody in the world that ever wanted to make a difference, be it in big things, or in the small things of daily life, be it good or bad, did so at the 'price' of loosing anonymity."

"What if I want to stay anonymous? Do I have to live on another planet to be anonymous?"

"You could still choose to stay anonymous, Alvas. But you would be an unknown entity and therefore excluded from any services. As far as the world goes, you don't exist."

"What a world. Where is my privacy?"

"Privacy will become one of the most fundamental human rights, Alvas. There just is no other way. The alternative is horror. The planetary society will have to deal with that. I hope they do so soon, and stop confusing anonymity and privacy."

"Still gives me the creeps, Novare. Especially since all this technology gets such a big push by military motives."

"That, Alvas, is the most interesting thing of all. Mr. J, I have a theory."

What theory is that, Novare?

"The military push to develop these digital technologies will eliminate anonymity, right?"

To a growing extent, yes.

"Therefore, in the long run, everybody around the planet will know more about many different people, cultures and nations. We will get closer to the different mindsets of people around the globe. Ignorance will turn into tolerance, and tolerance into respect. That will invoke an increased awareness of the futility of physical conflict to settle mental differences. Today's military efforts to enhance digital technology will ultimately change its own institutions, and its own reasons for being."

"That's what I call a liberal's naivety attack, Novare. You're just not being realistic."

"That's my point, Alvas. What you call 'realism' today, is a bad memory tomorrow. Technology changes reality and reality changes the way we think. It's a feed-forward system."

Well said, Novare. And your point is taken, Alvas. Each one of you pointed at one end of my 'axle of change'. Novare points at a reality that is changing dramatically through technology because of the evolutionary expansion of the human mind. Alvas points at the resistance of the human mind to change its static view of the world. A resistance that is supported by the institutions of political, economic, and religious power. In my opinion, that will change indeed. The alternative is isolation and hence: physical conflict on a scale we have not seen before.

Military operations themselves are shifting from the physical to the mental domain. Slowly but surely, we are climbing out of the mud of physical 'interaction' (destruction and mutilation of physical property) and biological 'interaction' (destruction and mutilation of lives) to the higher level of mental interaction. Physical conflict resolved by war is transforming into mental conflict resolved by diplomacy. Digeality will bring us closer to exposing the non-sense of physical destruction of human bodies to change human mindsets. The human mind will continue to climb the hill called 'peace of mind'.

"Will that eliminate the need for physical military power, JJ? Or are you having another naivety attack?"

"Mr. J. is not that naïve, Alvas. As long as you are around, there will be a need for a military system, as we know it today. Right, Mr. J?"

That is right Novare, although you cannot blame it on Alvas alone. You do contribute your share.

"But I *can* at least try to change my mind; that'll change Alvas' mind as well."

"Hah! Naivety indeed!"

Bytes of Meaning

How often do we say "what do you mean?" in our business meetings and personal conversations? Not because we don't understand the speaker's grammar, but because we don't grasp the meaning of the words used. Human minds are able to derive meaning from the broader context of the conversation, from body language, and from knowing the 'mindset' of the person that spoke the words. Computers on the other hand, take us by our word, in a very literal sense. During my digital day (in the previous section), I had to rely on my digital twin to understand what I mean, when I say, "remind me to stop at my favorite store." Entering this sentence into my Internet-search-engine's 'search' box yielded no answer, while entering just the words "remind me" resulted in the listing of 600,000 websites. The Net does not deal in semantics, only in syntax. Millions of words can be found in a heartbeat, but the computer has no clue what they mean, or what I mean when I enter words.

The same is true for computers that communicate with each other. The interacting computers might have different ways for expressing the same product or transaction. Hence, computers need a way to define meaning. Different schemes and languages are under development to let computers know what something means. This 'something' can relate to words, text documents, images, sound bites, or complete movie-sequences.

When I ask *my* digital twin to make an appointment with *my friend Jim* at *my* preferred time and place, the software needs to understand what *I* mean, not just what I say. The computer science and artificial intelligence communities, as well as businesses are undertaking massive efforts to find solutions. The first step is the use of a language called XML that enables us to attach semantic descriptions to digitally stored information. Microsoft will release its new version of the Office suit of software to make use of this language in 2004.

The future Net, that recognizes meaning, is called the Semantic Web. It is envisioned to have knowledge about the meaning behind documents, knowledge about the functionality of appliances and different software components, and knowledge about the meaning of wishes that you or I utter. Tim Berners-Lee, inventor of the hyperlinked domain of the World Wide Web, attaches an evolutionary importance to this future Net:

"The semantic web is not "merely" the tool for conducting individual tasks that we have discussed so far. In addition, if properly designed, the Semantic Web can assist the evolution of human knowledge as a whole."[29]

This would give the Net, or my digital twin, the ability to generate new memes on their own by looking for meaning in the concepts that now stay hidden in the syntax of documents across the Net. Is the Net indeed going to generate ideas of its own? Will we understand those ideas? Or does this vision lead to a harrowing dystopia of digital dictatorship?

"It doesn't care what I do and who I am and where I sit." These are the words of Axel Waibel, the director of Carnegie Mellon University's Interactive Systems Laboratories. His words are aimed at his computer that is not smart enough (yet) to be able to communicate intelligently with him. The computer's inability to know who he is, and whether he is speaking to the computer or somebody else in the room, is frustrating, and the lab's research is aimed at eliminating this hurdle.[30] The researchers explain how computers can, and should be observant of our human needs, and anticipate our wishes, desires and demands. Again, whether we consider this type of research of extreme importance to enhance the social enterprise, or whether we find it non-sense, the technology is coming, and the technology will not only change our lives, but will change our mind. The meaning of technology that connects human thought and digital knowledge, or the physical me and my DT, on a higher level of understanding goes deeper than mere 'convenience', because ultimately this type of research will lead to a seamless connection of human thinking and digeality. The most 'user-friendly' computer interface is no interface at all. We are working to do just that, with direct neural connections to hearing aids and visual prosthetics. Waibel's research will just pave the road for things to come.

While current research focuses on connecting the 'bodies' of computers to the human body, and on synchronizing their minds, the future will be the connection of the human mind and the computers algorithm. As challenging as this might sound, it has nothing to do with 'computers that supercede the power of the human mind', because as powerful as computers are in solving problems, humans are more powerful yet, in creating new ones. Problems are like puzzles: you have to create them, before you can solve them. Creating new puzzles is part of looking into the future; it requires purpose, intention, and design. The creation of new puzzles is the domain of the expanding human mind.

A Digital Noosphere?

Technology got ahead of us. Way ahead of us. Thousands of organizations around the world, Government Agencies, Universities, Non-Government-Organizations, and Businesses are creating islands of digeality. We only notice the physical renditions of these patches of digeality. We click online to order a book, and greet the Federal Express agent at our door to receive the book. We 'attach' a picture to our email to surprise a friend. We don't perceive the astonishing streams, rivers and lakes of manipulating bits and bytes that connect the events at the beginning and the end of such transactions. The moment we plug that connector into the phone line, or the moment we switch on that wireless transmitter, we connect ourselves to trillions and trillions of digital algorithm and renditions that literally span the globe, being available everywhere at the same time.

Every organization, and every connected individual, creates new knowledge and adds it to digeality. algorithms to design nano-size structures or new bio-technological creatures are developed and added every day. New ways to explore the universe or to investigate the quantum behavior of particles increase in power continuously. Cultures and cults, religions and political platforms, display their rationality or their emotions with the force of dramatic multi-media presentations, each one of them works on different premises and with different intentions. All of them create their own 'seashells', in relative isolation, mostly with the well-intended purpose of enhancing human existence, some with ill-intended goals. Few look at, or even realize the expanse of the tapestry they are helping to weave. Most carry out their activities with a mindset that is centuries old, and has long lost its validity. No wonder that many claim that 'technology has taken over evolution from nature', if they fail to accept the reality of mental evolution itself. We apply the same old economical, political and social 'principles' that served us well in the past and fail to adapt them to a different worldview that science has established over the course of the last century and a half.

"I'm confused, I'm frustrated, and my stomach aches, JJ, but I got the message. Something big is going on all right. How in the world are we ever going to deal with that?"

"Look at nature to answer that question, Alvas. It took life a while before it tamed the forces of matter and energy. We take it for granted that we keep the trillions of atoms in our bodies from falling prey to entropy. Now, our minds are trying hard – maybe not hard enough –

to gain control over the biological forces that evolution has programmed into our genetic codes. Memes have been ruling the world all along, but now they start touching and mingling. This will not result in unification, but in the creation of new memes. It reminds me of bacteria, Mr. J."

Bacteria?

"Remember, when the earth started to get covered by a dense layer of bacteria, in the waters of the sea, the surface, even in the atmosphere?"

Sure, but that was billions of years ago, Novare.

"Exactly. First, they developed in isolation, and produced a large variety of strains, each one adapted to a particular environment. When their populations grew, and their density became big enough, they started to meet. That is when things started to happen. Bacteria merged and cooperated to form eukaryotes, the cells of all of life."

What are you trying to tell me, Novare?

"Digeality is the new world in which minds can, and do meet. Like bacteria meeting biologically and form new organisms, digeality facilitates the merger of knowledge, and the creation of new ideas."

"A little far-fetched, Novare. Are you implying that digeality is to the future of the human mind what the bacterial biosphere was to the future of life?"

"That is exactly what I am saying, Alvas. A Noosphere, a layer of knowledge, is building. It is a digital Noosphere. With its emergence, the human mind is still in the process of crossing Mr. J's third threshold: the threshold to the domain of knowledge. Do you agree, Mr. J?"

As long as we clearly see that this digital Noosphere is not a distant goal in the future, Novare. It manifests itself through digeality indeed, but it is the soil for new ideas and new perspectives, not a horizon that signifies a destiny. Digeality is here to stay.

Digeality is a continuum of digital renditions and digital algorithm that produce renditions. Just like the space-time continuum of the universe, digeality has no absolute dimensions. In it, time and space can be manipulated at will, and digeality permeates everything and anything with the speed of light. Cosmic evolution created our first reality: the physical reality, as we know it. Biological evolution created the second reality: the mental reality of living organisms. This reality manifests itself through the sensory mechanisms of living creatures. The mental reality is the way living creatures experience the world as they form their own 'image' of the physical world. It allows living creatures to interact with that physical world, from bacteria to fish, and from fish to human beings. With digeality, human nature has created a third reality: the digital reality. Digeality is the representation of the human mental reality, of *human knowledge*. The emergence of the physical reality of the universe marked the beginning of *existence*. The emergence of the mental reality of life added a new dimension to existence, and marked the beginning of *being*. The emergence of the digital reality of human thought added a new dimension to being, and marks the beginning of *knowing*.

A new beginning: the beginning of human 'being' as the amalgamation of matter and energy, life, and mind. With the emergence of digeality, all three domains of evolutionary achievements are merging. Human beings are finalizing the process of crossing the third threshold. We are submerging ourselves in the domain of knowledge, with uncertainty and fear, because we have yet to learn to be choreographers of alternative futures. *Co*-choreographers with the rest of nature, that is.

To answer the question of this section: 'A digital Noosphere?' I suggest that digeality is a Noo*verse* rather than a Noo*sphere*, because of its universal reach. While digeality allows us to deepen our grasp of reality as it *is*, it is a new foundation for *reaching out* into the unknown territory of space, of time, and of understanding. However, the Nooverse not just a new domain of thought. It is the platform from which we will launch new physical realities that will change the lives of future generations. Digeality is the *drawing board, the laboratory, the production facility, and the executive boardroom* where new futures are designed, tested, and produced. Digeality is the meeting room of knowledge, and is the mirror that reflects all of human knowledge to all of us.

Chapter 10: The Knowledge Domain

"We have only recently emerged from the biological to the psychosocial area of evolution, from the earthly biosphere into the freedom of the Noosphere. Do not let us forget how recently: we have been truly men for perhaps a tenth of a million years – one tick of evolution's clock: even as proto-men, we have existed for under one million years – less than a two-thousandth fraction of evolutionary time. No longer supported and steered by a framework of instincts, we try to use our conscious thoughts and purposes as organs of psychosocial locomotion and direction through the tangles of our existence; but so far with only moderate success, and with the production of much evil and horror as well as some beauty and glory of achievement."

Julian Huxley[1]

The Story within

In part I, we visited some of the landmarks of scientific thinking, and saw the incredible changes that the last century and a half has brought us. The scientific view of reality changed from the permanency of a universe of galaxies, stars, planets, plants, animals and human beings to one that is constantly changing. We discussed the widening gap between the scientifically accepted view and the common understanding of the vast majority of the human population. Many of those that do accept evolution as the 'correct scientific framework' nonetheless act as if *change* is just the label for a cyclic human enterprise. In the meantime, technological change accelerates, continuing the evolution of scientific knowledge and magnifying the gradient between the momenta of technological and mental change. I called this a-synchronicity between the momenta the 'axle of change'.

In part II, we watched evolution at work. Three songs *of* the universe, *from* the universe, and *by* the universe; three themes, separated by thresholds at which entities emerged with radically new properties. These properties turned out to be different in *kind*; they gave the individual 'creatures' new degrees of freedom for their behavior. In ten episodes, evolution evolved as an amazing expanding symphony that created its own players, and instruments, writing its own composition. Until Homo sapiens came along, and changed the rules.

A Cocoon of Knowledge

Finally, nature separated the composing faculty from its integrated biochemical structure-function, and rendered this faculty in the form of the human mind. With that amazing innovation, we acquired the ability to co-write the script for the future of evolution. Is that what we are doing? Huxley's opinion about our use of this new faculty is obvious from the quote above: "so far with only moderate success." If you stop and think about it, that is not a big surprise. For thousands of years, we have considered ourselves as the 'absolute' anchor – at the center – of the universe. We needed all of our mental faculties to make it through life and fulfill our evolutionary duty to produce offspring. Only later, a growing number of human beings could afford to spend their time outside of the realm of day-to-day survival; they started *thinking*.

Philosophy was the first disciplined approach to explaining reality, and it spawned off all the natural sciences in the same sequence in which humans focused their attention on parts of the natural world. In parallel, philosophy spawned off mathematics that gave us the language to transform mental creations into a physical rendition of symbols. The sciences kept humans busy, well into the twentieth century, to find explanations for the physical world around them. Mythology, philosophy, and later, theology took care of explaining human affairs. They provided or reinforced the explanatory framework for people to operate in. Science had the reverse engineering of the world as its sole priority, before it turned its attention on the engineer itself. As we have seen in chapter 3, it is only a little over a century ago, when William James exclaimed "This [psychology] is no science; it is only the hope of a science."

With the recent advancements in neuroscience, cognitive science, information technology and psychology and the confluence of these sciences, we are getting the first glimpse at the mechanisms of the human brain and its potential for change. We have been seeking to explain the world we live in, and now we are seeking to explain the seeker. We are still far from explaining nature and evolution, but the intelligence embedded in the evolution of nature and the evolution of human knowledge about nature are converging. Evolution has produced a human mind that is, as we start to realize, capable of influencing the course of evolution. The physical and mental state of our planet testifies to that. Once again, we can now sketch the cocoon that symbolizes the history of the universe. Once again, nature created a new way of expanding its own horizons. This time, it added the domain of knowledge to the domain of life that sprang from the cosmic domain. Once again, the vector 'E' in the diagram of illustration 43 represents the expansion of the universe. Now, that thinking living humans inhabit that universe.

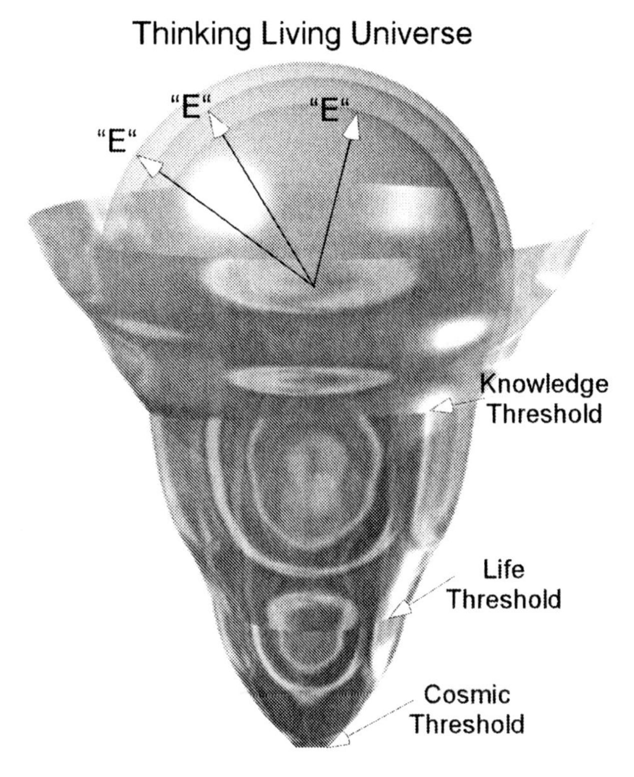

Illustration 43: The Domain of Knowledge. The cocoon continues to open up.

The surface of the inner sphere represents the 4-dimensional universe, the next bigger sphere presents the universe including life, and the outer one presents the universe, life and human knowledge, including all the products of the human mind. Now, the expanding universe contains all of the knowledge, and all of its physical renditions produced by human beings. Airplanes traversing the skies, the expressions of war and peace, your morning headache after the party, and the prayers we pray: they are all part of, and present *in*, the surface of the continuously expanding sphere.

We know that the universe itself continues to expand; evidence tells us that it is even expanding at an accelerated pace. We do not know what is happening to the expansion of life in the universe. Are we alone? Or is life expanding elsewhere? Either scenario might seem exciting to some and frightening to others. On the planet earth, of course, we know what happened. We do many things to decrease the diversity of life. Eighty percent of species live in rainforests, and the rainforest just one of the ecosystems that we offer on the altar of 'investor interests'. The universe, however, delivered life, and started transforming its lifeless atoms into living ones. On the macro-scale of galaxies, the universe never reversed its direction, at least not so far. I cannot help but thinking that the universe will stay on its course of expanding life throughout the solar systems and galaxies. One might therefore say, as Bill Joy did "the future doesn't need us."[2]

On the other hand, it does not seem unreasonable to assume that *we might want the future*. If this is so, then the future does need us as well. Why would the future need us? The answer is simple, really. Evolution itself gave us the capacity to create knowledge. We used that knowledge to reverse-engineer nature; we did so with enormous speed, and astonishing results. Our own knowledge expanded the choices we can make. Now, our choices range from preventing any future from happening altogether, all the way to creating new forms of life on other planets, and everything in between. That is why the future needs us: if we want a future at all, we hold the power to make that choice.

The cocoon that symbolizes the history of evolution has a direction: the direction of expansion, as I defined it in the previous chapters. In the domain of knowledge, the expansion manifests itself as increased knowledge and the increased diversity and versatility of the physical renditions of that knowledge. Many characterize this expansion as *increased complexity*. Complexity is certainly a feature of the expansion, but it does not explain evolutionary expansion. We have no physical metric that allows us to measure the expansion 'E'. Scientifically, we can only use such concepts as the anthropic principle to explain the direction of expansion. The time might have come for a science that has evolution itself, from the Big Bang to the digital age, and beyond, as its subject. Such a science might look at such things as entropy or 'information content' as possible candidates for the measurement of 'E'. For now, we can only answer the question 'why does the future need us?" with "because we are here."

Where Are We?

"Because we are here" is not as trivial a statement as it may sound. As an answer to the question "why does the future need us," it places us in an interesting situation, because of the intimately related question "why are we here?"

Science only gives us timeless answers, as it tries to explain how particles form atoms, atoms form molecules, molecules form cells, and cells form organisms, and so on. Nothing that relates to a possible contribution of Homo sapiens to the *future*. At one end of the scientific spectrum, or maybe 'above' it, we find philosophy. Not much help there either, as it tries to explain the present in the ontological (what *is*, the nature of *being*), or the epistemological (what we know, the nature of knowing) sense. Nothing there either, about Homo sapiens' contribution to the future.

At the other end of the scientific spectrum, or 'below' it, we find mathematics, which provides the ultimate tools for the rendition of our knowledge.

Although mathematics is the vehicle for the re-creation of reality on computers, it can only render what we already know, or at least, imagine. It doesn't help us answering the question about our role in the future either. Beyond mathematics, we are back to philosophy, thus bending the spectrum of sciences into a circle.

Religion gives us answers like 'serve God in this life, and thus earn a place in the hereafter'. While this is a reassuring proposition to secure a future for myself, it does not help much to figure out whether Homo sapiens has a role in shaping the future of the universe, and if yes, what that role is.

Herein lies hope for the future, where science and religion can meet, but it does not help us finding answers for today. With all of our knowledge, we are stuck in the past and the present, while the future continues to create itself. More importantly, we are still stuck in a worldview of a static and permanent universe, while evolution continues. That leaves us with a hypothesis, at best, or a conjecture as the next best. This hypothesis consists of four parts:

- *Human intelligence approaches the intelligence of nature, and the intelligence of nature's evolution.*
- *The Human mind is the co-author of the future, and the future of evolution.*
- *Digeality, as an extension of the human mind, is the foundation for the future of evolution, as it is the drawing board, the laboratory, the production facility, the meeting room, and the knowledge infrastructure for the creation and execution of alternative futures.*
- *As our minds explore alternative futures and execute upon them, the configuration of our mindscape, and hence, the way we think, changes. Therein, and in the physical manifestations we produce consequently, lies the future of evolution.*

We have to remind ourselves that the human mind is part of nature, and as such, part of evolution. It is nature that proceeds on its course, now with the human mind not only as its scout, but also as an active co-author of its script. In the human mind, nature has turned itself inside out, revealing some of its mysteries, as if we are destined to cross its own path of expansion and take over our share of the responsibility. We are peeking over the edge of evolution's own expansionary curve, as the amphibians peeked over the edge of the lakes and seas, when they proceeded to leave the water, as Huxley said on the occasion of the centennial anniversary, in 1959, of Darwin's publication of the origin of Species:

> *"Turning the eye of an evolutionary biologist on the situation, I would compare the present stage of evolving man to the geological moment, some three hundred million years ago, when our amphibian ancestors were just establishing themselves out of the world of water. They had created a bridgehead into a wholly new environment. No longer buoyed up by water, they had to learn how to support their own weight; debarred from swimming with their muscular tail, they had to learn to crawl with clumsy limbs"*[3]

The time has come to turn from recreating the past to creating the future. It's what we do anyway, but maybe a full awareness of our creative power can help make, or create different choices. It is in this sense that I interpret George Dyson's journey into the history of intelligence, in *Darwin amongst the machines*, that he sums up as follows:

> *"Evolutionists, who appeared to have dislodged mind and intelligence, are discovering that evolution is an intelligent process and intelligence an evolutionary process, rendering the separation less distinct. Technology, hailed as the means of bringing nature under the control of our intelligence, is enabling nature to exercise intelligence over us. We have mapped, tamed, and dismembered the physical wilderness of our earth. But, at the same time, we have created a digital wilderness whose evolution may embody a collective wisdom greater than our own. No digital universe can ever be completely mapped. We have traded one jungle for another, and in this direction lies not fear but hope. For our destiny and our sanity as human beings depend on our ability to serve a nature whose intelligence we can glimpse all around us, but never quite comprehend."*[4]

Even if we do not fully 'understand the intelligence of nature', as Dyson puts it, part of nature's intelligence is a set of algorithm that produce evolution itself, and the three major thresholds to the different domains of existence, living, and being. Now, nature includes human nature, and the intelligence of nature includes human intelligence.

Evolutionary Framework, Part 3

In part II of this book, we have watched the performance of evolution in three 'acts'. In the first two acts, we have watched the evolution of the universe and the evolution of biological life.

Act 1 started with a profusion of energy, and ended with the completed physical infrastructure of the universe, ready to produce life. Act 2 started with the first living cell, and ended with the appearance of the great apes, our evolutionary ancestors. Both acts showed the expansion of the *physical reality* that surrounds us. The physical reality is nature's way of rendering itself into perceivable phenomena. We called these renditions *natural, physical* or *external* renditions.

At the beginning of act 3, the physical reality of *nature*, as we know it today, was completed. Since the appearance of Homo sapiens, although the evolution of nature continued, no natural entity was produced that was new *in kind*: no new evolutionary threshold appeared in human history. The physical reality was the *first reality* to be established by nature. It manifests itself as the universe, the biosphere, and all the human artifacts that followed later.

With the appearance of Arr, the chains that tied the animal's mentality to its own bio-chemical functionality were finally cut. From now on, the second reality, the mental reality of one living creature, the human being, started to play an increasingly influential role: it is the reality as it is stored in our mindscape.

The slumbering 'flickers' of thought that higher animals had, are now 'freed' from the organic structures that harnessed them. With that liberation of thought, the human's mentality took over the lead-role of evolution, at least on our planet. With the first independent thought, creative human mentality started to write its own scripts. The virtual reality of individual and collective human knowledge was born.

In the thousands of years that followed, human beings rendered that knowledge into new physical realities. This virtual universe of human knowledge entered a new phase of acceleration, of its depth and its breadth, with every new medium invented by human creative thought. The invention of the alphabet, the printing press, the movie-picture, the phone, are but a few of the plethora of milestones that mark the curve of accelerating human knowledge. New technology increased the modalities of expression and communication of knowledge, in the arts, in science, and every other expression of human activity.

The Third Reality

However, one important and fundamental ingredient was missing in all the technologies invented up to the middle of the 20th century. We developed the technology to manufacture machines of great complexity, but had no means to create a functioning rendition of the knowledge that went into building those machines. We did not have the language to describe four-dimensional phenomena, be it a functioning machine, a moving weather system or the planetary movements in the solar system, and we did not have the technology to execute such language. Until Turing came along and showed us how to create both, the language and the machinery. The language was the language of bits and bytes, that machinery was the computer.

For the first time in human history, we found a way to render four-dimensional human knowledge into a physical reality. Sound, speech, images, movies and text are samples of the ways in which we render the universe of our knowledge, expressed in bits and bytes, and give it physical and perceivable presence. For the first time we were able to turn the imaginable into the reality of the digital world. To stress the reality of its physical presence on the one hand, and to distinguish it from the 'traditional' physical reality surrounding us, I called it *digeality*. Digeality is now emerging as the *third reality* that is permeating and shaping our lives.

Living organisms could emerge and diversify because nature had separated design (the genotype) from implementation (the phenotype). In fact, the design (in the form of RNA and/or DNA) existed first, and nature found ways to manufacture living organisms from it. In this way, the genotype is a representation of the phenotype, and the phenotype a representation of the genotype. With the advancement of higher animals, the phenotype 'got ahead' of the genotype. The 'slate' of the nervous system and brain had an expanding 'blank' portion. The mental reality started to inhabit a world that was not fully pre-programmed into the genes. Animals acquired behavior that could only be transmitted by *communication* to the next generation.

After centuries of experimentation, (human) nature found a way to once again separate design from implementation. That separation is accomplished by rendering knowledge into a physical reality. Just like the genotype is the physical system that represents the structures and the functions of a living organism, similarly, digital programs (we can call them digital 'memotypes') are the physical systems that represent the structure and functions of the artifacts that we create, including cars, bridges, business-to-business processes or molecular nano-machines. The genotype, through variation of individual genes, determines the structure and the functionality of future living organisms. Likewise, through variation of algorithmic parameters in digeality, human mentality has the ability to determine the structure and functionality of future artifacts and even living organisms.

This is not a futuristic view, and it is not a metaphor, but the reality of today. Once again, nature produced a new type of algorithm. This time the renditions produced by the human mind, become the (digital) algorithm that produce new artifacts, thus changing the physical reality we live in.

Digeality however is an emerging phenomenon, one that has just started to manifest itself. Likewise, the technologies that allow the physical realization of new digital designs, such as new-materials-technology, nanotechnology and biotechnology, are at their very early stages. With digeality, Richard Dawkins' memes acquire the meaning of external renditions that have their own physical implementation. I do not agree, however, with the viewpoint that memes continue the helpless tradition of the genes as the random victims of nature's tradition of trial and error. In *The Meme Machine,* Susan Blackmore makes the case for the illusion of a (conscious) free will, arguing that memes act, on the level of the evolution of thought, in the same fashion as genes do in biological evolution:

"Critics of the analogy between genes and memes often argue that biological evolution is not consciously directed, whereas social evolution is. Even proponents of memetics sometimes make the same distinction, saying for example that 'much cultural and social variation is consciously guided in a way that genetic variation is not'... The whole point about evolutionary theory is that you do not need anyone to direct it, least of all consciously. When human beings act, our actions have effects on memetic selection, but this is not because we were conscious. Indeed, the most mindless and least conscious of our actions can be imitated just as easily as our most conscious ones. Cultural and social variation is guided by the replicators and their environment, not by something separate from them all called consciousness."[5]

I have suggested that memes, unlike genes are never copied accurately, because of the way our mindscape absorbs new memes, and that this phenomenon is exactly a unique property of the mind, as opposed to the biochemical replication of genes. To be sure, as we can see from our 'pinhead' model, our minds alter genes in ways that we cannot influence consciously. The point is, however, that we can alter memes with intent. The lies that we sometimes concoct are good examples. We might even repeat them so many times that they become an intrinsic part of the 'knowledge' that we store in our mindscape. After a while, we might end up with the illusion that they are the truth. Whether this 'intent' has to do with consciousness depends on your definition of consciousness. That there can be a higher degree of purpose in modifying memes than in the mutation of genes is hard to deny, at least in my opinion.

My contention is that one of the products of evolution is an increasing solution space. I call it the *expanding decision space*, because the expansion of the versatility of individual entities goes hand in hand with the expansion of the solution space. The ability of individual entities to exercise the options available in the solution space is proportional to the size of that space. I equate this expanded versatility of individuals, and the expanded decision space to the expansion of what we call 'free will'. In this sense, evolution is a continuous expansion of the degrees of freedom of its creations. It seems that we are on a continuous journey of transformations that lead us from the helplessness of pre-determined destiny to the pro-activeness of finding our destination. Digeality marks an important crossroads in that journey. It marks the finalization of the independence of human thought. The journey that Arr began, is finding its completion.

I suggest then, that the physical reality, the mental reality, and now digeality as the third reality, are the constituents of an expanding space-time continuum in which the future takes place, as it evolves, and manifests itself. The reality of evolution itself determines the direction of this expansion.

The third reality is not a separate and independent reality. Just as the mental reality is a reflection of the physical reality, digeality is a reflection of the mental reality. Despite the enormous advances in the creation of 'artificial intelligence' algorithm, despite the fact that world-champion chess player Gary Kasparov lost against such sophisticated 'intelligent' machines, digeality has not gained 'independence of thought' as the human mind did, and it does not look as if this is about to happen.

Digeality is an extension of the human mindscape. As such, *digeality is an extension of human unconsciousness.* Digeality expands our range of mental impressions, and in making these impressions, our mindset will change, similar to our pinhead-model of the mind. Our mindscape contains renditions; we are never aware of the production of those renditions. Our mind works with the results of whatever biological calculations go on in our brains. Similarly, our minds use the results of any digital calculations to absorb them in our mindscape. Our mind uses all of these renditions to come up with novel thoughts. Chess-champion Kasparov should use 'Deep blue', the chess-playing computer he lost against, as a mental extension to play against Deep Blue. The results might be interesting.

The expansion of the mindscape increases the opportunity for the creation of novel thoughts, and it changes the way the mind produces new thoughts. Our mindscape-window has now new territory to cover, and faces new ways to emerge from an expanded territory that changed its shape and texture. In this way, *digeality expands our consciousness.*

Illustration 44 captures the intricate relationships between the three realities that constitute the time-space continuum of the future. All three are different renditions of the world, as they emerged at each evolutionary threshold. Each of them contributes to the structure and the function of every living or non-living entity in the realms of its influence, each of them modifies each one of the others, and each of them, through interaction with the other realities, modifies itself.

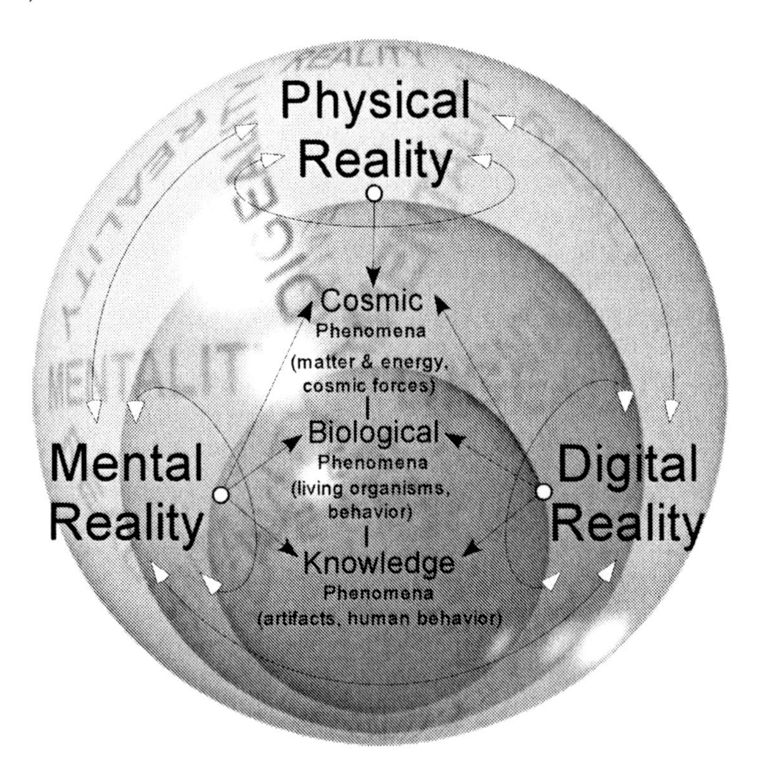

Illustration 44: Digeality: the third emerging reality in history.

This framework of realities determines the contours of what is imaginable, what is possible, and gives direction to what is preferable. If we leave one of them out, we continue to drift on the current of a future that doesn't need us. If we accept the reality of all of them, we accept co-authorship of our own future. That is the decision we face. It is a decision that nature will make one way or the other. It will be a conscious one, made by our mind, or it will be an unconscious one that emerges from the Darwinian mud of biological 'survival'. No doubt, nature will survive. The question is whether we will be amongst the surviving species, and if we are, what our definition of 'well-being' will be.

At this point, I should make a confession. I have no doubt that the human mind, once it emerged, has an evolutionary purpose. I am not suggesting that evolution *produced* the human mind with a purpose, although there is no reason why this could not possibly be so, if, and when, we discover the nature of the longitudinal algorithm that explains the direction of expansion. After all, the much touted 'randomness' is just one of the more effective ways to try out every imaginable possibility within a certain solution space. 'Randomness' is not equivalent to 'arbitrariness'. As we have seen, the solution-space was an expanding self-organizing tapestry of interrelated entities throughout evolution. This tapestry formed the ever more sophisticated test-bed for new randomly chosen variations of design parameters.

Each new addition changed the tapestry, and with it, provoked the emergence of a new goal: the new entity could establish a harmonious relationship with the tapestry, or it could not. If the new entity was successful, it raised the sophistication of the tapestry, and raised the bar for the next entity to be successful. Beautiful strategy, if you think about it! The only drawback is time. Many setbacks might occur, a tapestry could degrade, and catastrophes could wipe out complete strategies, which is what happened in nature indeed. Computer simulations based on such 'random' schemes are too time-consuming to be cost-effective, even with the fastest computers available today. Hence, mathematicians developed 'optimization methods' that do the trick in a shorter time. However, these methods force the user to define specific goals. Not so with nature. Nature had, quite literally, 'all the time in the world' to try out every possible design and test the design in the real world.

To avoid a conflict with the current scientific worldview, we might start looking for new applications of anthropic principle, until science changes its paradigm to include, and explain, the direction of evolution. My conviction of a direction stems from the emergence of human thought itself. With the independence of thought, we now have the capacity for (conscious) choice, and even the capacity to create new choices. Now, we have the ability to define purpose, goals, and objectives. More importantly yet, we can now *define* the values that we deem important to human well-being. For, as evolution continues, the human mind becomes the arrowhead of its direction. Now, we 'just' have to aim that arrowhead. How did we get here? Simple: evolution brought us here. By accident? Maybe, but it was an accident that got us where we are today, facing the choice to change our mind. We can accept the reality of evolution, and the interwoven tapestry of physical, mental and digital realities, or we can ignore these realities. The choice has consequences that nature (without human mentality) never faced in its fifteen billion year history. From science to religion, from politics to economics, and from family life to professional life: the choice influences the operational paradigm for our social infrastructure, and hence, the organizational structure of our social institutions.

What Evolution is

I can now extend my definition of evolution into the domain of knowledge. Let me correct that. I want to call it a perspective only, because it does not explain the process of evolution itself. The perspective does provide an interpretation of the continuity of the process as the expansion of knowledge.

Evolution is the process that continuously transforms knowledge (algorithm), encapsulated by physical manifestations (renditions) of reality, into novel renditions, in the direction of expansion.

In the cosmic domain, the relationships between the properties of matter and energy defined the algorithm, the 'recipes for behavior." These algorithm forged the creation of higher order entities with new emerging properties, like the properties of a range of new atoms. Relationships between the properties of new entities constituted the algorithm that, again, led to new entities, e.g. molecules. This process continued to form the diversity of galaxies, and the diversity of molecular compounds, until, at the threshold of life, enough knowledge was contained in some small local area of physical interacting compounds to harbor the design-elements for life.

In the domain of life, the properties of different compounds formed the relationships that forged different 'design elements' to form higher order organisms, e.g. the first prokaryotic cell. The new emerging properties of different prokaryotes established new relationships between them, and the resulting algorithm led to different forms of behavior (metabolic functions and interactions with other organisms) and new properties for the resulting higher order organism, e.g. the eukaryotic cells. By amalgamation of many different physical entities, the new organism harbored the knowledge for its own design, the knowledge for its own replication, and the knowledge to manufacture a new organism through its own algorithmic machinery. Variation of parameters in the design blueprint led to differentiation, and the production, reproduction and cooperation of different cells.

In multicellular organisms, the knowledge for the cooperation between different cells and organs was rendered as a separate algorithmic machinery: the nervous system and the brain. Execution of such an algorithm resulted in a specific mode of behavior. I called the totality of the knowledge harnessed by a living organism the mental reality of the organism, from the knowledge embedded in the genotype, upward to the final rendition of the phenotype and the palette of possible actions of the organism in its environment. In higher animals, this mentality expanded to include new algorithm that could be learned by actions or by example.

In the domain of knowledge, nature had created a separate algorithmic machinery that could capture new algorithm independently from any immediate physical behavior. This machinery used the rendering mechanisms from the senses that the organism already possessed to give the new knowledge a physical existence. At first, the expression of these new knowledge-constructs, or memes, was fragmentary, and sequential in nature. Finally, with digital technology, the human mind created an external machinery for the rendition and execution of algorithmic knowledge.

This begs the question: 'what is next'? Many foresee that same digital machinery taking over from human nature. In my opinion, they fail to see that current digital technology is 'just' the extension of the human mind. We are about to accept responsibility for the future, and digital technology is the extension of our mental platform to make that possible. The machines will have to wait, because we are *not* on the final approach to a fourth evolutionary threshold, yet. There is a twisted kind of comfort in the idea of machines taking over. In that way we can continue our complacent attitude that permeates the world today, and we don't have to 'change our minds'. No such luck. We are alone at the forefront of evolution, and solely responsible to expand our knowledge, so that we can render new futures, and take the next step on the road to changing destiny into destination.

We have accepted the fact that genes expressed their knowledge in the physical renditions that we call organisms, 'to ensure the survival of the genes', according to Dawkins. Variations of the genes allowed them to create the arms, legs, and eyes that warranted their survival. We take for granted the extensions that our minds created to expand our physical reach, strength and dexterity. Digeality, which *does* expand our mental reach, depth and dexterity, will wake us up to a new reality.

The Expanding Thinking Living Universe

All of the interacting realities as depicted in illustration 44 are the subject of evolutionary expansion, as depicted in illustration 45. We can now finalize the properties of expansion as they manifest themselves in the domain of knowledge, which consists of the mental and the digital realities as the representation of the world around us.

In the domain of knowledge, new entities are the knowledge constructs that our minds produce from the background chaos present in our individual and collective mindscapes. Precisely because memes are transformed, even between two individual minds, there is plenty of chaos to go around. The emergence of digeality added a new source of chaos by providing each individual, and every community of individuals with a virtually unlimited amount of new 'mind-impressions' in the form of websites, software ap-

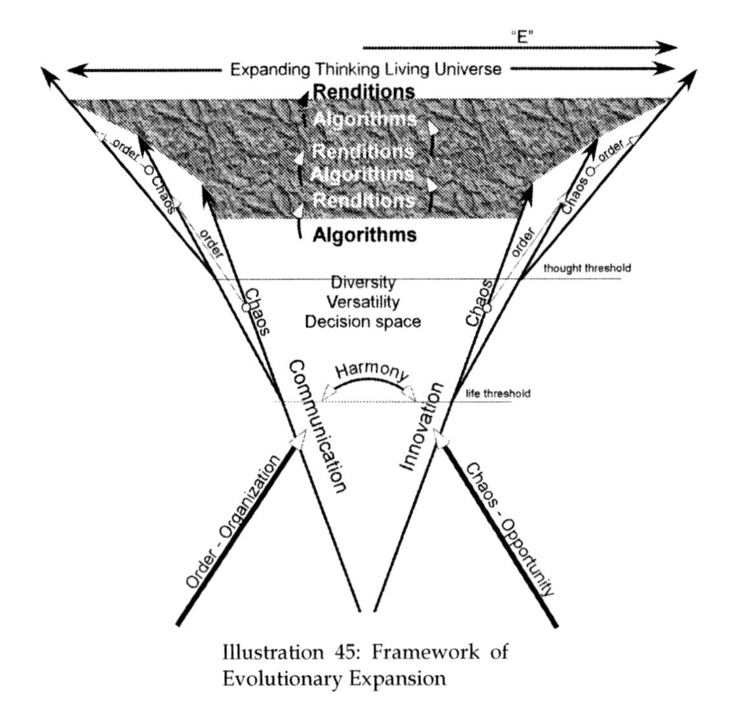

Illustration 45: Framework of Evolutionary Expansion

plications, and communication tools. The chaos comes to us as movies, songs, tunes, news, emotional outbursts of individuals, email, chat rooms, virtual-reality games, gambling, dating services, bulletin boards, and so forth, expanding our mindscape to unbearable proportions, but at the same time changing our mind by offering totally different perspectives and opportunities for the creation of novel thoughts of our own. I therefore maintain that in the domain of knowledge, as in the cosmic domain and the domain of life:

The drivers of evolution are the creation of chaos and the creation of order.

Looking at the world today, it seems that chaos did get ahead of order. Therein we might find the reason for the attempts to 'restore order' with every means we have, like the tools of war when the tools of dialogue fail, instead of using the opportunity for new, higher structures of order. However that might be: The continuation of evolution is dependent on the maintenance of harmony between chaos and order.

When we fail to use the tools of the domain of knowledge, we fall back on the tools of the domain of life: physical destruction. We have yet to learn to use the power of our mental tools, and the new tools of digeality. They afford us the possibility to create new ways to structure the physical reality of the world, and, in my opinion, they will. It might take a generation or two, but the new chaos of human thoughts, expressed in digeality represents new opportunities for order and will spark and fuel two processes that determine the way in which the world evolves. These processes are *innovation* and *communication.*

> *Evolution creates new, individual entities and environments through innovation, and establishes and maintains their integrity through communication.*

Each evolutionary domain - the cosmic domain, the life domain, and the knowledge domain - provides the substance that individual entities in each higher domain need to create and maintain the internal order of their structure-function, and the external order of their behavior. Inorganic matter provides the soil from which living organisms emerged and organized themselves. Organic matter provides an elaborate chain of food from which the variety of living organisms draws the substances to sustain themselves. The biologically 'chained' mindscape of the human mind provides the necessary chaotic substances, in the form of impressions that originated from the genotype and from life's experiences, for the creation of independent (conscious) information-constructs that we call thoughts. They gain their independence in the form of the 'mindscape-window'.

The products of evolution are individual entities and a networked hierarchy of environments, often called eco-systems. The properties of these products are *diversity, versatility,* and *decision space.*

Diversity. In the cosmic domain, we observed the growing diversity of physical entities that emerged through the ages. In the life domain, we observed diversity as the amazing proliferation of living species, and the diversity of individual creatures within species, as defined by variations in the genotype, the phenotype, and extensions of the phenotype. In the knowledge domain, we can now observe the incredible diversity of thoughts. The domain of the 'memes' might be the most diverse yet. With the emergence of digeality, this worldwide variety of thought is now accessible by every human being (although that 'every' still lies in the future). This accessibility rings the bell for the next round of increased diversity, and not, as some might think, the convergence of thought. That is because it is the human mind, and with that, it is nature at work here. That means evolution; which means expansion.

Versatility. As I have introduced, versatility has three defining components: *awareness, processing capability* and *dexterity.*

Awareness. In part two, we have seen how nature continuously improved the technology that individual entities use to perceive the world around them. With the higher animals, and human beings, nature had given creatures the ability to form a complete mental 'picture' of the environment they lived in. With the emergence of independent thought, a new world of awareness opened up. Human minds can reflect on their own thoughts and modify them, thus creating new representations of the world. This produced the world of symbolic representations and philosophical, scientific, artful, religious and political renditions of the world developed.

New mental concepts emerged, and from them, humans created the technologies for the expansion of their own awareness. Microscopes and telescopes, communication technologies and a host of media led to the continuous expansion of awareness.

With digeality, traditional limitations of human awareness disappeared. We can now build digital renditions of phenomena that remained inaccessible to us by older technologies by varying their space and time dimensions at will, as if adjusting our own scale. We can travel through blood vessels as if we had the size of an atom, or we can travel through space like the starship enterprise in a Star Trek movie. Digeality gives us the possibility to study, design, and manufacture physical systems on any scale, from the scale of atoms in nanotechnology, and the scale of single genes in biotechnology, to the scale of planets and stars for designing space stations and livable habitats on remote celestial bodies.

Processing capability. In the cosmic domain, physical systems are the algorithm; they are the expression of the relationships between the properties of the components of the system. The algorithm and the executing machinery are one and the same system. Starting with the intelligence captured in the genotype, nature created a variety of separate mechanisms for the execution of algorithm that sustained the life and the behavior of organisms. In the domain of knowledge, a variety of technologies transferred the execution of algorithm outside the mental reality of the human mind, from mathematical calculus carried out on clay tablets or pieces of paper to the slide rule, the pocket calculator and the computer. With digeality, specifically through the Internet, all the processing power in the world is being connected, thus connecting the knowledge that emerges on the planet. This is how digeality increases the processing power of every individual, and every community of individuals.

Dexterity. Nature continuously invented new technologies to expand the physical dexterity of its creatures. The same is true for the creature's mental dexterity. Nature gave animals just enough processing capabilities to deal with the features of new organs. It would be of little use to transform the fins of fish into the legs of land animals if the animal had no way of 'knowing' how to use them. With the domain of knowledge, mental dexterity entered a new era. Mental dexterity leaped ahead of what was needed to 'merely survive'. Nature 'approved' of this new feature and did not select against it. Some of you might still wonder why nature produced this new feature, and allowed it to go forward. Human mental dexterity would produce the world we live in, and would produce alterations of nature itself. Is that what nature wants? Or does nature not care? It had produced more useful technologies for the last fifteen billion years, so there must be some 'rational' for creating this 'surplus' in mental dexterity. If we don't accept such a rational, we might have arrived at the point where we should create one, and define a purpose.

We applied our new mental dexterity to expand our physical dexterity. We equipped ourselves with trains, planes and automobiles; with forklift trucks, cranes and screwdrivers. Later, we went to work on the expansion of our mental dexterity and invented 'digital'. Digeality makes digital dexterity available to everyone, thus expanding the foundation of our common knowledge, and creating a massive digital foundation for further expansion of that knowledge.

Decision space. At several points throughout the book, I have suggested that evolution produces an ever-expanding domain of possibilities. The appearance of every new entity, and every new creature changed the environment it participates in. The interactions of the new creature with its environment added to the diversity and functionality of that environment. Expansion of the versatility of creatures further increased the range of possible interactions within the environment. Over a period of billions of years of evolution the environment of the planet presented an ever increasing range of options for new species to 'make a living', and be successful enough to reproduce itself. The increasing diversity and increased versatility also gave individual creatures more possibilities to interact with the environment.

With the higher and 'programmable' animals, when behavior could be taught and learned, the possibilities increased again. After crossing the barrier to independent thought, a new domain of possibilities opened up. This is the expanding decision space. It is a solution space, a range of options, no matter who or what makes the choice.

Digeality gives this decision space another boost by expanding the mind of every human being, and in doing so, changes it.

Changing our Mind

In his foreword to Susan Blackmore's *The Meme Machine,* Richard Dawkins recognizes, and acknowledges the dual use of the word meme. On one hand, it can have the meaning of an algorithm; on the other hand, it can be a rendition. By admitting this dual interpretation of the meme, Dawkins cleared a major objection, namely the infidelity of memes to replicate accurately, against the meme theory:

"I believe that these considerations [to consider the meme as algorithm in stead of a rendition] greatly reduce, and probably remove altogether, the objection that memes are copied with insufficient high fidelity to be compared with genes."[6]

How do you copy the algorithm hidden in a rendition? Sure enough, algorithm like a recipe for apple-pie are easy to render, say on paper, or in the form of words. This type of recipe might have a high fidelity for being copied accurately. Conveying your idea about a new product to your boss might not be that easy, which you'll find out when you're asked, "Could you run that by me again?" or "What in the world do you mean?"

Susan Blackmore solves the meme-copy-infidelity problem by pointing at increasingly sophisticated technologies to do the copying, and by shifting from copying renditions to copying algorithm. Technologies range from verbal reproduction to digital ways of copying memes:

"Note that this evolutionary process has made memetic-copying mechanism more similar to genetic ones. One of the great worries for memetics was the Lamarckian[7] *'inheritance of acquired characteristics." We can now see that with further developments of meme-copying technology the tendency is, just as it presumably was for genes, towards a non-Lamarckian mechanism – that is, copy-the-instruction not copy-the-product. The precise way it is done will always be different for memes and genes but the basic evolutionary process is the same."*[8]

What Are We?

According to Blackmore and Dawkins, human beings are not only 'gene replicating machines', but also 'meme replication machines'. In their view, the evolution of human ingenuity will produce technologies that assure the *correct* replication of memes. The rest is simple. Darwinian selection mechanisms will select preferable memes and combine them into complete 'memotypes' that define the mentality of our children and grandchildren. That is how biological evolution did it, and that is what mental evolution will do. Right? Not quite.

Many neo-Darwinists think of a biological organism as a vehicle to assure the propagation of genes. The gene is primary, and the body is – just – the vehicle. For me, in relation to biological evolution, this perspective is as good as any. After all, metaphors are the tools to construct explanations, including scientific ones. Human beings arrived on the scene anyway, no matter what perspective you fancy. However, now, this perspective of the 'selfish gene' is being transposed to the domain of knowledge, or human mentality. Daniel Dennett sums it up best, as a slogan, in *Consciousness Explained*: "A scholar is just a library's way of making another library."[9] This slogan refers to the Darwinian view on the human body as the vehicle for the transmission of genes. In the same way, the scholar is just a vehicle for the transmission of memes. It is true; a human body doesn't produce or change genes. However, the human mind does produce and does change memes. I find the Darwinian mindset that promotes the helplessness of the human mind troubling at best and dangerous at worst.

Does this mean that I advocate a different perspective on the evolution of biological life? No, it does not mean that, as might be clear to the reader of part 2 of this book. It *does* mean that I have a different perspective on the *totality* of evolution, as follows from my suggestions in the previous chapter. In my view, each new evolutionary domain added a different and more sophisticated mechanism for variation and selection.

In the *cosmic domain*, 'genotype' and 'phenotype' were equivalent. Both were one single rendition produced by matter and energy. Variation and selection occurred in one action that succeeded, or did not succeed, through physical and chemical interactions of matter, to form new entities. The process could be completed in an instant, or take a long time, but most of the times it took place under violent circumstances.

In the *life domain*, the genotype was separated from the phenotype. Each has its own physical rendition, and its own mechanisms to execute the underlying algorithm. At first, the genotype determined, and created the algorithm that the phenotype could execute. We give these algorithm names like 'instincts' or 'habits', and we include the resulting behavior in the 'extended phenotype'. Variation takes place in the genotype, without touching the 'final product'. This was an incredible advancement over the violent physical and chemical clashes of entities in the cosmic domain. One advantage is a decreased 'waste' of final 'product', because variation of the design might itself lead to invalid designs that would never produce a phenotype. More importantly, the compatibility of genes across individual creatures meant a far greater potential for the creation of novel designs. Nature had produced a mechanism that could take parts of existing blueprints and recombine them to produce new designs. The mechanism that produces the final product (the phenotype) is compatible with any new combination.

It would take humans 100,000 years to reinvent such universal production mechanisms in the form of robots that line up to form 'industrialized' production lines of a variety of different cars. The software industry is still struggling to develop 'universal machines' that can produce new computer algorithm from the design of a new application. Thus, the separation of genotype and phenotype, and the separation of variation and selection was a step forward in the evolutionary process. The process *itself* became more intelligent. Communication between genes – the formation of a new sequence – led to innovative designs. If the design was feasible (meaning the universal mechanism was able to manufacture a phenotype), the organism was produced. The communication (interaction) of the new organism with its environment determined the degree of harmony with the environment. If the new instrument contributed to the symphony, it continued playing. If not, it did not produce offspring. Sometimes, the new instrument found employ in a different orchestra, like plants that only flourish high up in the mountains. If the new organism was successful in the environment, it contributed to the innovation of the environment itself, thus changing the rules for the next species to appear on the scene.

Then, Homo sapiens, as the only species ever, *entered the domain of knowledge*. Equipped with a faculty to create autonomous knowledge constructs, Homo crossed a threshold of a significance that (at least) equals the significance of the threshold to the domain of life. With the crossing of this threshold, the gene lost partial control over the content of the autonomous thought-products. As Steven Pinker explains, the genes remained in control of many thought-producing algorithm, the basic topology of the mindscape, and of many pre-configured behavioral algorithm. However, the mind now had a new degree of freedom; the freedom to create new thoughts. In addition, the mind acquired a new degree of freedom to modify its way of thinking (e.g. the Hebbian plasticity of the brain). With Homo, the genes *lost their hegemony over the (human) brain*. We all know, with all our emotions and feelings, how hard the genes try to determine our thoughts, but at times, thoughts of our own making prevail.

Back to the notion of memes. Memes are the knowledge-constructs that I call 'independent thoughts'. Independent thoughts are, in my terminology, *mental renditions*. If we communicate these memes to the outside world, we create renditions of our thoughts. These renditions are *also* memes, according to Dawkins, the inventor of the concept. In my terminology then, memes are both, mental (or 'internal') renditions or physical (or 'external') renditions.

Therein lies the difference between genes and memes, and therein lies, in my opinion the true evolutionary importance of memes. Memes are renditions, not algorithm. They might be renditions of algorithm, but they are *always* renditions. Memes travel between human minds and such communication is always the exchange of renditions. We never observe or experience the execution of any algorithm. Renditions are a medium, the medium becomes the message, in the sense as explained by McLuhan and Shlain. Hence, the rendition will be 'subjectively' processed by any human brain. In addition, the message, which is the rendition, will 'massage' the receiving brain and change its way of thinking. Such is the fate, and the virtue, of memes. Genes are *replicators* that produce exact copies of themselves in the right environment. A sequence of genes *is* the algorithm, and as the medium, it *is* indeed the message. There is no further content hidden within it. A copy of it does the same thing as the original when executed by its environment. A sequence of genes is the absolute 'bottom-line' algorithm for life; it does not represent anything else. High fidelity reproduction is their 'purpose' in life; biological reproduction, and thus biological evolution *depends* on it. Biological innovation happens by a variation of the genes, a process that the genes have no control over. Compared to genes, memes are not *replicators* but *imitators*.

On one hand, memes mold the mindscape in ways precisely as Dawkins and Blackmore describe. If the mindscape is submitted to a whole package of related memes, like a package that represents a certain scientific discipline, or a religious way of thinking, the impressions will mold the mindscape in certain ways. If this happens over longer periods of times, or isolated from the reception of different memes, a certain view of the world will emerge, and, moreover, a certain way of generating views of the world. They are also right in saying that certain memes are better at imitating themselves than others. The meme 2+2=4 has been imprinted into our minds with such repetition and 'precision', that there is hardly any difference between its imitations in different minds.

On the other hand, as I have discussed using the pinhead model, the mindscape changes the meme. This change is dependent on the momentary four-dimensional 'topography' of the mindscape. This capacity for change is not a 'defect' of the brain. Quite the opposite, 'reading something different' in a transmitted meme or a collection of memes, is an amazing source of innovation in the human mind.

Are we a meme machine? My answer is yes, but with an important difference from the common 'memetic' interpretation. We are meme-*creators*, not copiers. Many will argue that it would be a tough life, if memes cannot be communicated accurately. They are right. The low copying fidelity of memes is the very basis for cultural and personal diversity. Some lawyers might be out of a job if memes were high fidelity replicators, because there would be no room for interpretation. We surely try our best to build 'common bases' of memes that can be replicated accurately. Science, religion, cultures, languages, law & order, military organizations, companies and families depend on a common set of memes. We usually call such 'tapestries' of memes the 'common sense' of such communities. It is the foundation for their behavior. Because we are meme-creators rather than replicators, humanity shows the rich diversity on our planet. With digeality, we might be able to define a global set of 'equality memes' that guarantees a new set of human rights and a new minimum set of common values. Does this mean that we move towards a unification, a 'socialism of the mind?'. Are we therefore converging to a common lifestyle or behavior? This assumption is painfully wrong, because it ignores the creativity of human imagination and the power of digital (and other) technologies.

There is plenty of room at the top of a sound base of expanded human rights, a base that transforms entitlements into the standard toolset of every human being, just like eyes and ears, and lungs and fingers. Because of the 'creativity' of the human mind to modify memes, the diversity of memes available to the collection of the 6 billion minds on the globe is considerably larger than the number of available genes. This means that the room for expansion, on top of a growing base of 'common memes' is virtually infinite. In this way, a growing number of individuals, and a growing number of communities can display, create, define or choose (whichever you prefer), a growing diversity of *lifestyles*, or if you prefer, 'extended phenotypes'. These lifestyles are expressions of their own 'memotype'.

Not so, says Susan Blackmore. Technology should help to make memes more reliable replicators. Digital technology makes that possible indeed. You can burn precise copies of your CD's or DVD's. You can buy precise copies of the program (the algorithm) that does the burning of CD's for you. The effect of this high fidelity copying of renditions and of algorithm multiplies the proliferation of renditions (like CD's) and, as we know all too well, the Internet adds another multiplication factor. What about algorithm?

Most of us don't care about the algorithm. We are interested in the renditions these algorithm produce. In the case of an algorithm for producing apple-pie that I received in an email from my mother, I'll be interested in following it precisely so that my apple-pie turns out exactly like hers. When I buy an application program (an algorithm) for my PC, be it a text processing, voice recognition or computer aided design application, I am interested in its capability to create a multitude of *different* renditions. Again, the net-result of the exact replication of algorithm is a proliferation of renditions, or memes.

Digeality is not only the ubiquitous proliferator of memes; it is also, as an extension of the human mind, the multiplier of the *diversity* of memes. On top of that, digeality has a selection mechanism for memes all of its own, as we can experience from our junk-email, and the seemingly arbitrary hyperlinks that we follow on the Net. Herein lays another reason why digeality is a veritable third reality that exercises its influence on the future of evolution.

With the introduction of independent thought, nature added a new feature to its process of evolution. Nature added the meme and the capacity to create new ones by learning and by modification. Learning is nothing else than the creation of mental renditions, or 'mental memes'. Modification is the creation of new memes by manipulating existing memes or the relations between them. With this feature, nature exponentially expanded our options for exhibiting different lifestyles. With this expanded decision space, we took another step in the direction of *choice*. Also, in my opinion, digeality, by extension of our mind, represents a step on the road to true self-consciousness. When we reach that destination, we might be able to finally explain what consciousness is.

Memes are the most powerful products of nature yet. With memes, we can create new types of vehicles, new types of buildings, new types of power generating plants, even new atoms and molecules, and, as I already mentioned, we now have memes that can create new genes.

Memes, as I have interpreted them, and the meme-creating faculty of the human mind, add a new layer of intelligence to the process of evolution. Digeality complements the mentality of the human mind, similar to the way the wheel complements the walking ability of our legs.

If the genome is the engine that propagates humanity into the future, if our bodies are the vehicles that get us there, then our minds are the steering wheel.

Where do we Go from Here?

Human beings entered the domain of knowledge. This changed the rules of evolution itself. In River out of Eden, Richard Dawkins makes it utterly clear that, in the evolution of life, there is no feedback from the phenotype on the genotype. He calls it "Wrong, utterly wrong! Genes do not improve in the using; they are just passed on, unchanged except for very rare random errors."[10] Darwin is king in the domain of life. The world would be a strange place indeed, if our children would acquire our learned habits, acquired illnesses, and our way of thinking developed over the time before we conceive them.

> *"Time and again, cooperative restraint is thwarted by its own internal instability. God's Utility Function seldom turns out to be the greatest good for the greatest number. God's Utility Function betrays its origins in an uncoordinated scramble for selfish gain."*[11]

I submit that this so called 'selfishness' has guided evolution to continuously increase knowledge, to constantly increase the diversity within environments and relentlessly increased the versatility of its individual creatures. The result has been a tapestry of astonishing complexity that maintains its harmony in ways that we can truly call miraculous. A planet that doubled its population in just a little over a half century without totally destroying itself is miraculous too. Innovation originates in individuals. This is true for individual physical entities in the cosmic domain, it is true for the life domain through the mutation of a gene in an individual, and it is true in the domain of knowledge through the creation of new memes. If you want to call the pressure that leads to individual change 'selfishness', fine. The result however has been a weave-world in which 'every body' finds a place to 'be well'. There is no absolute metric that allows the measurement of 'common good'. It is a property that can only be seen relative to the environment of the individual entity, similar to the relativity of space and time.

Many want to carry the baggage of Darwinian explanations into the domain of knowledge, and apply it to the evolution of the mind. That is when the problem arises. It results from a view on the emergence of human beings as merely a biological descendent from the great apes with a 'twist' that twist being a halo that emanates from the brain, halo that we call 'mind'. This view, in my opinion, is simply wrong. Darwinian thinking fully acknowledges the fundamental difference between the cosmic domain and the life domain, but does not so for the difference between the life domain and the cosmic domain.

In *River out of Eden*, Dawkins describes the evolution of human beings as a continuity of biological evolution that passes 10 thresholds on its journey. The first of Dawkins' thresholds is the 'replicator-threshold': "the arising of some kind of self-copying system in which there is at least a rudimentary form of hereditary variation, with occasional random mistakes in copying." The last threshold, Dawkins calls the 'Space threshold': "After radio waves, the only further step we have imagined in the outward progress of our own explosion is physical space travel itself: Threshold 10, the Space Travel Threshold."

"Where do we go from here?" is the question asked in the heading of this section. My suggestion is to go back to the threshold from the domain of life to the domain of knowledge, and acknowledge the evolutionary difference between the two. I suggest then, to leave Darwinian thinking behind at the doorstep of this threshold, and give credit to the creative power of the human mind. Then, we'll take it from there.

In the cosmic domain, entities evolved by *formation*. Physical and chemical processes repeated themselves, based on the information embedded in the properties of particles and energy, thus re-constructing the same physical entities, over and over again. Innovation occurred by the formation of new combinations of matter and energy.

In the life domain, entities formed by *reproduction*. Biological processes used the knowledge embedded in the genes to produce the same organisms. Innovation occurred by variation of the design-knowledge embedded in the genes.

In the knowledge domain, entities form by *learning*, and learning is the creation of memes. Mental processes use the knowledge embedded in our mindscape - a tapestry of memes - to produce different memes. Innovation happens by conscious or unconscious recombination of memes. Intention and focus drive the 'variation' of memes, not randomness. Randomness can play a role, but it is certainly not the sole driver. This changes the way evolution happens.

Sure enough, each domain also uses the mechanisms of the previous domains. Biological reproduction is dependent on repetition: execution of a cell's copying algorithm relies on the accurate repetition of physical and chemical processes to recreate the biological unit, atom by atom, and molecule by molecule. In the same way, mental learning depends on the biological processes that produce thoughts, and as such, depends on the fidelity of reproducing the same biological machine, from generation to generation.

Darwinian tools are the square pegs that don't fit the round holes of evolution in the knowledge domain. Learning is creation. Learning and creation are the hallmarks of the domain of knowledge. Once we acknowledge that, we can go forward. If we don't, we might as well get rid of our prefrontal lobes, live the 'happy lives' of zombies, and leave the future to chance, or to computing machines with an excellent random-variation-generator.

We are like Russian dolls, because we incorporate all three mechanisms of re-creation: repetition, reproduction and learning. Each new layer added a new dimension to creativity, in a very literal sense. Each one expanded the decision space for individual creatures. It is what nature did. Nature raised us above biological existence into the mentality of 'being'. If we accept the newly acquired freedom, we accept responsibility for the futures that we co-create with nature.

The emergence of the third reality completes the process of waking up the human mind to its true potential.

As a global tapestry of knowledge, it will give us a different perspective on planetary issues, and make us reach out into the universe. Dawkins' 'selfish' genes produced the value foundation of all human beings. 'Selfish' memes will be the 'leading thoughts' that determine the nature of the expansion of that foundation. These leading thoughts will be carried by the thought leaders of the future. Digeality presents unprecedented opportunities for the creation of new ideas and new realities, and thus raises the bar of responsibility for future leaders of every type of community. If these leaders close their eyes to evolution, the new ideas will carry the seeds of peril and coercion. However, if they change their minds and acknowledge the power of evolution and the human co-authorship of the future, we can look forward to the most fascinating, and most satisfying times humanity has ever experienced.

I am astonished by the reluctance, or ignorance, especially by religious institutions to accept the reality of evolution. This ignorance is a major cause of the friction in the 'axle of change' that I described in part 1: the friction between our (scientific) explanation of reality and the mental paradigm we operate in that reality. The friction is enormous, the axle about to break. Therein lurks another answer to the question "where do we go from here?"

The first step has to be a change of mind, because if we refuse to accept the reality of evolution, we miss the opportunity to shape our future, and the future will shape us. Nature has brought us here. If God is behind all it, then, with the addition of the domain of knowledge, God handed us the steering wheel. Evolution has no brakes, but now, it does have a steering wheel. The road into the future is a slippery one; therefore, the use of brakes would be hazardous, as sharp jerks of the steering wheel would be. Digeality is our new companion, because it provides all the passengers with knowledge and an expanded capacity to learn as the landscape changes around us. It allows us to peek ahead of us, because it can translate our imagination into the third reality, and we can ask that reality "what if?" Such is the digital spirit. Digeality gives us the tools to start minding our own future in ways never possible before. Digeality gives us the colors and the brushes to paint new realities; it gives us the sheets, the tunes, and the rhythms to compose new symphonies; the words and the grammar to write new poems; and opportunity to make mistakes before they become reality.

Chapter 11: Minding the Future

"One of the most difficult social developments is changing a culture, but this is what must be done to prepare for continuing globalization. The world must be seen as a collectivity – the uni-diversity that it is, in fact, but not yet fully in perception. The change in perception must begin, however, with individuals. Such a change cannot be forced from the top...
...But if there is to be sufficient peace in the world to permit opening economies wider to global markets and competition, such a mind change will have to occur."

Jack Behrman[1]

Framework for the Future

Imagine that you are a futurist, equipped with all of the modern mathematical tools for trend-analysis, but without the insights into evolution that we gained in the last century-and-a-half. Imagine, sitting at the edge of the world just before, or after the occurrence of any of the three evolutionary thresholds.

Would you have predicted the emergence of galaxies and solar systems from the ball of fire right after the Big Bang? Or the emergence of parrots, daisies and goldfish when observing the hot planet earth? While watching primates in the trees, a futurist might conclude that, since they came from the water onto land and from land to the trees, the next 'logical' step would be to fly! Would we predict that the monkeys would develop into beings, capable of creating autonomous thoughts? That these beings could 'take' these thoughts and turn them into suits-and-ties and space shuttles?

Before defining a 'framework for the future', it is important to remind ourselves that the future is unpredictable. In fact, the unpredictability of the future is an inherent part of the process of evolution itself, as most of the 'longitudinal algorithm', the algorithm that describe evolution along the history of the expanding universe, remain a mystery.

"Just a little while ago, you said that our intelligence was approaching the intelligence of evolution itself, JJ. What gives?"

Good point Alvas. I'll try to answer that. On the one hand, evolution did become more predictable than ever before.

"Can you explain, JJ?"

Let's look at the cosmic domain first. Although the big question remains, whether the universe will accelerate its expansion, stop the acceleration, or even start to compress in the future, we have a good scientific 'handle' on what is happening today, and what *might* happen in the future. We did not have such insight just eighty years ago.

"I got that."

Then, let us proceed to the domain of life. There, we can see a similar situation. Although we don't know what nature thinks of next, and what the effects would be, we do understand how biological evolution works. Therefore, we do have a profound understanding of what is *possible*, or at least, what is *imaginable*.

"You mean, we know that changing genes or their sequence will result in new organisms, with a new form and new functions?"

Exactly. In addition, we have control, fortunately or unfortunately, over the evolution of nature.

"Like in making species disappear?"

It's not only negative, Alvas. We have been changing the genetic makeup of living organisms for over a century. Medicine, for example the development of antibiotics, is impossible without genetic modification. The important thing is that we can do it, and that we are just at the beginning of what biotechnology can do.

"Same for nanotechnology?"

Same thing. We now have the digital tools to design, render, analyze and study new 'nano' and 'bio' structures. As I said before, our memes have now control over genes and other molecular structures. In the sense that we have that control, and possess the new tools of digeality (and other technologies), human nature has more 'predictive' power than ever before: we now have the capacity to create new artifacts, and new forms of life. With that, we are part of a future that we helped orchestrate.

"Therefore, the future becomes more predictable!"

Slow down, Alvas. As I said, on the one hand, you are correct. On the other hand, however, as we have seen throughout the book, evolution created an ever-larger decision-space. With our entrance into the domain of knowledge, with the independence of thought, the expansion of that decision-space accelerated. Birds cannot leave the atmosphere, but we can. Living organisms cannot change their own genetic makeup, but we can. Planets cannot build cars, but we can. The introduction of new possibilities brings new uncertainties, new vulnerabilities, and a new potential for non-linear effects of the changes we make.

"Sounds like one step forward, two steps back, JJ."

I am not sure that you should describe it in terms of 'forward of backward', Alvas. What it does mean however is that the new tools of digeality not only give us new possibilities to change the future, but that we will be dependent on those tools to manage that future. This is where 'approaching the intelligence of evolution' comes in. We will be at the helm of the future of evolution. The knowledge domain is the new wildcard and trump card at the same time, if you will.

"Scary."

Sure is, Alvas. That is why I suggest my 'framework of the future'. It doesn't imply the predictability of the future, as the future always 'predicts the present', never vice versa. The framework, however, reveals the contours of the decision-space, and its most important properties. Maybe that way, we can better judge the directions we should *not* take.

My framework for the future has three components, all of which we have discussed at length throughout the book:

- *Evolution is the process that shapes the future*
- *The human mind is the co-author of the future*
- *Digeality is the knowledge foundation for imaginable futures*

The future will show whether my framework carries the weight, I attach to it. The following is a brief description of all three components, placing the elements already discussed in previous chapters into *one* evolutionary context.

Evolution

'*Evolution is the process that shapes the future*': the first pillar of the framework for the future. The existence of evolution seems to be the most scientifically grounded proposition of the three, at least for the domains of cosmic evolution and the evolution of biological life. Ironically, the denial of evolution as the explanation of the emergence of humanity is the strongest barrier against a worldwide embrace of evolution's expansionary power. So-called creationists of several religious (and cultural) denominations, knowingly or unknowingly, do not accept evolution's existence, and hence, its consequences. I am convinced that this will change in the coming generation (or two); digeality itself will deepen and distribute our understanding of evolution, and religions will re-focus their attention from the medium of the written word to the meaning of the message encapsulated by those words, disentangling medium and message. Just imagine what will happen when evolution turns out to be the undisputable explanation of reality in the eyes of religious institutions. If evolution is reality, then the same God has created that reality, and, therefore, made human beings the co-creators of the future.

The following table summarizes all of the discussed properties that characterize evolution.

Illustration 46:

Properties of Evolution

Process of evolution	**algorithm & renditions** **chaos & order** **communication & innovation** **harmony** **relativity**
Properties of evolutionary products: individual entities and environments	**diversity** **versatility:** ➢ *awareness* ➢ *processing* ➢ *dexterity* **decision space**

I have defined evolution as a perpetual spiral of algorithm and renditions; a spiral that manifested itself in three major stages that I called the cosmic domain, the life domain, and the knowledge domain. Each new domain starts with a 'threshold'. This doesn't mean that there cannot be continuity from one stage to another; it only means that we do not know what happened at those thresholds. With the modified Hubble telescope, we can peek into the past to just millions of years away from the Big Bang, the threshold to our existence. New telescopes and deep-space probes, with the help of digital technology, might – in the not too distant future - shed light on the threshold itself. Similarly, Miller-like experiments, molecular genetic research and space probes might – again with the help of digeality – find an answer to the question 'what is life?', and shed light on the second threshold. Finally, neuroscience, cognitive psychology and artificial intelligence research might shed light on the mysterious threshold to knowledge, and advance our insights into the mind-body mystery, the plasticity of the human brain, and the human mind. Each domain produced its own individual creatures, first physical entities, then living organisms, and finally thought constructs, or memes, or - generically – knowledge, and each domain produced its own environments, defined by the individual entities and their interactions.

Each domain emerges from the previous one, and adds on to the previous one. Ultimately, each domain might be explained as new interactions of the substances from the previous one. When such explanations are found, many questions will be asked in new ways: is there a soul? What is the soul? Where is the soul?

I have no answer to these questions. My bias is that we will be able to explain the apparent discontinuities between the domains, and turn the thresholds into continuities. It is also my heart-felt opinion that this will require a conceptual shift in the scientific worldview, and that it involves physical phenomena that we have not yet discovered. For the time being, the thresholds remain mysteries.

I defined the most important property of creatures (including physical entities) in each new evolutionary domain as the autonomy that the creatures acquired from the forces that ruled the behavior in the previous domain. The entities in the cosmic domain acquired a new degree of autonomy from the forces that ruled the 'world' before that time (if there was such a thing, e.g. a 'multiverse'). Living organisms acquired a new degree of autonomy against the background of the thermodynamic forces of the universe, while at the same time harnessing those forces in every molecule of their bodies, and using them as a necessary means to maintain order and integrity. With the appearance of Homo, thought acquired a new degree of independence of the biological processes that govern the brain and the nervous system. Just as the organic body harnesses and needs 'cosmic' matter and energy to maintain itself, the human mind harnesses and uses the biological processes *and* the physical properties of matter and energy to do its work.

I suggested *innovation* and *communication* as two fundamental characteristics of the process of evolution. Communication is nature's way to create *order*. In the cosmic domain, communication manifests itself as the interactions between entities – consisting of matter and energy – in both the small world of quanta, and the big world of galaxies and solar systems. In the life domain, communication manifests itself as the interactions between living organisms, and between organisms and entities from the cosmic domain. As the evolution of life evolved, so did the organs for non-physical interactions between living organisms, and a new form of non-physical interaction – of communication – emerged. Let us remind ourselves, however, that every form of communication is an interchange of matter and energy between two living creatures, at least as far as we know. Also, we do not know about all of the possible forms of such communication; that's why we don't know *how* ospreys manage to come home to the same nest every spring. Finally, in the domain of knowledge, human nature found a way to communicate knowledge. After many attempts at the development of new media, human minds invented digital technology, by far the most complete way for communicating knowledge yet.

Innovation is nature's way to create novel patterns out of the background *chaos*. In the cosmic domain, that chaos is maintained by the perpetual increase of entropy in the universe. In the domain of life, chaos is maintained by the background of matter and energy, and by the organism itself by recycling matter and energy that the organism needs for the maintenance of its own order. Random recombination of DNA molecules provides the necessary chaos for the creation of novel organic genotypes and hence, phenotypes. In the knowledge domain, the constant stream of experiences provides the brain with a chaos of self-modifying memes. Digeality increases the potential for 'mindscape-impressions' exponentially. One of the most mystifying features of long-term evolution is the maintenance of a finely tuned balance between order and chaos. Nature's harmony, at least so it seems, will continue with us, or without us. The difference between the past and the future is: we do have more choices today, and we have the ability to make those choices.

The creation of order and of chaos drive the evolution of the universe. Order is predictability and stability; order is repeatability and reproducibility; order is meaning and communication; and order is status-quo and conformism; boredom and stagnation; and ultimately order is death. Chaos is unpredictability and confusion; chaos is every possibility knowable and unknowable; chaos is opportunity and surprise; and chaos is emptiness and diffusion; ultimately, chaos is death too. Order in itself, as well as chaos in itself cannot sustain any system. It is the interplay between the two, and a finely tuned balance between them, that produces innovative expansion: in the universe, in life, and in human thinking. Chaos and order are the two edges of a deep and mysterious gap that runs like a canyon through space and time. Nature manages to bridge that gap, and keep the two in harmony. Human nature seems to stumble in its use of that ability. No nation, no society, no culture, no corporation, and in today's world, no planetary society can survive without order and chaos, both continuously expanding, while maintaining harmony between the two. Innovation is our destiny. If we embrace innovation as the unknown world of expanding opportunities, and if well-being is our destination, then the road we take is our choice, no, the road is of our own making. Through communication and innovation, the rich tapestry of diversity arises, providing the background for yet new forms of communication and innovation. Expanded versatility of individual entities and of environments emerges. Expanded diversity results in an ever-greater solution-space for the potential actions of individual entities. Increased versatility of individual entities and creatures elevates the *solution-space* to an increased *decision-space*.

The last attribute of the process of evolution is relativity: the absence of any dimension that provides an absolute metric that allows us to 'rank' events or experiences. Einstein showed us the relativity of time and space, Darwin showed us that the success of an organism can only be measured relative to its environment, and relativity takes the form of subjectivity in the domain of knowledge. I must reiterate once again that we include all of the manifestations of human thought, including feelings and emotions in our use of the word 'knowledge'. Objectivity simply does not exist, but is only a collectively 'accepted truth', 'common sense', or an agreed upon standard. All of them, even scientific theories, are subject to change, in time, and in geographical space of the globe.

The attributes of the products of evolution: increased diversity, increased versatility and an increased decision space, serve as qualitative and necessary criteria for evolutionary expansion. Accepting them has far-reaching consequences for the way we manage the world as we go into the future. Rejecting them has equally far-reaching consequences, because that amounts to a conscious human attempt to halt evolution. Only elimination of humanity itself can, at least temporarily, achieve this monstrous goal.

The Mind

'The human mind is the co-author of the future': the second pillar of the framework for the future. Although the human mind has only just begun to be a serious subject of scientific study, we know that the mind has its basis - solely or largely - in the human brain and nervous system. As such, the mind is the 'executive authority' that makes the choices, consciously or unconsciously, from the vast array of intertwined algorithm that 15 billion years of evolution have etched into the structures of molecules, organs and communication-networks of the human body.

What signifies the difference between the human mind and the animal mind, in my opinion, is not what we commonly call 'consciousness', and it is not 'awareness' (I defined awareness as 'the acknowledgement of existence' by one entity, or creature, of another), but the autonomy – the independence from the biological background – of thought. That is what elevates human beings (exclusively) into the domain of knowledge.

As higher animals developed 'room to spare' in their expanded brains, the new algorithms their brains produced, led to immediate action and reaction. Bodily action was the only way for animals to express their thoughts and emotions. This 'room to spare' expanded exponentially with the appearance of Homo, and other changes in the brain (e.g. Pico's PIM's that we discussed in chapter 8) enabled the autonomy of human thought.

We do know about the richness of algorithm and renditions, embedded in the human body, its nervous system and its brain. Millions of years of evolution have left their algorithmic imprints in the human 'condition'. C.G. Jung refers to this common set of algorithm and renditions hidden in the recesses of our mind as *the collective unconscious* that nature has bestowed on *all* of us:

"On one side its contents points back to a preconscious, prehistoric world of instinct, while on the other side it potentially anticipates the future – precisely because of the instinctive readiness for action of the factors that determine man's fate."[3]

Human beings are the 'Russian Dolls' of their common evolutionary experiences; as such, our emotions and feelings were once the 'signposts' for the behavior of our ancestors. Thinking isn't new, but the process of thinking attained tremendous new power with the appearance of human beings: the power to separate the products of thinking from the bodily expressions of that thinking.

"As I said before, JJ, thinking is nothing new."

"We heard it, Alvas. Stop putting the spotlight on yourself. It might expose you as the source of a lot of misery in the world."

"Novare, I just..."

That's enough guys. I had thought that the two of you started to realize that one cannot do much at all without the other. Remember Jung's words:

"Conscious and unconscious do not make a whole when one of them is suppressed and injured by the other. If they must contend, let it at least be a fair fight with equal rights on both sides. Both are aspects of life. Consciousness should defend its reason and protect itself, and the chaotic life of the unconscious should be given the chance of having its way too – as much of it as we can stand."[4]

Jung proceeded to say, "This means open conflict and open collaboration at once." Maybe you can display some more of the 'open collaboration' side of the equation.

"Sorry, JJ."

"I'll try, Mr. J."

Steven Pinker's *Blank Slate* is one of the most recent accounts of our common mental foundation. There is no such thing as a blank slate. However, there is 'room at the top': the mind possesses a vast amount of 'freely programmable' space in the human brain to learn, create, modify, and execute novel algorithm. In *The Blank Slate,* Pinker states as his Third Law of behavioral genetics:

"A substantial portion of the variation in complex human behavioral traits is not accounted for by the effects of genes or families."[2]

Pinker estimates that the ratio between genetically determined and acquired behavioral human expressions is around fifty-fifty. With recent research, we are only beginning to understand the plasticity of the brain, which manifests itself through phenomena like the Hebbian effect, or the ability of the brain to develop new neurons even later in life. I have illustrated the flexibility of the brain, and its fundamentally predetermined base structure – its genetic 'factory setting' – using my 'high-tech pinhead model'. I call this 'freely programmable space' of the brain, combined with its plasticity, the 'application space' of the mind. The genetically determined portion then can be called the 'operating system' of the mind. To be sure, I am not trying to make the human mind equal to our PC's. What a computer can do, a human mind can do too (be it that some calculation of some things might take the human mind more than a lifetime); however some things the mind can do, today's computers cannot do. Just tell your computer a new joke that comes along, and look at the computer's reaction. This flexibility of the brain, however, is only one of the ways in which our minds can change and expand.

No matter how hard we try, our minds only deal with *renditions*. We are not able to 'observe' the thinking process itself. The algorithm that we execute when we produce renditions, or retrieve them from memory, escape the spotlight of our attention. Beyond a certain threshold, we can even no longer hold one coherent thought construct in the focus of our attention. We need help, and resort to verbal explanations, written notices, or a circle through a sequence of thoughts in our mind. The upside, and therein hides the enormous potential, is that our minds are capable of observing, absorbing, and comprehending renditions that represent knowledge of great complexity. Our minds can 'look' at a chart, blueprint, balance sheet, or a movie and make use of it without grasping the mountain of knowledge necessary to produce the particular rendition of the image in its focus. Naturally, the processing and absorption of such renditions is dependent on the state of our mindscape (the positions of the pins and the algorithm in the pin-computers, etc. in our pinhead model) at that stage in our lives. Our mindset at that moment determines what we learn, how we learn it, and how it affects our worldview. The pinhead-model (illustration 47) introduced in chapter 8 serves as a metaphor for the plasticity of our minds that mold themselves into new shapes as we process the experiences in our lives.

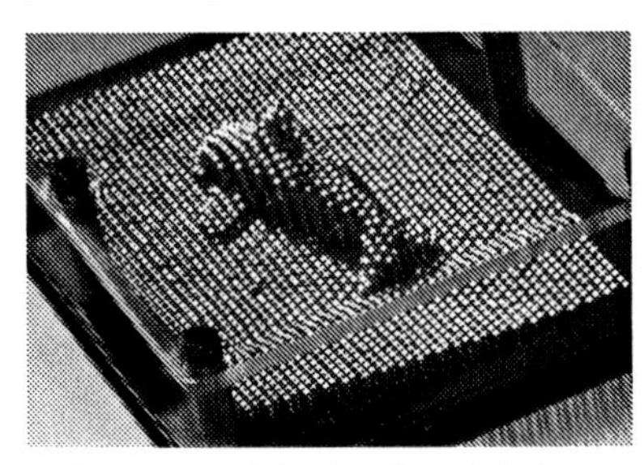

Illustration 47: Pinhead-model; the medium is the message and the 'massage'.

The important fact is that we do not have to comprehend an underlying algorithm to *make sense* out of a rendition presented to our mind.

Our minds can operate in meaningful ways on many levels of (hidden) complexity. From simple beginnings like the use of fire and the wheel, all the way to our ability to fly an airplane without having to understand the underlying complexity, we have demonstrated this ability of our minds, resulting in both good and bad effects on our social state of affairs.

In the sixties, my first step to become a 'computer programmer' was to learn the use of 16 'on/off' switches on a computer-panel. Each switch represented one 'bit', while the pattern of 16 bits constituted a single instruction to the computer. Over the years, 'computer languages' became available that allowed the expressions of functions for the computer to perform on higher levels of abstraction. When using the lowest levels of programming, more than 80% of human 'intelligence', and programming time, went towards the representation of a given problem in the particular language, not towards solving the problem itself. Today, virtually everybody is becoming a computer programmer. When we use spreadsheets, send an email, or 'download' our children's newest pictures, we send a set of enormously complex instructions to our PC's. All the algorithm are already present in the computer, or on the Net, and there is no need for the human mind to understand the complex functionality of those algorithm.

On the contrary, our minds are perfectly capable of dealing with 'higher level' problems, such as making a long-term financial projection, maintaining our bank account, or composing a photo-album from our children's pictures. Our minds are capable to handle 'knowledge about knowledge', in the form of higher-level renditions that are 'simple' representations of their underlying algorithmic complexity. Both, the application space of the brain and its ability to deal with levels of renditions, contribute to the mind's potential for expansion: we are equipped to lead the next stage of evolution, and prepared by nature to take on new responsibilities.

The human mind is where the future emerges, with or without conscious consideration, and with or without planetary collaboration. Digeality gives us a platform of unprecedented accessibility and communication of knowledge to enable such mental collaboration. Digeality has the power to lift the dreams of a common morality and a planetary purpose from the realm of the naïve to the domain of the mentally and physically possible.

The human mind is the place where the future of mental evolution has taken residence. The mind is the room that has the exclusive view on the future. It is the room where the totality of what we call human nature coalesces into that entangled web of feeling, knowledge and experience that we might call our 'soul': the ultimate guide to the unknowable territories of the future, where new manifestations of diversity and versatility are waiting to be discovered, created and explored. This 'room with a view' will lift us out of the primordial swamp of Darwinian passivity onto the road of mental creativity.

Digeality

'Digeality is the knowledge foundation for imaginable futures': the third pillar of the framework for the future. I have presented digeality as the third emerging reality in the history of the universe. Three stages of evolution, three thresholds that were crossed, and three emerging realities. One of those realities, digeality, is just starting to make its impact on every aspect of human endeavor. None of the realities is independent from the others; however, each reality acquired a new degree of autonomy relative to the previous one: a new degree of freedom for individual entities to act beyond the forces that ruled the previous reality, and an increased capacity for the creation of novel expressive algorithm and renditions. Each new reality builds upon and incorporates the features of the previous one; the new reality controls and 'tames' the forces and processes of the one it emerges from. In fact, each new reality emerges from the previous one by creating breakthrough novel products from the knowledge embedded in the structures of the previous reality, as symbolized in illustration 48.

The cosmic domain is the realm of the emerging physical reality through the processes of formation of atomic and molecular structures, and the repetition of these processes of formation.

As I have described it, the physical reality is the rendition (I called this type of physical rendition, as it manifests itself independent from our perception, *external* or *physical* renditions), of knowledge - or algorithm - embedded in that physical reality. It took the universe 3-7 billion years to form all of the ingredients, its multitude of different physical structures and physical forces necessary for the road that would lead to life.

Three Realities

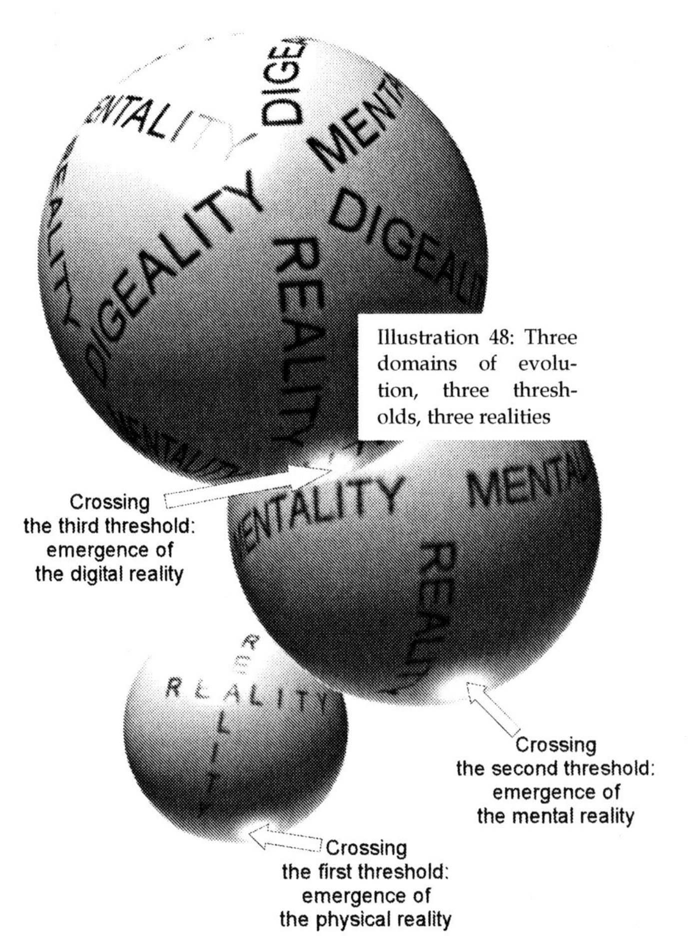

Illustration 48: Three domains of evolution, three thresholds, three realities

The life domain is the realm of the emerging mental reality. The new autonomy of organisms required an internal representation of the physical reality, resulting in a continuous quest for more sophisticated sensory organs and better mechanisms to construct an image of the physical reality, in which the organism had to make a living. Sensory organs that better matched the physical reality of the environment, improved dexterity to act in that environment, and increased (brain) power to 'make sense' of the mental representation resulted in expanded versatility of individual organisms, as well as expanded diversity and complexity of the environment itself. The mental reality is the rendition (*internal* or *mental* rendition) of knowledge *about* the physical reality as it manifests itself to the organism. In the domain of life, this manifestation ranges from the bio-chemical sensitivity of the first living cell to its environment to a sophisticated mental projection of the environment inside the brains of higher animals.

It took biological evolution about two billion years to start producing all of the mechanisms necessary to proceed on the road that would lead to the production of independent thought. Only after the eukaryotes invented multi-cellularity and the innovative power of sex, nature was on the road of the emerging mental projection of the physical reality, leaving behind its total reliance on the chemical interface between itself and its environment.

Finally, the domain of knowledge is the realm of the emergence of what I call digeality. For a hundred-thousand (at least) years, human beings experimented with a rich variety of 'media', all of them geared towards the rendition of human knowledge; all of them attempts to express and communicate the products of the human mind, including thoughts, emotions, intuitions and imaginations. The introduction of the phonetic alphabet and the invention of moveable type were the first steps on the road to express and communicate our knowledge.

None of them were 'complete' in their expressive power, and none of them enabled us to create thinking machines that could (re-) produce knowledge: media captured the products of knowledge, not the knowledge itself. Media were the renditions of the products of thinking, e.g. through verbal language or the printed word. The products could be reproduced artificially, but not algorithm that produced the renditions. A sophisticated representation of the mental reality of the human mind was impossible until the 'bit' was found to be the basic building block for such representations, and the programmable computer that could execute such representations, was invented. This is the historic importance of digeality. It marks an evolutionary milestone on the road towards the creation of new knowledge constructs and their physical manifestations.

Digeality in its current form is just the beginning of our ability to express algorithm. Different ways to render knowledge, different ways to express algorithm, and different ways to execute those algorithm will emerge. Today's computers and programming paradigms are based on old sequential, predictive, and non-self-adjusting paradigms of the 1940's, and will surely be superseded by new and more powerful ones. In the same sense, the interfaces between the mental and the digital realities, as sophisticated as they are, are still based on the paradigm of syntactic transformation of mental renditions to digital renditions. A new paradigm, based on the elimination of (manual) transformation of information, and on the communication of meaning, is on the horizon.

"Question, JJ. Or better: a minor pretzel."

Go ahead, Alvas.

"Remember my problem with gravity in part 1 of the book?"

I sure do.

"And my problem with relativity?"

I do.

"My problem is this: I can sense gravity, at least when the body didn't consume too much alcohol. By sensing gravity, I keep the body upright, you see. However, I cannot sense relativity. No matter how you try to explain it, I don't have a sense for it. I can't *feel* relativity. How come?"

"Simple, Alvas."

"Not to me, Novare, it isn't."

"You emerged on this planet, and for that matter, very close to its surface. You, as well as all living species, never needed a sense for the relativity of space and time. It's only relevant at large speeds and great distances. You never needed it, that's why you don't have it. It's that simple. In the same way, you don't have a 'sense' for the quantum behavior of particles either. Our bodies are too big to notice quantum behavior, and too small to notice relativity."

"Hmm. Are there any other things that do exist, but yet, I can't sense them?"

"Of course, Alvas. Don't you remember the 'dark energy' and 'dark matter' that Mr. J discussed in part 2? Everything we talk about in this book is based on the 10^{78} atoms that we can observe. These atoms only account for less than ten percent of the matter and energy that *has to be* out there. You don't have a sense for that either, at least as far as we know."

"Can you shed light on that, JJ?"

No, I cannot, Alvas, except for the following: when we talk about realities, we should not forget that all three realities, the physical reality, the mental reality and digeality are only renditions of what we *know*. We can render the imaginable, but we *cannot* render the unknowable; not in our minds, and not in digeality. Thanks for bringing that up.

"Welcome, JJ."

Thresholds to the Future

The framework for the future is based on the knowable, and therefore grounded in the domain of knowledge. The future might bring new discoveries that change our view of the world again. Until then, the suggested framework for the future provides a new perspective that leaves behind the static view of a permanent world. It opens up a radically new world of possibilities, both good and bad, and transforms the mindset we need to deal with the future. If we accept the pillars of the framework - the human mind, digeality, and the process of evolution - as the foundation upon which possible futures are built, we accept a new order of responsibility: the responsibility for the future of humanity, the future of the planet, and the future of the universe. Many might think that, because evolution plays its 'games' over such long periods of time, it would be outright stupid to waste our time on such farfetched issues. "We have enough to keep our own company, our own economy, or for that matter the world going," they might say, "we don't have time to worry about a thousand years from now".

These folks would miss the point completely. Such an attitude would express the ignorance of the historic moment in time, right now, and right here. History is at a cusp, no, a pivotal point. Humanity is in the painful process of waking up to its destiny, and to its power to find new destinations within destiny's widening contours. Just as the amphibians raised their heads above the waters of the oceans to view a new terrestrial world, we are just starting to lift our heads above the troubled waters of biological determination, and can glimpse the horizons of a new way of being, in which the mind prevails. If we miss this pivotal point in human history, and continue the illusion of a permanent world, we will end history itself, at least the history of humanity, as we know it. Thus, our own mindset can stop the show of our own human future.

Minds that Mind?

Here we are, at the beginning of the third millennium in the Gregorian calendar. A millennium that might well be characterized in future history-books as the millennium of the mind, and the millennium of the maturation of digeality: the millennium in which human nature passed an evolutionary threshold.

We finished the previous millennium with a century of consciously designed violence that broke all records in the history of the world. When I write these words, we continue the tradition of our attempts to settle mental divisions with the tools of physical elimination. Every politician, military officer, and every commentator makes clear that war is a 'necessary evil' and only justified as a 'last resort'. I cannot help but feel that this war will accelerate the birth of a gigantic change, economically, politically and socially. Humanity needs more proof that it is entering the domain of knowledge, where the rules of the game change and mental, not physical, interaction assures a sustainable future, which is the only future possible. The conflicts will, in my view, take away the last curtain of isolationism that veils the deepest chasm around the planet today: the chasm of knowledge. We can only bridge the divides of hunger, poverty, and all the other divides permanently, if we close this knowledge divide first and forever. Digeality provides the tools to make this possible. Efforts such as the 'digital divide project' by the United Nations is a beginning, but can only serve as a catalyst for the creation of a broad and massive emphasis on the dissemination of knowledge, and the renewal of education around the world.

War, in the midst of a maturing and mind-opening digital world, might be the necessary evil that finally lifts humanity into the domain of knowledge, and contribute to its own demise as a useful tool to resolve the conflicts of the future. I am deeply confident about the profundity of positive change in the not too distant future, because I am convinced of the power of human mentality, and its emerging digital extension.

Evolution is the underlying process that defines the future. No matter whether the reader agrees with my framework for the future, recognition of the process of evolution is a necessary condition for embracing change. Recognition alone is not enough, however. We will have to embrace its underlying expansionary power, if we want to play the role as co-choreographers of the future.

"That would be against Divine Intervention, JJ. Better watch your words."

"No so, Alvas. What if God has designed the world to follow an evolutionary path, and given the exclusive capacity to create knowledge to human beings?"

"Simple, Novare. In that case God would expect us to take up the responsibility for the continuation of evolution! At least if JJ's assertion that we are at this pivot in the history of evolution. If he is right, then we are changing from being the subject of evolution to becoming its author. We better start thinking of the consequences."

"That wouldn't hurt anyway, right? We are becoming too powerful for our own good, if we don't change our mind. Digeality advances molecular technologies at a fast pace. Soon, we will be able to design what we imagine, and to manufacture what we design. Right, Mr. J?"

Yes Novare. We are getting closer to submerging ourselves fully into the domain of knowledge, and closer to waking up to our own power. Just like the genotype can only express itself by developing a phenotype, the awakening of human knowledge was never complete until that knowledge started to manifest itself in digeality. Through it we can fully express human knowledge, advance that knowledge and create novel renditions. I agree, it's time to wake up and change our mind about the world we live in today, tomorrow and the day after tomorrow.

If we mind about the future, embracing evolution and its expansionary nature is fundamental, but it is not enough. If we keep believing that the human being is just another animal, just another random product of Darwinian evolution with or without a 'halo' around its head that we call 'mind', we are doomed. If we keep assuming that we are just another animal that might or might not survive its environment as it 'struggles for biological survival', without any conscious influence of its own, we fail to acknowledge the awesome power and adaptability of the human mind. On the threshold to knowledge, we acquired the independence of knowledge, an accomplishment that will prove to be as significant for our future as the new property of life was to the future of our planet.

If we embrace evolution, *and* embrace the potential of the human mind, we still miss the third pillar of the framework for the future. Do we mind about the future of science? Do we mind about what new forms of life we create using bioengineering? The future of human nature? Do we mind about how we 'grow' future living environments and our 'utensils' for living, using molecular technology? Do we mind about our own social interactions, the way we are governed, and our economy? Do we mind about the 'now-after' – the future of our great great grandchildren? Do we mind about creating new space and new time for them by vitalizing the universe and continue the transformation of dead materials into living organisms, and into thinking living organisms, the process that nature has started?

If the answer to *any* of these questions is 'yes', we have to embrace digeality for what it is: the drawing board for alternative futures, and an intrinsic part of our day-to-day-reality. No science, no engineering activity, and no social, professional, economic, political, and personal activity is, or will be possible *or* desirable without the digital reality. It will change – in fundamental ways we can hardly imagine – the way we live, and the way we 'make a living'. My little excursions into the digeality in chapter 9 are just the barely visible tips of one gigantic iceberg of change. We can choose to 'endure', with fear or amusement, the changes that our technologies rain down on us, or we can change our mindset and take control, and, within the framework of our evolutionary destiny, orchestrate our own future as best as we 'know how' today, and know even better tomorrow.

As we change our minds, we will retool our minds into minding sustainable long-term futures, and our minds will not stop, but guide the show of evolution.

Machines that Mind?

Another showstopper for human evolution, however, lingers behind the curtains. At least according to a vision promoted by the Artificial Intelligence community. According to proponents of this vision, intelligent machines will take over the evolutionary baton from the human mind, and they will do it soon. According to them, the emergence of Robo Sapiens is *imminent*. It is only a matter of a few decades. One of the most notable and renowned representative of the 'intelligent machine' vision of the future is Ray Kurzweil. To give you a flavor of his vision, I present a longer quote from a March 25, 2002 interview with Kurzweil:

> *"So technology itself is an exponential, evolutionary process that is a continuation of the biological evolution that created humanity in the first place. Biological evolution itself evolved in an exponential manner. Each stage created more powerful tools for the next, so when biological evolution created DNA it now had a means of keeping records of its experiments so evolution could proceed more quickly. Because of this, the Cambrian explosion only lasted a few tens of millions of years, whereas the first stage of creating DNA and primitive cells took billions of years. Finally, biological evolution created a species that could manipulate its environment and had some rational faculties, and now the cutting edge of evolution actually changed from biological evolution into something carried out by one of its own creations, Homo sapiens, and is represented by technology. In the next epoch this species that ushered in its own evolutionary process - that is, its own cultural and technological evolution, as no other species has - will combine with its own creation and will merge with its technology. At some level that's already happening, even if most of us don't necessarily have them yet inside our bodies and brains, since we're very intimate with the technology - it's in our pockets. We've certainly expanded the power of the mind of the human civilization through the power of its technology.*
>
> *We are entering a new era. I call it "the Singularity." It's a merger between human intelligence and machine intelligence that is going to create something bigger than itself. It's the cutting edge of evolution on our planet. One can make a strong case that it's actually the cutting edge of the evolution of intelligence in general, because there's no indication that it's occurred anywhere else. To me that is what human civilization is all about. It is part of our destiny and part of the destiny of evolution to continue to progress ever faster, and to grow the power of intelligence exponentially. To contemplate stopping that - to think human beings are fine the way they are - is a misplaced fond remembrance of what human beings used to be. What human beings are is a species that has undergone a cultural and technological evolution, and it's the nature of evolution that it accelerates, and that its powers grow exponentially, and that's what we're talking about. The next stage of*

this will be to amplify our own intellectual powers with the results of our technology.
What is unique about human beings is our ability to create abstract models and to use these mental models to understand the world and do something about it. These mental models have become more and more sophisticated, and by becoming embedded in technology, they have become very elaborate and very powerful. Now we can actually understand our own minds. This ability to scale up the power of our own civilization is what's unique about human beings."[5]

The reader might notice a resonance with some of the thoughts in the quote with perspectives presented in my book. Kurzweil grounds his vision in the expansionary power of evolution, and points to the emerging power - and influence - of digital technology. That is where the similarity with my perspective ends. A few notes will highlight the differences that have a defining impact for our human future.

First, the evolution of the human mind, and of human knowledge, is just starting, rather than ending. This evolution doesn't manifest itself in natural biological alterations of the human brain (although these are possible, and can now be artificially created), but in the potential for the conception, understanding, and creation of new 'software': the novel knowledge constructs of the future. With digeality, human nature finalized the creation of an 'extended phenotype', that concluded the crossing of the third threshold into the domain of knowledge. The genotype created the technology of the 'bodily' phenotype. Similarly, the 'memotype' of the human mind created digeality as the 'mental' phenotype. Kurzweil acknowledges this when he says:

"With DNA, evolution had an information-processing machine to record its experiments and conduct experiments in a more orderly way. So the next stage, such as the Cambrian explosion, went a lot faster, taking only a few tens of millions of years. The Cambrian explosion then established body plans that became a mature technology, meaning that we didn't need to evolve body plans any more...Homo sapiens evolved in only hundreds of thousands of years, and then the cutting edge of evolution again worked by indirection to use this product of evolution, the first technology creating species to survive, to create the next stage: technology, a continuation of biological evolution by other means."[6]

Digital technology is the manifestation of human knowledge, and other technologies (like nanotechnology and biotechnology) are the means of rendering that knowledge into novel physical creations. Technology is an important part of the future of evolution with an unprecedented impact on our physical reality. However, technology is an expression of underlying algorithm. The source from which these algorithm flow, or the algorithm that produce new algorithm yet, is that exclusively human faculty: the human mind. With the emergence of digeality, human nature finally found a way to express this knowledge, not only as a medium in the traditional sense, but as a new reality that can produce novel renditions from human knowledge.

Second, the emergence of digeality or technology in general, does *not* mark another – the fourth – threshold in the story of evolution. Digeality is 'merely' the rendition of human knowledge, while other technologies produce the *physical* renditions of that knowledge, similar to nature's way of translating the knowledge embedded in the genotype into the physical renditions of the phenotype. The perceived singularity is not the 'merger of human intelligence and machine intelligence', as Kurzweil puts it, but is rather the singularity marked by the emergence of digeality as the third reality, which is a rendition of human 'mentality'.

Third, throughout evolution, nature continuously added new features to existing creatures. In the course of its journey, nature crossed three thresholds. Each crossing not only gave a new thrust to the expansion of diversity, versatility and the decision space, but also – as we have seen – added a radically new *kind* of property that contributed to further expansion. With the first crossing nature added a new ('our') existence, with the second crossing nature added life, and with the third crossing nature added knowledge. As I suggested, the most important feature of each crossing was the increased autonomy of nature's creatures. Every new degree of freedom added to the previous ones. All the acquired degrees of freedom, expressed in the expanded versatility of individual creatures, characterized by the three properties of *awareness, processing power, and dexterity*, manifest themselves in only one creature produced by nature. That creature happens to be the human being. The 'singularity' marks the beginning of a new road of expansion, not an end. At the helm of that expansion is the total human being; the most complete creature nature has produced to-date.

Fourth, intelligence is mostly used in the sense of solving problems, and not the creation of problems, as I discussed already briefly. I suggest that our mind will stay ahead of computers because of its imaginative, its intuitive, and its creative talents. These talents allow us to deal with issues of high complexity, without having to deal with the complexity itself. Computers help us with complex knowledge, and our minds create knowledge about knowledge.

Fifth, this does not mean that (human) nature will not find, or create, a fourth threshold (let us remind ourselves that these thresholds are 'just' the gaps in our knowledge about nature's ways). I am suggesting that technology does not represent such a new threshold; technology does not produce anything we don't already know. On the contrary, technology is our means for rendering knowledge. I also suggest that a potential fourth threshold will be found on the same path of *addition* that nature has taken thus far. Accumulation is the name of the game, and the path of the future will require all of the features already present in the nature of the human being. Many Artificial Intelligence labs seem to be discovering this, according to Rodney Brooks, director of MIT's AI lab:

> *"We have turned labs that were used to assemble silicon and steel robots into labs where we assemble robots from silicon, steel and living cells. We cultivate muscle cells and use them as the actuators in these simple devices, the precursors of prostheses that will be installed seamlessly into disabled human bodies. Some AI (artificial intelligence) Lab faculty who study how to make machines learn have stopped building better search engines and begun inventing programs that can learn correlations in the human genome and thereby making predictions about the genetic cause of disease. We have turned rooms that used to house mechanical CAD (computer-aided-design) systems into rooms where we measure the cerebral motor control of human beings, so that eventually we can build neural prostheses for people with diseased brains. And our vision researchers, who used to build algorithm for detecting Russian tanks during the cold war, now build specialized vision systems to provide guidance during neurosurgery. Similar transformations are happening throughout engineering departments, not just at MIT, but throughout the world."*[7]

Maybe, finally, AI experts discover that kicking the old radio might bring its sound to life, but that kicking the wrong algorithm executed by the wrong machines is not worth the wait: it will not gain independence from the thermodynamic forces of the universe that way, as living organisms did; likewise, the machine will hardly be stimulated into producing thoughts that acquire independence from its underlying flow of electrons, as humans do. I am convinced that we will build a plethora of robots from silicon, steel, molecules, and living components. I am convinced that we will enhance our ability to repair and augment our bodies (including the nervous system and brain) beyond our current imagination.

My lifelong passion for the power of digital technology should protect me from being branded as a Luddite. I had the opportunity to watch digital machines come 'alive' with awe, pride and admiration. What I observed was the brilliance of the human mind, and the products of its thinking activity. Digeality will expand – as it already does – our minds, although personally, I wonder why some people talk about new hardware inside of my skull when wireless technology can connect my brain to every thinking machine on the planet. My digital twin can talk to everybody else's – and everything else's – digital twin and communicate to the physical me what I want, need or desire. That brings me to my last and most important note.

Sixth, while 'Robo Sapiens' taking over the role at the helm of the future might sound frightening, the impact of digeality as I described it, goes beyond the impact that intelligent machines could ever have. The idea of machines that are smarter than human beings might give us an eerie feeling; it is a soothing idea at the same time. After all, if the machines solve all of our problems, the life as a 'domesticated' creature might not be so bad: no responsibilities and everything is taken care of. We might even *think* that we are in charge. Who is to say, whether your dog thinks that he is taking *you* for a walk, instead of the other way around?

Not so with human beings; at least not until, if and when, a new threshold is on the horizon. Don't get me wrong. We will build a plethora of robots, non-living and living ones, and everything in between. Will the machines be intelligent? I think so. Will they be smart? In *Our Molecular Future, how nanotechnology, robotics, genetics, and artificial intelligence will transform our world*, Douglas Mulhall explains that they might be:

> *"We may need artificially enhanced intelligence to understand nature's mysteries and the complexities that we're unleashing upon ourselves. If we're too squeamish about that, let's remember that we have the beginnings of such enhanced intelligence, in the form of computers that help us run everything from cars to life support, and implants that replace defective parts of our bodies. Moreover, if we want to stop killing ourselves by the tens of millions in violent conflict, we're going to have to get much smarter, because our genes haven't been able to solve that by themselves. This means addressing at least one probability: that the day of Homo sapiens may be numbered, either by nature's disruptive technology or our own. In a blink of the geological eye, we may be surpassed by our own creations or annihilated by nature's. Right now, we are not ready."*[8]

I am in total agreement about the potential virtues, and the perils of technologies. I am in agreement with the mindless way in which we develop and apply some of our technologies. However, I am in total disagreement with the Mulhall's conclusion of being 'surpassed by our own creations'.

Being smart has to do with *intuition* and *imagination*, driven by *intention, purpose* and the amazing power of *curiosity*. In this sense, machines are *not* smart, while humans are - or at least can be - smart. *Intelligence* is the foundation of knowledge that defines the solution space in which smart people make imaginative *and* intelligent choices. Intelligence has the flavor of order: the knowable and the known, the predictable, the expectable and the dependable. Smartness on the other hand, has the flavor of the unpredictable, of opportunity, the unknown, and the flavor of innovation. No change, no innovation, no scientific breakthrough, no groundbreaking engineering insight, and no new expressions of art take place without this smartness. Smartness is creativity.

Digeality provides both, the accessibility to new renditions of intelligence, and new expressions of smartness. It expands the human mindscape and the human mind-window, providing the order of new knowledge structures and the chaos of new renditions.

All of human knowledge, the expressions of intelligence and of smartness will be embedded in the new reality of the digital world, for everyone to see, hear and touch them, and to use them for the creation of new futures.

All we have to do is to change our mind about our own evolutionary history and the evolutionary potential of the human mind. We have to stop using our common static framework for the future and discover that such a framework is just the rim of a rearview mirror, pulling us back to the mud-wrestling match of Darwinian thinking.

Digeality presents us with a forward projecting mirror that reflects the evolutionary nature of nature: the nature of the universe, the nature of biological life, and the nature of human knowledge. The digital reality connects us and gives all of us the insights to make conscious changes for the future, as it reflects and projects our knowledge forward into the unknown. That is the spirit of the emerging digital reality. That is the digital spirit. It is the projection of the human mindset as it views reality.

Our minds are at the helm of the future of evolution, surrounded by the rest of nature: the known and the unknown, the knowable and the unknowable. We should row our boat with the oars of the knowable, and be humbled by the mysteries of the unknown and the unknowable. Our mindsets determine the direction of the boat.

Chapter 12: The Digital Spirit

Illustration 49: The digital spirit: the spirit of the universe, of life, and of human thought coalesce into one new reality.

"I believe in human nature. It captures the natural morality, one that is not necessarily different at all, from a religious one. Humans are born with the capacity for empathy, for compassion towards others. Like spite and anger, compassion and love are part of our nature. They are in our genes.... We are capable of love, because evolution made us into social beings."

James Watson, February 2003.[1]

Digital War: the Road to Peace.

As I write these words, the colors of military camouflage cover the mountains of Afghanistan and the deserts of Iraq. The first chapter of the book started with war and so does the last chapter. Why? Because the camouflage colors not only cover mountains and deserts, but they also cover millions of TV screens and computer displays around the world. In the 20th century, most technological breakthroughs came from military applications. Technological innovation was the driving force of the cold-war arms race. It had the highest priority, and attracted more money than any other application of technology. As that tradition continues into the 21st century, military applications are at the forefront of technology. War is once again the focus of my attention, because:

War is going Digital.

The 'state of affairs' in any 'theater of operation' can be viewed and manipulated from any place in the world, as long as a satellite or a phone plug is in sight. 'Embedded' journalists send their digital movies real-time around the world, making it part of digeality, spanning the globe, for everybody to fear, 'enjoy', or disregard. The CNN anchor could hardly conceal his excitement as the first images of tanks were rolling through the Iraqi desert. The images are still wobbly and blurry, a limitation that is just a matter of time. The journalist in the scene points out a tank and gives the name and rank of its commander, 'surprising' the parents at home. Commentators call military commanders the 'remote conductors' of the military 'orchestra'. Precision-guided missiles compare the landscape they cross with the digital landscape in computer-memory. The physical missile that carries the warhead is guided by its digital twin that carries the knowledge about the target. The war became more digital than ever before in the time span of just ten years. Owen Cote, professor and associate director at the MIT *Security Studies Program* explains:

> *"In the first [Gulf] war, only about ten percent of the U.S. airplanes were capable of using precision weapons. Now, every airplane has the capability to launch precision weapons – laser-guided bombs and satellite-guided bombs navigated by the GPS signal. Ten years ago they were pretty much using gravity bombs, not all that different from WWII and Vietnam,"*[2]

The increasing use of digeality makes me wonder: how long will human soldiers still be necessary to carry arms to the place of deployment, or to drive the vehicles that carry the arms? Obviously, this must be just the beginning of digital warfare? Where would that lead to?

Today, we are still looking at the war through the straws of narrow bandwidth, limited viewpoints, and biased commentary. That is about to change. Already, unmanned airplanes with digital eyes and ears verify their location in real time and fly under the control of a joystick that maneuvers the plane's digital twin on a computer-screen. Tanks talk to airplanes and ships that can look through the digital eyes of either one, at the click of a mouse. Human soldiers, combat-vehicles, airplane-carriers, all have their digital twins that listen to the commands of the commander's digital twin. Tanks can see through the eyes of helicopters ahead, and the eyes of airplanes above. It doesn't matter whether its day or night, the sensitive digital eyes can see ahead, or look back.

The digital eyes and ears transfer the images and sounds of war into a wave of photons that engulfs the globe. We just have to tap into the wave from our couches to pass judgment and provide cheap advice. Those bored by a lack of action can be assured: it will get better. As Owen Cote explains,

> *"The whole issue of advanced networking and communications really becomes essential when you start talking about mobile targets. It's possible that this war will demonstrate the first really big major successes against mobile targets. What we may see here is that networks of persistent sensors and precision weapons will give us capabilities against mobile targets that resemble the capabilities we now have against fixed targets – which is, basically, if we find you and decide we want to attack you, we're going to. We can't do that yet with mobile targets, but we'll be demonstrating in this war some nascent capabilities in that area."*[3]

In the future, soldiers wear helmets that guide the machine-guns in their hands to the right targets, thus replacing human judgment by more precise computer algorithm. Digeality will have an infinite number of 'sensory organs' around the globe, and can zoom in on any country, any building, any bridge, any vehicle, and any movement on troops, or any individual soldier at any time. The orchestration of digital twins will produce the necessary instructions to each individual soldier and every piece of equipment. Cameras and microphones embedded in the soldier's gear will display the local terrain and surrounding enemy, so that the news on TV becomes even more personal. This is not senseless cynicism, but points to a milestone in the road to digital conflict resolution.

Human soldiers have two important roles. They carry the weapons to the area of combat, and they use human judgment in their immediate environment in the 'theater'. Both roles are on the road to elimination by digital technology. Digitally controlled 'smart' weapons and digitally guided weapon carriers will know and recognize their trajectories and their targets. Conventional machine-guns that spray a wave of bullets will be seen as a waste of money. Digital algorithm, the products of simulation and programming, will make the need for human judgment on the battlefield the exception rather than the rule. That would leave the war in the hands of a few human conductors of an orchestra of digital twins. No logistics for the physical movement of troops, no supply lines to protect: just a clean 'surgical' procedure that eliminates the enemy.

What happens, however, when the enemy has the same digital capability? Where are the living soldiers to eliminate? Who is the enemy? Civilians that have the wrong mindset? Or their leaders, remotely performing a similar digital orchestra to eliminate the attacker? The answer seems obvious: digeality itself has to be attacked! How? Surely, the physical facilities that harbor the digital nerve centers should be targeted. What if these nerve centers are spread around the world, networked by the redundant channels and relay centers of an Internet, and able to move thousands of kilometers in a blink of an eye? What do we do then?

Shut the Internet down? That would shut the world down. The answer is *software*. 'Special (software) forces' will penetrate the enemy's digital territory. Intelligent virus-agents will break down the firewalls that protect the digital brains of the enemy, and destroy the enemy's digital twin. As simplistic as this scenario might sound, it points in the direction that we are already taking. Also, this scenario serves as a metaphor for the path that many human activities will take: the path from atom to bit. This is the direction that nature maintained over its evolutionary course that lasted fifteen billion years; the course from a world of matter and energy to a world of knowledge; from the violence of nuclear reactions that shaped the past, to the power of knowledge that will shape the future.

A handful of atoms harness enough energy to divert the planet earth from its orbit around the sun. A truckload of atoms could peel a continent away from the planet, catapult it to outer space and reshape it to become a new moon. The evolution of the cosmos required a maximum of physical violence. Life conquered physical violence, and even conquered entropy. Living organisms rendered physical forces harmless, and turned them into a source of useful energy. With the emergence of autonomous thought, human nature started on a road of tempering the forces of biology. Cultures, religions and political systems are powerful expressions of our quest for mental harmony, designed to control and transcend the forces of instinctive biological behavior. Physical processes resign to the power of biological life, and biological life resigns to the power of knowledge. The language of knowledge is spelled in bits and bytes that spread the power of knowledge around the globe. If the victory of knowledge over physical violence is the spirit of evolution, then the victory of bits over atoms is the spirit of digeality: the digital expansion of the human spirit; the digital spirit.

It's all in the Mind

According to James Watson (in the quote at the beginning of this chapter), who, together with Francis Crick, unraveled the mystery of DNA in 1953, human nature is 'in the genes'. In the same interview with the German magazine Der Spiegel, the interviewer asks whether *human rights* might be written in our genes. Watson's answer points an arrow right at the core of the perspective in the book:

> *"I don't believe in human rights. I believe in human duties and responsibility. What would these human rights be? Where would they come from? We don't have a right to food and health. We can wish for something like that, it would be useful. However, to have a right means to earn it. Rights require a contract between people. Instead, they say, rights come from God, unchangeable and absolute. That is wrong. Rights are in no way part of human nature."*[4]

Watson's words sharply point out the dilemma we face: Do we view human nature as the sum of biological activities? If we continue to do so, we ignore the changing reality of our planet, and continue our games of social mud wrestling. Or do we accept the evolution of the human mind as the continuation of biological evolution? If we do, human rights *become* a natural part of human nature. After all, the possession of arms and legs is not an entitlement; they are the tools that nature provided to *all*. Equality amongst human beings is defined by the common set of tools provided to every human being; no questions asked.

My framework for the future places mental evolution at the helm of human nature. Mental evolution is coming to life, while biological evolution fades to the background, continuing its journey to the baton of its own clock. *The gene has lost its dominance, the meme is taking over.* Variation and selection take on the added meaning of intention and purpose, design and acceptation, or rejection. Humanity has arrived at a singularity indeed. That singularity marks the end of the *Darwinian* evolutionary reign and the beginning of *mindful* evolutionary expansion.

Two billion years ago, bacteria established biological globalization, and after that, millions of new species emerged through the combination of knowledge embedded in a variety of bacteria. Mutations, mergers, joint ventures and acquisitions produced new knowledge and new technologies that spread around the globe as new genotypes and new phenotypes.

Over the course of a century, human activity established the physical globalization just as the bacteria did, two billion years ago. On top of that, in just a few decades, we expanded that physical globalization through digital technology. We seem to be slow to discover the mental globalization that is taking place because of the maturing digital reality.

That is precisely where humanity finds itself today. Memes span the globe, our minds start to connect, the differences and the similarities confront each other. Human knowledge and human emotions emerge from the innermost sanctum of the human minds that span the globe, to form a space-time continuum of human thought. The thoughts are human, the renditions digital. The messages are composed by fifteen billion years of human evolution; the medium is new, and digital. Evolution created the mind, the mind created the medium and paints it with knowledge, and the medium will, in turn, change the mind. In an ever expanding spiral of mental activity, medium and message will contribute to the expansion of human nature; an expansion in which human 'rights' are the mental extension of human nature.

Biological evolution springs from new knowledge (genotype) that manifests itself in the biological technology of a new species (phenotype) as it interacts with its physical environment. Mental evolution springs from new knowledge (memotype) that manifests itself in new technology (extended phenotype) as it interacts with its mental environment.

Biological expressions manifest themselves as the dance of atoms and molecules. Mental expressions manifest themselves as the dance of thoughts and emotions. Digeality renders the dance in bits and bytes. This digital knowledge continuum is the soil for the expansion of human nature. This expansion, like any evolutionary expansion, starts by spreading new useful tools throughout the population, thus expanding the common base of the species. The expanded common base then becomes the foundation for new knowledge and new technologies that can blossom from it.

Today, the common base of human tools is a narrow one indeed. More than half of the world population has hardly any tools at all. Deep chasms separate the worlds of 'developed nations' and the rest. Hunger, poverty and disease are but a few of the names for the tunes that emanate from these chasms. All of these chasms share a common void: the void of knowledge, the biggest divide of all. Therein lays our hope, because digeality is spanning the world, with total disregard for national or cultural borders, carrying human knowledge to every corner of the planet and back. That new reality provides algorithm and renditions for everybody to see, and to act upon. Digeality opens up the world and makes its rich diversity transparent and accessible. Digeality doesn't discriminate, it exposes arrogance and ignorance as the alibis for inaction of the 'ones that know', and it starts filling up the knowledge voids of the 'ones that don't know'. As Hazel Henderson points out in her book *Beyond Globalization*, globalization is more than the narrowly defined monetary growth of global economic enterprise. Digeality is the colorful rendition of the globalization of mentality: a powerful new foundation for political, social and economical innovation, where "Broad participation by citizens, employees, the poor, and marginalized groups - so greater democracy, equity, and transparency - [that] are requirements for re-shaping the global economy,"[5] can become a reality.

A mental globalization has started that reaches beyond war and peace, beyond economic expansion, and beyond borders of any physical kind, as it expands the common base of human nature that in turn provides the foundation for innovation. This *mental globalization* is the mind-altering meaning of digeality: it is the *digital spirit*.

Towards Peace of Mind

"I'm not sure I like that transparency, Novare. I'll lose my anonymity."

"You have been anonymous for millions of years, Alvas. It is time for you to come out. After all, that is why I am here. Yes, you will lose your anonymity. But you will win recognition. You will be counted, Alvas. Better yet, you count!"

"I'm not so sure about that, Novare."

"You're just scared, that's all. You are all of evolutionary history, Alvas. You are the one with all the experience, but kept it inside. I, on the other hand, can make your experience come to live, and use it to produce something new, and change the world."

"So you are the present?"

"I am the gate between the past and the future, Alvas. The present does not exist. Either we live in the past, or we are creating a future. Can't you see? That's why we need each other. The past gives me the substance from which to create change that is why I need you. Change is the only way to tell the difference between the past and the future. I am that change, Alvas."

"What if I don't *want* change?"

"Change, Alvas, is life, because it is the only thing we are aware of. Nothing you can do about that, except to die."

"You sound real scary, Novare. I don't know where to find the courage to face that reality of yours."

"You have all the courage you need, Alvas. After all, you survived millions of years of attacks on your life. What you need is direction."

"Is that the direction of evolution, as JJ explained it? His framework for the future?"

"Indeed. Only one thing is still missing."

"What's that?"

"Together, you and I can find the right direction. What's missing is a purpose."

"You mean, like defining a destination?"

"More like an intention, Alvas."

"Now you make me curious. And more scared. Leaving my anonymity makes me vulnerable."

"Vulnerable? The biggest threat to your security doesn't come from the outside, but from inside, Alvas. Don't worry, I'll help."

"Aren't you scared then?"

"Of course I am."

"Why do it then? Being counted, becoming 'somebody'?"

"Because I mind the future, Alvas. I want to help shape it, and I want to be part of it."

"OK then. I have my own reason though."

"Which is?"

"It's that curiosity. It would burn me up, if I stayed behind."

"That's as good a reason as any."

"But curiosity killed the cat, Novare."

"And it landed a man on the moon, Alvas."

"Aren't you curious, Novare?"

"Of course I am."

"Are you curious, JJ?"

If both of you are, I have no choice.

"Just tell us, Mr. J. *What is our intention*?"

That is the best question you ever asked, Novare. And the most difficult one to answer, because we'll have to define it.

We have crossed the threshold of knowledge. Our minds are at the helm of a continuum of three entangled realities. These realities will change our mindset. The nature and the depth of change depend on the answer to that very question: ...What is our intention?

Notes

Prologue: Minding the Future

1. Savage, Marshall T. (1994) *The Millennial Project*. Little Brown & Company, p17.
2. Mello, Anthony De, (1992) *Awareness, The Perils and Opportunities of Reality*. Doubleday, Image Books, p5.

Part I: GENERATION OF CHANGE

1. The Charter of the United Nation was signed on June 26, 1945, in San Francisco, at the conclusion of the United Nations Conference on International Organization, and came into force on October 24, 1945. The Statute of the International Court of Justice is an integral part of the Charter. Source: www.un.org/Overview/Charter/preamble. Html.
2. Schrödinger, Erwin, (1944, 1967) *What is Life? & Mind and Matter*. Cambridge University Press, p1.
3. Peter C. Bishop, Ph.D., University of Houston-Clear Lake, Studies of the Future.

Chapter 1: Mindset

1. Albert Einstein, from an address on the occasion of the fifth Nobel Anniversary Dinner at the hotel Astor in New York, December 10, 1945. Source: Einstein, Albert, (1954) Ideas and Opinions. Wings Books, p115.
2. Source: Weelen, Paul, (1995) *Limburg bevrijd*, (the liberation of Limburg). Van Geyt, Ljubljana, p112.
3. Ray Kurzweil, author of *Spiritual Machines*, renowned scientist and entrepreneur, inventor of speech recognition and speech synthesis, and advocate of so-called 'strong artificial intelligence', which says that computers can achieve and surpass human intelligence and consciousness within the 21st century.
4. Edward de Bono's 'lessons' on thinking, as explained in his book *Lateral thinking*.
5. Excerpts from ex US President Jimmy Carter's Nobel Peace prize acceptance speech, Oslo, December 10, 2002. Source: cnn.com.
6. Sir Charles Sherrington, (1940) *Man and his nature*, Cambridge University Press. Quoted from: Schrödinger, Erwin, (1944, 1967) *What is Life? & Mind and Matter*. Cambridge University Press, p121.
7. Coates, Joseph (2002) *Where Is Bin Laden*. Source: www.wfs.org/jcoates.htm.

Chapter 2: Changing Minds

1. Sources: www-gap.dcs.st- and ac.uk/~history/Mathematicians/Ptolemy.html, and Russell, Bertrand. (1945) *The History of Western Philosophy*. Simon & Schuster, p108ff.
2. The material on Heraclites, and quotes from his philosophy, are from: Russell, Bertrand. (1945) *The History of Western Philosophy*. Simon & Schuster, p42-47.
3. Source: www.gap.dcs.st-and.ac.uk/~history/Mathematicians/Copernicus.html.
4. Article by Peter Landry, on: www.blupete.com/Literature/Biographies/Science/Bruno.htm.
5. Launch of the spaceshuttle that released Galileo in 1989. Introduction of launch and illustration from the Introduction to *A journey to Jupiter*. Source: galileo.jpl.nasa.gov/mission/mission.html.
6. From an article by J.V. Field, see : www.gap.dcs.st, and .ac.uk/~history/Mathematicians/Galileo.html.
7. Source: illustration and text for Jupiter mission: galileo.jpl.nasa.gov/news/release/press021029.html.
8. Picture courtesy Joachim Amkreutz, information source: www.endex.com/gf/buildings/ltpisa/tpnews/1999/ltpsn121899.htm.

9. Paul Boutin, science writer for Wired magazine, on October 15, 2001 on his website paulboutin.weblogger.com/2001/10/15.
10. Source: www.sci.esa.int. At a post-launch conference held between Kourou, Paris and ESOC Darmstadt, the Director General of ESA Antonio Rodota said, "The launch of XMM represents a success for Europe. We are very proud of our satellite, and of the Ariane-5 launcher.... The construction of this new spacecraft has led us to develop new technologies which place Europe today at the forefront of space science."
11. Portions of bibliological information from: www.gap.dcs.st- and .ac.uk/~history/Mathematicians/Newton.html.
12. The Isaac Newton Group of Telescopes (ING) consists of the 4.2m William Herschel Telescope (WHT), the 2.5m Isaac Newton Telescope (INT) and the 1.0m Jacobus Kapteyn Telescope (JKT). The ING is located 2,350m above sea level at the Roque de Los Muchachos Observatory (ORM) on the island of La Palma, Canary Islands, Spain. The operation of the site is overseen by an International Scientific Committee, or Comité Científico Internacional (CCI). Source: www.ing.iac.es/PR/images_index.html.
13. Bibliographical information from: www.gap.dcs.st and .ac.uk/~history/Mathematicians/Newton.html.
14. Source: www.ing.iac.es/PR/images_index.html. Picture credit: courtesy Javier Méndez, Isaac Newton group of telescopes.
15. From an article, based on an interview with Didier Queloz, published on the European space agencies website at: sci.esa.int/content/news/index.cfm?aid=28&cid=1902&oid =30133. Queloz is a member of ESA's Scientific Advisory Group for its Darwin planet-search mission.
16. Source: sci.esa.int/content/searchimage/.
17. More information about NASA's *Terrestrial Planet Finder* program on NASA's website at: www.jpl.nasa.gov/stars_galaxies/planet_hunting/location.html.
18. More information on ESA's *DARWIN project* at: sci.esa.int/home/darwin/.
19. From Charles Darwin's Voyage of the Beagle, his own account of his travels. This can be found at www. literature.org/authors/darwin-charles/the-voyage-of-the-beagle/.
20. Biographic information on Darwin from: 1) www.lib.virginia.edu/science/parshall/darwin.html, 2) www2.lucidcafe.com/lucidcafe/library/96feb/darwin.html and 3) Mayr, Ernst, (2001) *What Evolution Is*. Basic Books, p10ff.
21. Durant, Will. (1968) *Van Socrates tot Bergson*. Salamander Books. [Translated from Will Durant, The story of philosophy, 1939], Vol I, p93, 94.
22. Mayr, Ernst, (2001) *What Evolution Is*. Basic Books, p9.
23. Source: www.groups.dcs.stand.ac.uk/ ~history/Mathematicians/ Einstein.html.
24. It is more precise to say that no *information* can travel faster than light. Several experiments have been carried out were particles like photons travel faster than light. Certain global phenomena can exhibit speeds faster than the speed of light, like the Big Bang itself, which must have expanded faster than the speed of light. Such phenomena cannot, however, be used to carry any messages.
25. QUBE is my imaginary programming language of the future. It might be based on the characteristics of quantum computing.
26. Courtesy European Space Agency. Source: www.ing.iac.es/PR/science/galaxies.html.
27. The times given in the example are not exact, because the speed of the spacecraft itself is not taken into account. The mathematical formulas that establish the relationship between different reference frameworks, called 'Lorentz transformations', were developed by physicist Conrad Lorentz in 1897. Einstein, unaware of their existence, developed them himself.
28. Such a simulator does not exist today, but will be technically feasible in the near future, when computers have enough power to simulate the complexity of the universe.
29. Courtesy of James Amkreutz, the author's 9-year-old nephew.
30. Picture used with the kind permission from Dr. Pavlos Akritas of the Solvay Institutes, Brussels, Belgium, solvayins.ulb.ac.be.

31. Feynman, Richard P., (1985) *QED*. Princeton University Press, p5.
32. Ibid, p9.
33. Source: www-groups.dcs.st-and.ac.uk/~history/Mathematicians/ Schrodinger.html.
34. www-groups.dcs.st-and.ac.uk/~history/Mathematicians/ Heisenberg.html.
35. Hawking, Stephen W., (1988) *A brief History of time*. Bantam Books.

Chapter 3: New Beginnings

1. Huxley, Julian (1992) *Evolutionary Humanism*. Prometheus Books. [Originally published in 1964, as *Essays of a humanist*. Harper and Row], p29.
2. Source of information at: sm3b.gsfc.nasa.gov/mission-updates/mission/mar9.html.
3. Courtesy NASA at: sm3b.gsfc.nasa.gov/mission-updates/mission/mar9.html.
4. The speed with which a galaxy was moving toward or away from us was relatively easy to measure due to the Doppler shift light. Just as a sound of a racing car becomes lower as it speeds away from us, so the light from a galaxy becomes redder. With a sensitive spectrograph, Hubble was able to determine the red shift of light coming from distant galaxies.
5. J.B.S.Haldane, *DAEDALUS, or, science and the future*, Transcribed by: Cosma Rohilla Shalizi, at: www.santafe.edu/~shalizi/Daedalus.html.
6. Mayr, Ernst, (2001) *What Evolution Is*. Basic Books, p41.
7. Ibid, p43.
8. Hunt, Morton, (1993) *The story of psychology*. Anchor Books, p145.
9. Ibid, p2.
10. Quoted from Fukuyama, Francis. (2002) *Our Posthuman Future*. Farrar, Straus & Giroux, p161.
11. John Brockman, "*The third culture*," at: www.edge.org/3rd_culture/index.html and Roger Kimball in "the two cultures today," published in The New Criterion Vol. 12, No. 6, February 1994, at: www.newcriterion.com/archive/12/feb94/cultures.htm.
12. *A Brief History of Computing*, By Jack Copeland, at *www.cs.usfca.edu* and www.AlanTuring.net/turing_archive/pages/Reference%20Articles/BriefHistofComp.html.
13. Information on Turing and computers can be found at www.turing.org.uk/turing/index.html. This site is maintained by Alan Hodges, author of *Alan Turing, the enigma*, considered one of the best books on Alan Turing.
14. A Brief History of Computing, By Jack Copeland, at *www.cs.usfca.edu*.
15. J J O'Connor and E F Robertson at: www-groups.dcs.stand.ac.uk/~history/Mathematicians/Turing.html.
16. Hodges, Andrew, (1999) *Turing*. Routledge, p35.
17. Dyson, James. [editor Robert Uhlig] (2001) *A History of Great Inventions*. Caroll & Graf Publishers.
18. Source: www-history.mcs.st-andrews.ac.uk/history/ Mathematicians/Hollerith.html
19. Hodges, Andrew, (1999) *Turing*. Routledge, p36.
20. Kurzweil, Ray. (1999) The age of spiritual machines, when computers exceed human intelligence. Penguin Books.
21. Information on Teilhard's life and work from: 1) King, Ursula, (2000) *Spirit of Fire, the Life and Vision of Teilhard de Chardin*. Orbis Books, 2) Chardin, Teilhard, (1955, 1975) *The Phenomenon of Man*. Harper & Row.
22. Source: www.rice.edu /fondren/woodson/mss/ms50/#bio
23. Sir Julian Huxley in his foreword to: Chardin, Teilhard, (1955, 1975) *The Phenomenon of Man*. Harper & Row. The foreword was written in 1958.
24. United Nations Development Programme, (2002) *Human Development Report 2002, Deepening democracy in a fragmented world*. Oxford University Press, p1. Report available at: www.undp.org/hdr2002/complete.pdf
25. Bhalla, Surjit (2002) *Imagine There's No Country: Poverty, Inequality, and Growth in the Era of Globalization*. Institute for International Economics, ISBN: paper 0-88132-348-9, available at: www.iie.com/publications/publication.cfm?pub_id=348. Surjit Bhalla is managing di-

rector of Oxus Research and Investments, an economic research, asset management, and emerging-markets advisory firm based in New Delhi, and a former economist with the World Bank, Brookings Institution, Rand Corporation, Goldman Sachs, and Deutsche Bank.

26. Erwin W. Muller, who received his doctorate from the Technical University of Berlin in 1936 at age 25 under Gustav Hertz. Working at Penn State University, he used a "field-ion microscope," a new device based on the field emission microscope to obtain the pictures. Source: Regis, Edward, (1995) *Nano: The Emerging Science of Nanotechnology: Remaking the World-Molecule by Molecule.* Little Brown & Company, p24-44. The Greek philosophers Leucippus and Democritus were the first ones to come up with the idea of an "atom," the Greek word for "indivisible, a smallest particle of matter. Different materials, that had nothing observable in common, must have some underlying components, that make them different, they surmised. Not until 1799, when the French chemist Joseph-Louis Proust noticed that he could mix very specific quantities of copper, oxygen and carbon to form copper-carbonate, that serious consideration was given to the existence of this smallest particle, the atom. In June 1827, Robert Brown observed the erratic motion of pollen grains, dissolved in water: the notion of Brownian motion was born. Almost a century later, Brownian motion, all but forgotten, Albert Einstein specified an experiment to prove the existence of atoms. In 1908, French experimental physicist Jean Perrin carried out the experiment, enough to prove the existence, however, he was not able to actually observe individual atoms.
27. Castells Manuel, (2001) *The Internet Galaxy, Reflections on the Internet, Business, and Society.* Oxford University Press Inc., New York, p282.
28. Are We Spiritual Machines? Ray Kurzweil vs. the Critics of Strong AI. Originally, published in June 18, 2002 in by the Discovery Institute. Published on KurzweilAI.net on June 18, 2002.
29. Huxley, Julian (1992) *Evolutionary Humanism.* Prometheus Books. [Originally published in 1964, as *Essays of a humanist.* Harper and Row], p81, 82.
30. Rushworth M. Kidder in: *Three Women's Moral Courage: Why We Care,* ethics newsletter, January 2003, at: www.globalethics.org/newsline/members/currentissue2.tmpl #01060316042521.
31. McLuhan, Marshall, (ninth printing, 2001, first edition 1964, MIT edition, 1994) *Understanding Media.* The MIT Press.
32. Bill Joy's influential article *Why the future doesn't need us* in WIRED magazine, Issue 8.04, April 2000. Bill Joy is cofounder and Chief Scientist of Sun Microsystems.

Part II: SONGS FROM THE UNIVERSE

1. Albert Einstein, *The world as I see it,* Originally published in Forum and Century, Vol. 84, pp. 193-194, the thirteenth in the Forum series, "*living Philosophies.*" As published in Einstein, Albert, (1954) *Ideas and Opinions.* Wings Books, p11
2. Hawking, Stephen W., (1988) *A brief History of time.* Bantam Books, p 174.
3. King, Ursula, (2000) Spirit of Fire, the Life and Vision of Teilhard de Chardin. Orbis Books, p226.

Chapter 4: *Act One, "To be or not to be"*

1. From "*The world as I see it*" by Albert Einstein, Originally published in Forum and Century, Vol. 84, pp. 193-194, the thirteenth in the Forum series, "*living Philosophies.*" As published in Einstein, Albert, (1954*) Ideas and Opinions.* Wings Books, p9.
2. Rees, Martin, (2001) *Our cosmic habitat.* Princeton University Press, p179.
3. Einstein, Albert, (1954) *Ideas and Opinions.* Wings Books, p290.
4. Hawking, Stephen W., (2001) *The Universe in a nutshell.* Bantam Books, p202. The expression 'anthropic principle' was coined in 1974 by cosmologist Brandon Carter. Information on the anthropic principle can be found at: www.anthropic-principle.com/primer.html.

5. Moravec, Hans, (2000) *Robot*. Oxford University Press. In Moravec's view, future robots will possess intelligence that surpasses that of human's. Moravec calls these robots, the inventions of the human mind, our 'mind-children'.
6. Hawking, Stephen W., (2001) *The Universe in a nutshell*. Bantam Books, p78.
7. Rees, Martin, (2001) *Our cosmic habitat*. Princeton University Press, p134. According to Rees, Roger Penrose thinks that the inflation theory is a "fashion the high physicists have visited on the cosmologists."
8. Hawking, Stephen W., (2001) *The Universe in a nutshell*. Bantam Books, p35.
9. Penrose, Roger, (1994) *Shadows of the Mind*. Oxford University Press, p420.
10. Source: article by Frank L. Lambert, Prof Emeritus, Occidental college and scientific advisor to the J. Paul Getty Museum, see: http://www.2ndlaw.com
11. For information on entropy see also: www.infoplease.com/ce6/sci/A0817435.html
12. Penrose, Roger, (2000) *The Large, the small and the Human mind*. Cambridge University Press, Canto Edition, p48.
13. Rees, Martin, (2001) *Our cosmic habitat*. Princeton University Press, p130. To the time of 1 second, as mentioned in the quote, Rees gives the following comment: If the postulated starting point is much earlier, then the inferred precision would have been still greater: far larger numbers, indicating far more impressive-seeming fine-tuning, are quoted by other authors.
14. Tipler, Frank, (1995) The Physics of Immortality: Modern Cosmology, God and the Resurrection of the Dead. Anchor Books.
15. Coates, Vary T., *Technological glimpses of the future* (No 10), Futures Research Quarterly, Fall 2002, Volume 18, Number 3, World Future Society, p62.
16. For information on the properties of all atoms, see the award-winning interactive, and fun periodic table at: www.webelements.com
17. Particles can have different electric charges (neutral, positive or negative), or they can have a different angular magnetic momentum and magnetic moment, called 'spin'.
18. hubblesite.org/newscenter/
19. Liebes, Sidney, Sahtouris, Elisabeth, Swimme, Brian (1998) *A walk through Time*. John Wiley & Sons, Inc, p18.
20. Hamel, Gary, (2000) *Leading the revolution*. Harvard Business School Press, p57.
21. Main source: Rees, Martin, (2001) *Our cosmic habitat*. Princeton University Press. If we add up all the mass of all the matter that we can observe in the universe, the gravity that holds everything together could not occur. There is simply not enough (observable) matter to cause this gravity. Hence, we assume the existence of more "stuff" (which accounts for about 90% of all matter!), that we cannot "see." We actually assume two components to this missing stuff. One is "dark energy" and the other one "dark matter." Today, it is suspected that dark matter particles have no electrical charge and that they have not been detected because they pass right through any ordinary material. Dark Energy is the energy, also called "vacuum energy," that exercises the anti-gravity repulsion force that drives galaxy-clusters in the universe outward. According to theory, this dark energy accounts for 76% of mass-energy (remember the Einstein equivalence between mass and energy according to e=mc2), dark matter for 30%, and only 4% for "ordinary atoms," half of which we can "see and touch"; the other half free-floating throughout the universe.

Chapter 5: The Cosmic Domain

1. Eddington, Sir Arthur, (1933) *The expanding universe*, Cambridge University press, p126.
2. Dyson, Freeman, (2001) *Disturbing the Universe*. Basic Books, p251.
3. Rees, Martin, (2001) *Our cosmic habitat*. Princeton University Press, p 180.
4. Russell, Bertrand. (1945) *The History of Western Philosophy*. Simon & Schuster, p46.
5. Einstein, Albert, (1954) *Ideas and Opinions*. Wings Books, p292.
6. Ibid, p290.
7. Merriam Webster unabridged collegiate dictionary at: unabridged.merriam-webster.com
8. Seth Loyd, *The computational Universe*, published on edge.com, 10/24/2002.

9. Pico, Richard M. (2002) Conscious Mind in Four Dimensions, biological relativity and the origins of thought. McGraw Hill, p13.
10. Richard M. Pico's book *Conscious Mind in Four Dimensions, biological relativity and the origins of thought*, is one of the inspiring sources for my approach to 'domains of evolution'.

Chapter 6: *Act Two, "Making a Living"*

1. Used with kind permission from Gregory G. and Mary Beth Dimijianhttp. For a selection from their extraordinary work see: www.dimijianimages.com or www.photoresearchers.com.
2. Deutsch, David, (1997) The Fabric of reality. Penguin Books, page ix.
3. Stuart Kauffman, What is Life?, published in: Brockman, John (2002) The next fifty years, science in the first half of the twenty-first century. Vintage Books, p128.
4. For an account of the evolution of the early earth, see: Liebes, Sidney, Sahtouris, Elisabet, Swimme, Brian (1998) *A walk through Time*. John Wiley & Sons, Inc.
5. Picture use, Courtesy Joachim Amkreutz.
6. Pico, Richard M. (2002) Conscious Mind in Four Dimensions. McGraw Hill, p35, p42-43. His account for the 'bubbles' is, like many other ideas about the origin of life, a hypothetical one. "We will employ this model of the primordial protocell as a hypothetical vehicle for the transition from pre-life to life. Thus, we can begin to imagine how a unique collection of molecules with their interrelated chemical reactions and products contained within the walls of phospholipid membranes, may have formed the first living cells .... Such naturally occurring spheres [bubbles] may have formed the basis of the first independent protocells, semiclosed volume spaces that could have survived in aqueous solutions of the primordial Earth. The phospholipid protocell, or a similar structure, would have provided a charged boundary, a barrier that created within its walls the first internal organic environment by capturing a tiny volume of the local aqueous solution. Thermodynamically, this can be viewed as a semiclosed (nonequilibrium) system that would have permitted the exchange of energy as well as matter through its semipermeable membrane of phospholipid strands, while being held in a three-dimensional spatial formation by the interaction of phosphate charge with the surrounding water molecules."
7. During these turbulent times, a new breed of green and purple bacteria produced the remarkable invention of photosynthesis, just a few hundred million years after the first cell appeared. Using hydrogen sulfide as their source of hydrogen, they absorb sunlight to synthesize organic molecules to build their bodies. However, they released sulfur as waste and released it into the atmosphere. Then nature changed the rules again. A new species, called cyanobacteria developed the capability to use water as the source for hydrogen, producing *oxygen* as waste. The best-known form of photosynthesis is the one carried out by higher plants and algae, as well as by cyanobacteria and their relatives, which are responsible for a major part of photosynthesis in oceans. All these organisms convert CO_2 (carbon dioxide) to organic material by reducing this gas to carbohydrates in a rather complex set of reactions. Electrons for this reduction reaction ultimately come from water, which is then converted to oxygen and protons. Energy for this process is provided by light, which is absorbed by pigments (primarily chlorophylls and carotenoids). Chlorophylls absorb blue and red light and carotenoids absorb blue-green light, but green and yellow light are not effectively absorbed by photosynthetic pigments in plants; therefore, light of these colors is either reflected by leaves or passes through the leaves. This is why plants are green. Source for information on photosynthesis: Wim Vermaas, *An Introduction to Photosynthesis and Its Applications*, Professor, Department of Plant Biology, and Center for the Study of Early Events in Photosynthesis Arizona State University, photoscience.la.asu.edu/photosyn/education/photointro.html.
8. Sources for information on the life and functionality of prokaryotes: Gould, Steven Jay, edt. (2001) *The Book of Life*. W.W. Norton & Company, p42, and Liebes, Sidney, Sahtouris, Elisabet, Swimme, Brian (1998) *A walk through Time*. John Wiley & Sons, Inc, p34.

9. For a bio on the illustrious scientist and political activist John Haldane, see: www.spartacus.schoolnet.co.uk/SPhaldane.htm. For one of Haldane's famous publications, DAEDALUS *or Science and the Future*, A paper read to the Heretics, Cambridge, on February 4, 1923, see: www.santafe.edu/~shalizi/Daedalus.html.
10. Haldane's and Oparin's theories differed in their assumptions about the basic materials (Haldane: carbon dioxide, ammonia and water vapor, Oparin: methane, ammonia, hydrogen and water vapor) used to create the first organic materials (amino acids), the material that provided the necessary carbon (Haldane: Carbon dioxide, Oparin: Methane) and the energetic forces that ignited the chemical reactions.
11. For a brief, informative summary of recent discoveries relating to the possible origin of life, see: www.chem.duke.edu/~jds/cruise_chem/Exobiology.
12. James Watson and Francis Crick, *Molecular Structure of Nucleic Acids: A Structure for Deoxyribose Nucleic Acid*, Volume 171 of the British journal Nature, April 25, 1953. Working with nucleotide models made of wire at the Cavendish Lab at Cambridge University, Watson and Crick attempted to put together the puzzle of DNA structure in such a way that their model would account for the variety of facts that they knew described the molecule. Once satisfied with their model, they published their hypothesis.
13. Image and caption source: www.ornl.gov/hgmis.
14. The explanation of 'genetics 101' is based on information from various sources, but mainly the explanations in: Grace, Eric S., (1997) *Biotechnology unzipped*. Joseph Henri press, p26 - 29.
15. The initial work on amino acids in Murchison was done in the laboratories of NASA Ames research center, and led to the first convincing evidence of amino acids of extraterrestrial origin. More than thirty years later analysis of Murchison meteorite samples continues to reveal exciting new results. These include the finding of an excess of certain amino acids and the finding of extraterrestrial helium. See: www.ast.cam.ac.uk/AAO/local/www/jab/astrobiology/murchison.html.
16. Source : www-curator.jsc.nasa.gov/curator/antmet/marsmets/alh84001/.
17. Source: www.chem.duke.edu/~jds/cruise_chem/Exobiology/sites.html.
18. Source: National academy of Sciences, Steering committee on Science and Creationism (1999) *Science and Creationism: A View from the National Academy of Sciences, Second Edition*. The National Academy Press. Electronic version free at: books.nap.edu/books/0309064066/html/index.html.
19. The website for the *Origin-of-Life Prize* is www.us.net/life. Readers are advised that this site invites serious scientific contributions only.
20. Pico, Richard M. (2002) Conscious Mind in Four Dimensions. McGraw Hill, p 50.
21. Schrödinger, Erwin, (1944, 1967) *What is Life? & Mind and Matter*. Cambridge University Press, p69.
22. Ibid, p79.
23. Ibid, p74.
24. Information on the appearance of EK's based on Gould, Steven Jay, edt, (2001) *The Book of Life*. W.W. Norton & Company, p43, 44, 45, and Mayr, Ernst, (2001) *What Evolution Is*. Basic Books, p42.
25. Gould, Steven Jay, edt, (2001) *The Book of Life*. W.W. Norton & Company, p44.
26. Description of Eukaryotes based on: David J. Patterson, School of Biological Sciences, University of Sydney, Australia, and Mitchell L. Sogin, The Josephine Bay Paul Center in Comparative Molecular Biology and Evolution Marine Biological Laboratory Woods Hole, USA: *Eukaryotes (Eukaryota), Organisms with nucleated cells*. The following description of characteristics of eukaryotes is quoted from this article. "The eukaryotes are distinguished from prokaryotes by the structural complexity of the cells - characterized by having many functions segregated into semi-autonomous regions of the cells (organelles), and by the cytoskeleton. The most evident organelle in most cells is the nucleus, and it is from the presence of this organelle that the eukaryotes get their name. Most cells have a single nucleus, some have more (some have thousands) and others like red blood cells of ourselves

have none - but they can be shown to derive from cells with nuclei. Nuclei contain most of the genetic material of a cell - with other elements of the genome located in mitochondria and plastids (if those organelles are also present). The nucleus is bounded by a membranous envelope. The nuclear envelope is part of the endomembrane system that extends to include the endoplasmic reticulum, dictyosomes (Golgi apparatus) and the cell or plasma membrane that encloses the cell. The envelope is perforated by nuclear pores, which allow compounds to pass between the nucleus and the surrounding cytoplasm. Some protists have more than one kind of nucleus - using one to retain a copy of the genome for purposes of reproduction, and another in which some genes have been greatly amplified, to regulate activities. Within the nucleus, the genes are located on a number of chromosomes. The total amount of DNA in a nucleus measuring less than one hundredth of a millimeter across may stretch to over a meter. When not in use this is kept within a nucleus measuring only a few microns across by being bundled up in superhelical arrays. The cytoskeleton is comprised of a rich array of proteins. The major ones are tubulin (which forms microtubules) and actin (forming microfilaments) and a myriad of interacting proteins, which effect movement or create the skeletal architecture of cells. The cytoskeleton provides shape for the cell and support for membranous organelles. It also provides anchorage for motility proteins which transport materials within the cell and cause deformations which bring about the movements of the entire cell - or organism." This article can be found at: tolweb.org/tree?group=Eukaryotes&contgroup=Life#about.

27. Dawkins, Richard, (1995) *River out of Eden, A Darwinian View of Life*. Basic Books, p46.
28. Symbiosis is the process in which unrelated organisms come together and form a stable association - as do lichens (algae and fungi) and coral reefs (coelenterates and dinoflagellate algae). This idea was promulgated at the turn of the 20th century (Mereschkowsky, 1910) and promoted later in the same century by Lynn Margulis (Margulis, 1970). Symbiosis has been and still is an important driving force in the evolution of eukaryotes. Through this mechanism, complementary metabolic capabilities, life cycles, and competences of different organisms have been brought together to create an amalgam that is greater than the sum of the parts. The close proximity of partners in symbiosis creates opportunities for coevolution of genomes.
29. Gould, Steven Jay, edt, (2001) *The Book of Life*. W.W. Norton & Company, p44, 45.
30. Ibid, p45.
31. Mayr, Ernst, (2001) *What Evolution Is*. Basic Books, p42. The appearance of the eukaryotes does not signal end of the prokaryotes. They remain abundant in our modern world. Their biomass is estimated to be the same as all of the eukaryotes; they live as parasites on organisms and on organic waste matter. An amazing fact is the constancy of prokaryotes: one third of the early fossil species are indistinguishable from their modern counterparts, according to Mayr.
32. As the reader might guess, *a history of the unconscious*, at least the one suggested by Alvas, does not exist. He mentions it here to draw attention to his own emerging existence, making sure the reader understands that he already existed at the time, and suggesting the beginning of some form of internal processing capability of organisms: the rudimentary beginning of a brain. The information about jellyfish, however, is accurate.
33. Permission to reproduce courtesy of digitalbiology.com. This company produces 3-dimensional 'digital displays' of biological life, for musea and exhibitions. For more information, see: www.digitalbiology.com/ and www.paleoindustrial.net.
34. Gould, Steven Jay, edt, (2001) *The Book of Life*. W.W. Norton & Company, p56.
35. Antrhopods belong to the phylum of *antrhopoda*. They are invertebrate animals with jointed limbs, the body divided into segments. Best known are the insects and the spiders, but many other, less well-known animals belong to this phylum.
36. Reprinted with kind permission from Gregory G. and Mary Beth Dimijianhttp. See: www.dimijianimages.com or www.photoresearchers.com.
37. Liebes, Sidney, Sahtouris, Elisabet, Swimme, Brian (1998) *A walk through Time*. John Wiley & Sons, Inc, p155.

38. Ibid, p190.
39. Gould, Steven Jay, edt, (2001) *The Book of Life*. W.W. Norton & Company, p219.
40. Ibid, p219.
41. Picture and caption source: http://www.cwu.edu/ ~cwuchci/ washoebio.html.
42. Gould, Steven Jay, edt, (2001) *The Book of Life*. W.W. Norton & Company, p214, 215.
43. For more information, see: http://www.nads-sc.uiowa.edu/director.htm.
44. Penrose, Roger, (2000) *The Large, the small and the Human mind*. Cambridge University Press, Canto Edition.

Chapter 7: The Life Domain

1. Reprinted with kind permission from Gregory G. and Mary Beth Dimijianhttp. See: www.dimijianimages.com or www.photoresearchers.com.
2. Chardin, Teilhard, (1955, 1975) *The Phenomenon of Man*. Harper & Row, p138.
3. Here are Frank Tipler's claims: "I have presented and defended my Omega Point Theory at length in my book *The Physics of Immortality* (Doubleday, 1994).... As science, the Omega Point Theory makes five basic claims about the universe: (1) the universe is spatially closed (has finite spatial size and has the topology of a three-sphere), (2) there are no event horizons, implying the future c-boundary is a point --- the Omega Point, (3) Life must eventually engulf the entire universe and control it, (4) the amount of information processed between now and the final state is infinite, (5) the amount of information stored in the universe diverges to infinity as the final state is approached." It should be noted that Tipler sees life transforming itself into intelligence near the very 'end', when he says: "Thus life (which near the final state, is really collectively intelligent computers) almost certainly must be present *arbitrarily close* to the final singularity in order for the known laws of physics to be mutually consistent at all times." See: www.math.tulane.edu/ ~tipler/ summary.html.
4. Deutsch, David, (1997) *The Fabric of reality*. Penguin Books, p179.
5. Ibid, p181.
6. For information on the different kingdoms, see: www.abdn.ac.uk/~nhi708/classify/kingdoms.html
7. Mayr, Ernst, (2001) *What Evolution Is*. Basic Books, p162.
8. Gould, Steven Jay, edt, (2001) *The Book of Life*. W.W. Norton & Company, p39.
9. Ibid, p39.

Chapter 8: *Act Three, "Creating Knowledge"*

1. *The Thinker*, sculpture by Auguste Rodin. The following short description can by found at www.rodinmuseum.org/index.html. "The Great French sculptor Auguste Rodin (1840-1917) brought monumental public sculpture into the 20th century and established a new sculptural freedom which continues to haunt our imaginations. His stated aim was to be faithful to nature; he steadfastly refused to idealize his subjects, creating instead an unprecedented combination of outer realism and psychological insight."
2. McLuhan, Marshall, (ninth printing, 2001, first edition 1964, MIT edition, 1994) *Understanding Media*. The MIT Press, p65.
3. Mayr, Ernst, (2001) *What Evolution Is*. Basic Books, pp 240-245.
4. www.culture.fr/culture/arcnat/chauvet/en/index.html On Sunday, December 18, 1994, Jean-Marie Chauvet led his two friends, Éliette Brunel and Christian Hillaire, on the Cirque d'Estre toward the cliffs.... They gathered up the essential tools, hesitated for a moment, and then returned to their discovery.... They explored almost the entire network of chambers and galleries.... They discovered hundreds of paintings and engravings. Their life is changed.... On December 29, 1994, an expedition, led by the discoverers, was undertaken. Direct dates obtained in 1995 have added an unexpected dimension to the discovery. Three samples taken from charcoal drawings of two rhinoceroses and one bison have yielded dates between 30,340 and 32,410 BP (before present). Considering the statistical

margins of error, this means that the paintings were made at the very ancient date of approximately 31,000 years ago, within an interval of 1,300 years.

5. Pico, Richard M. (2002) *Consciousness in Four Dimensions.* McGraw Hill, p172ff.
6. Ibid, p177.
7. Ibid, illustration based on Pico's book.
8. Ibid, p199.
9. Goldberg, Elkhonon, (2001) *The Executive Brain.* Oxford University Press, p ix.
10. Searle, John S., (1997) *The Mystery of Consciousness.* The New York Review of Books, p5.
11. Pico, Richard M. (2002) *Consciousness in Four Dimensions.* McGraw Hill, p199.
12. Edelman, Gerald M., Tononi, Giulio, (2000) *A Universe of Consciousness; how matter becomes consciousness.* Basic Books, p19.
13. Pinker, Steven. (2002) *The blank slate,* The modern Denial of Human Nature. Penguin Group.
14. Ibid, p421.
15. Damasio, Antonio, (1999) *The Feeling of What Happens.* Harcourt Brace & Company, p312.
16. Ibid, p 322.
17. Crick, Francis. (1994) *The Astonishing Hypothesis.* Touchstone, p3.
18. Ibid, p199.
19. Dawkins, Richard, (1976) *The selfish gene.* Oxford University Press.
20. Ibid, p192.
21. Ibid, p192.
22. McLuhan, Marshall, (ninth printing, 2001, first edition 1964, MIT edition, 1994) *Understanding Media.* The MIT Press, p7.
23. Shlain, Leonard, (1998) *The Alphabet versus the Goddess.* Penguin Books, Ltd. For information, wee also: http://www.artandphysics.com. In addition to being an author, Shlain is also Chief of Laparoscopic Surgery at California Pacific Medical Center in San Francisco and Associate Professor of Surgery at UCSF. He was a pioneer in the field of video-assisted laparoscopic surgery and presently holds five patents for surgical devices.
24. McLuhan, Marshall, (ninth printing, 2001, first edition 1964, MIT edition, 1994) *Understanding Media.* The MIT Press, p17, 18.
25. Johnson, Steven, (1997) Interface Culture, How new technology transforms the way we communicate. HarperCollins, p145.
26. Nicholas Negroponte in the forward to: Downes, Larry and Chunka Mui, Chunka. (1998) *Unleashing the Killer App.* Harvard Business School Press, pxi.
27. From a transcript of a seminar given by Nicholas Negroponte on June 6, 1976 at the University of Technology, Eijndhoven, the Netherlands. Transcript by the author.
28. Negroponte, Nicholas, (1995) *Being Digital.* First vintage books edition, p230.
29. Dyson, Esther. (1998) *Release 2.1.* Broadway Books, p340.
30. Levy Pierre, (1997, 2001) *Cyberculture.* University of Minnesota Press, p148.
31. Screenshot from a scene generated by the computer-video-game 'SimCity'. Courtesy: www.simcitycentral.net/screenshots.shtml".
32. Gartner's Predictions for 2002" can be found at: (www3.gartner.com/1_researchanalysis/focus/predictions2002.html).
33. *Self-Reliance and Creative Destruction,* by Prof. Bryan Caplan, Dep't of Economics,George Mason University, Virginia, USA. Source: www.gmu.edu/departments/economics/bcaplan/davis2.htm.
34. Mandel, Michael J. (2000) *The coming Internet depression.* Basic Books, p17.
35. Levy Pierre, (1997, 2001) *Cyberculture.* University of Minnesota Press.

PART III: MEMORIES OF THE FUTURE

1. United Nations Development Programme, (2002) *Human Development Report 2002, Deepening democracy in a fragmented world.* Oxford University Press, p14. Report available at: http://www.undp.org/hdr2002/complete.pdf.

2. Zey, Michael G., (2000) The Future Factor. The Five Forces Transforming our Lives and Shaping our Human Destiny. McGraw-Hill, p221.

Chapter 9: Digeality

1. Kaku, Michio, (1998) Visions: How Science Will Revolutionize the 21st Century. Bantam Books, p355.
2. Funding for the Sloan Digital Sky Survey (SDSS) has been provided by the Alfred P. Sloan Foundation, the Participating Institutions, the National Aeronautics and Space Administration, the National Science Foundation, the U.S. Department of Energy, the Japanese Monbukagakusho, and the Max Planck Society. Source: http://www.sdss.org/ digital sky Project.
3. See: space.jpl.nasa.gov/.
4. See: www.spaceholdings.com/contact_us.php & http://www.starrynight.com/.
5. See: www.syz.com/DU/index.shtml. Readers that are curious about tourism in space can check out www.spacefuture.com/tourism/tourism.shtml.
6. Technology Research News February 21, 2003: www.technologyreview.com/ offthewire/3001_rnb_022103_2.asp.
7. NSF PR 02-92 - November 14, 2002, Using Computers, Scientists Successfully Predict Evolution of E. Coli Bacteria, see: http://www.nsf.gov/od/lpa/news/02/pr0292.htm. Readers that like to play the 'evolutionary game' should visit www.alife.fusebox.com. The site is run by artists who "see these algorithm as a starting point for a new artistic exploration where the interactivity is not only between the user and the computer program but within the computer system itself."
8. See: www.dllab.caltech.edu/.
9. Richard Dawkins in *Son of Moore's Law*; essay published in: Brockman, John (2002) *The next fifty years, science in the first half of the twenty-first century*. Vintage Books, p155, 157.
10. Regis, Edward, (1995) Nano: The Emerging Science of Nanotechnology: Remaking the World-Molecule by Molecule. Little Brown & Company, p47.
11. Ibid, p48.
12. See: avst.larc.nasa.gov/downloads/Morphing_Prp_SPIE2002_Final.pdf.
13. *Reshaping aircraft*. Technology Review, March 2003, p27.
14. *Harnessing Quantum Bits*, Computers that can simultaneously process information in numerous alternate realities are less theoretical than you might think. MIT Technology Review, March 2003.
15. Business 2.0, August 2002, p53.
16. Ibid, p57.
17. See: www.darpa.mil/grandchallenge/rules.htm.
18. See: www.nads-sc.uiowa.edu.
19. *Simulating surgery*, Technology Review, March 2003, p26.
20. See: grin.hq.nasa.gov/ABSTRACTS/GPN-2000-001541.html.
21. *Futures Research quarterly*, summer 2002, Vol 18, Number 2, originally published in *the Financial Times*, Monday, March 25, 2002.
22. See: www.media.mit.edu/wearables/.
23. See: www.cs.uoregon.edu/research/wearables/.
24. See: wearables.cs.bris.ac.uk/.
25. See: www.wearable.ethz.ch/.
26. See: www.darpa.mil/iao/index.htm, http://www.darpa.mil/iao/TIASystems.htm.
27. See: www.darpa.mil/iao/HID.htm.
28. See: www.darpa.mil/iao/index.htm.
29. *The semantic web*, by Tim Berners-Lee, Scientific American, May 2001 issue.
30. *The Observant Computer*, Technology review, April 2003, p67.

Chapter 10: The Knowledge Domain

1. Huxley, Julian (1992) *Evolutionary Humanism*. Prometheus Books. [Originally published in 1964, as *Essays of a humanist*. Harper and Row], p81.
2. Bill Joy's, *Why the future doesn't need us* in WIRED magazine, Issue 8.04, April 2000.
3. JH2: Huxley, Julian (1992) *Evolutionary Humanism*. Prometheus Books. [Originally published in 1964, as *Essays of a humanist*. Harper and Row], p80.
4. Dyson, George B. (1997) *Darwin among the machines*. Perseus Books, p228.
5. Blackmore, Susan, (1999) *The Meme Machine*. Oxford University Press, p239.
6. Ibid, page xii.
7. Lamarck developed a theory of evolution, around the same time as Darwin, with a fundamental difference: Lamarck assumed the inheritability of characteristics that an organism acquired during its lifetime. Today, that theory has been refuted.
8. Blackmore, Susan, (1999) *The Meme Machine*. Oxford University Press, p214.
9. Dennett, Daniel, (1991) *Consciousness Explained*. Back Bay Books, p202.
10. Dawkins, Richard, (1995) *River out of Eden, A Darwinian View of Life*. Basic Books, p2.
11. Ibid, p151-160.

Chapter 11: Minding the Future

1. Jack N. Behrman, Luther Hodges Distuingished Professor Emeritus, University of North Carolina's Kenan-Flagler Business School, in *Moral Buttresses and Obstacles in the Globalization Process*, published in *Futures research Quarterly*, Volume 18, Number 4, winter 2002, p57.
2. Pinker, Steven. (2002) *The blank slate, The modern Denial of Human Nature*. Penguin Group, p373, 380.
3. Jung, C. G., (second edition, 1968, tenth printing, 1990) *The Archetypes and the Collective Unconscious*, Princeton University Press, p279.
4. Ibid, p288.
5. *After the Singularity*, A Talk with Ray Kurzweil, John Brockman, editor of Edge.org, interviewed Ray Kurzweil on the Singularity and its ramifications. Originally published on March 25, 2002 on Edge.com. Interview can be found on www.kurzweilai.net.
6. Kurzweil, Ray, *The Intelligent Universe*. Originally published on Edge.com, Nov. 7, 2002. Published Dec. 12, 2002. on: www.kurzweilai.net.
7. Rodney Brooks, *The Merger of Flesh and Machines*, in: Brockman, John (2002) The next fifty years, science in the first half of the twenty-first century. Vintage Books, p187. Rodney Brooks is director of the Artificial Intelligence Laboratory and Fujitsu Professor of computer science at the Massachusetts Institute of technology (MIT).
8. Mulhall, Douglas (2002) Our Molecular Future, how nanotechnology, robotics, genetics, and artificial intelligence will transform our world. Prometheus Books, p312.

Chapter 12: The Digital Spirit

1. James Watson, co-discoverer (with Francis Crick) of the double-helix structure of the DNA molecule, in an interview with the German magazine *Der Spiegel*, February 2003.
2. *The New War Machines*, interview with MIT professor and security expert Owen Cote discusses "smart" munitions and other tools in the high-tech U.S. arsenal. Interview published in technology review, March 2003.
3. Ibid.
4. James Watson, co-discoverer (with Francis Crick) of the double-helix structure of the DNA molecule, in an interview with the German magazine *Der Spiegel*, February 2003.
5. Henderson, hazel, (1999) *Beyond Globalization*. The Sarov Press, p21.

Bibliography

1. Anton, Philip S., Silberglitt, Richard, Schneider, James, (editors, 2001) *The Global Technology Revolution*. RAND.
2. Armstrong, Karen, (2001) *The battle for God*. Ballantine Books.
3. Barnett, Lincoln, (1952) *Einstein und das Universum*, Fischer Bucherei, (Translated from *The Universe and Dr. Einstein*, 1949).
4. Bergson, Henri, (1911, 1998) *Creative Evolution*. Dover Publications, Inc.
5. Bertman, Steven, (2000) *Cultural Amnesia*. Preager Publishers.
6. Bhalla, Surjit (2002) *Imagine There's No Country: Poverty, Inequality, and Growth in the Era of Globalization*. Institute for International Economics, ISBN: paper 0-88132-348-9, available at: www.iie.com/publications/publication.cfm?pub_id=348
7. Blackmore, Susan, (1999) *The Meme Machine*. Oxford University Press.
8. Bono, Edward, de, (1971) *the use of Lateral Thinking*. Pelican books.
9. Brockman, John (1995) *The Third Culture, Beyond the Scientific Revolution*. Simon & Schuster.
10. Brockman, John (2002) *The next fifty years, science in the first half of the twenty-first century*. Vintage Books.
11. Buchanon, Patrick J., (2002) *The Death of the West*. Thomas Dunne Books.
12. Canton, James. (1999) *Technofutures*, How leading-edge technology will transform business in the 21st century. Hay House, Inc.
13. Castells Manuel, (2001) *The Internet Galaxy, Reflections on the Internet, Business, and Society*. Oxford University Press Inc., New York.
14. Chandler, Keith, (2001) *The Mind Paradigm*. Iuniverse.com.
15. Chardin, Teilhard, (1955, 1975) *The Phenomenon of Man*. Harper & Row.
16. Coates, Joseph, Mahaffie, John B., Hines, Andy, (1997) *2025, Scenarios of US and Global Society Reshaped By Science and Technology*. Oakhill Press.
17. Coates, Joseph, (2002) *The next thousand years*, World Future Society Conference, July 2002.
18. Coleman, Daniel, (2000) *Working with Emotional Intelligence*. Bantam Books.
19. Crick, Francis. (1994) *The Astonishing Hypothesis*. Touchstone.
20. Damasio, Antonio, (1999) *The Feeling of What Happens*. Harcourt Brace & Company.
21. Dennett, Daniel, (1991) *Consciousness Explained*. Back Bay Books.
22. Dawkins, Richard, (1989 edition, originally published 1976) *The selfish gene*. Oxford University Press.
23. Dawkins, Richard, (1995) *River out of Eden, A Darwinian View of Life*. Basic Books.
24. Deise Martin V., Nowikow Conrad, King, Patrick, and Wright Amy (2000) *E- Busines*. Price Waterhouse Coopers LLP.
25. Dertouzos, Michael, (2001) *The Unfinished Revolution*. Harper books.
26. Deutsch, David, (1997) *The Fabric of reality*. Penguin Books.
27. Downes, Larry and Chunka Mui, Chunka. (1998) *Unleashing the Killer App*. Harvard Business School Press.
28. Drucker, Peter F., (2001) *The Essential Drucker*. HarperCollins Books.
29. Durant, Will. (1968)Van Socrates tot Bergson. Salamander Books. [translated from Will Durant, The story of philosophy, 1939]
30. Dutcher, Jim and Jamie, (2002) *Wolfs at our door*. Pocket Books.
31. Dyson, Esther. (1998) *Release 2.1*. Broadway Books.
32. Dyson, Freeman, (1997) *Imagined Worlds*. Harvard University Press.
33. Dyson, Freeman J., (1999) *The Sun, The Genome, and The Internet*. Oxford University Press.
34. Dyson, Freeman, (2001) *Disturbing the Universe*. Basic Books.
35. Dyson, George B. (1997) *Darwin among the machines*. Perseus Books.
36. Dyson, James. [editor Robert Uhlig] (2001) *A History of Great Inventions*. Caroll & Graf Publishers.
37. Edelman, Gerald M., Tononi, Giulio, (2000) *A Universe of Consciousness; how matter becomes consciousness*. Basic Books.

38. Eddington, Sir Arthur, (1933) *The expanding universe,* Cambridge University press.
39. Einstein, Albert, (1954) *Ideas and Opinions.* Wings Books.
40. Feynman, Richard P., (1985) *QED.* Princeton University Press.
41. Fukuyama, Francis. (2002) *Our Posthuman Future.* Farrar, Straus & Giroux.
42. Gates, Bill, (1999) *Business at the speed of thought.* Warner Books, Inc.
43. Gates, Bill, (1995) *The Road Ahead.* Penguin Books.
44. Gilder, George, (2002) *Telecosm.* Touchstone Books.
45. Gleick, James, (2002) *What just happened,* Random House.
46. Goldberg, Elkhonon, (2001) *The Executive Brain.* Oxford University Press.
47. Gödel, Kurt, (originally published 1931, 1992) *On formally undecidable propositions of principia mathematica and related systems.* Dover Publications.
48. Goldberg, Bernard, (2002) *BIAS.* Regnery Publishing, Inc.
49. Goswami, Amit, (1995) *The Self-Aware Universe.* Tarcher/Putnam.
50. Gould, Steven Jay, (2002) *The Structure of Evolutionary Theory.* The Belknap Press of Harvard University Press.
51. Gould, Steven Jay, edt, (2001) *The Book of Life.* W.W. Norton & Company.
52. Grace, Eric S., (1997) *Biotechnology unzipped.* Joseph Henri press.
53. Hagen, Steve, (1999) *Buddhism, Plain & Simple.* Broadway Books.
54. Hamel, Gary, (2000) *Leading the revolution.* Harvard Business School Press.
55. Harris, Marvin, (2001) *The Rise of Anthropological Theory.* AltaMira Press.
56. Hawking, Stephen W., (1988) *A brief History of time.* Bantam Books.
57. Hawking, Stephen W., (2001) *The Universe in a nutshell.* Bantam Books.
58. Hearnshaw, L. S., (1987) *The Shaping of Modern Psychology.* Routledge and Kegan Paul.
59. Henderson, hazel, (1999) *Beyond Globalization.* The Sarov Press.
60. Hesselbein, Frances, Goldsmith, Marshall and Beckhard, Richard (editors), (1996) *The Leader of the Future.* The Drucker Foundation.
61. Hodges, Andrew, (1999) *Turing.* Routledge.
62. Hofstadter, Douglas R., (1979,1999) *Godel, Escher, Bach: an Eternal Golden Braid.* Basic Books.
63. Hunt, Morton, (1993) *The story of psychology.* Anchor Books.
64. Huxley, Julian (1992) *Evolutionary Humanism.* Prometheus Books. [Originally published in 1964, as *Essays of a humanist.* Harper and Row].
65. Ions, Veronica, (1999) *The History of Mythology.* Quadrillion Publishing, Limited.
66. Irwin, Robert, (1999) *Buying a Home On the Internet.* McGraw-Hill.
67. James, William, (1948) *Psychology.* World.
68. Johnson, Steven, (1997) *Interface Culture, How new technology transforms the way we communicate.* HarperCollins.
69. Johnson, Steven, (2001) *Emergence.* Scribner.
70. Jung, C. G., (second edition, 1968, tenth printing, 1990) *The Archetypes and the Collective Unconscious,* Princeton University Press.
71. Kaku, Michio, (1998) *Visions: How Science Will Revolutionize the 21st Century.* Bantam Books.
72. King, Ursula, (2000) *Spirit of Fire, the Life and Vision of Teilhard de Chardin.* Orbis Books.
73. Korten, David C. (1999) *The post corporate world.* Berrett-Koehler Publishers Inc and Kumarian Press Inc.
74. Kristof, David and Nickerson, Todd W., (1998) *predictions for the next millennium.* Andrews McMeel Publishing.
75. Kurzweil, Ray. (1999) *The age of spiritual machines, when computers exceed human intelligence.* Penguin Books.
76. LeDoux, Joseph, (2002) *Synaptic Self.* Viking, Penguin Group.
77. LeDoux, Joseph, (1996) *The Emotional Brain.* Touchstone.
78. Levine, Rick, Locke, Christopher, Searls, Doc and Weinberger, David, (2000) *The Cluetrain Manifesto.* Perseus Books.
79. Levy Pierre, (1997, 2001) *Cyberculture.* University of Minnesota Press.

80. Levy, Pierre, (1997, 1999) *Collective Intelligence*. Perseus Books.
81. Liebes, Sidney, Sahtouris, Elisabet, Swimme, Brian (1998) *A walk through Time*. John Wiley & Sons, Inc.
82. Lightman, Alan, (1993) *Einstein's Dreams*. Warner Books, Inc.
83. Mandel Michael J. (2000) *The coming Internet depression*. Basic Books.
84. Marx Hubbard, Barbara, (2001) *Emergence*. Hampton Roads Publishing Company.
85. Maslow, Abraham H., (1961, 1998) *Maslow on Management*. John Wiley & Sons.
86. Maslow, Abraham H., (1968) *Toward a Psychology of Being*, second edition. Litton Educational Publishing.
87. Maslow, Abraham H., (1970) *Motivation and Personality*. 2nd edition, Harper and Row.
88. Mayr, Ernst, (2001) *What Evolution Is*. Basic Books.
89. McLuhan, Marshall, (2001, first edition 1964) *Understanding Media*. The MIT Press.
90. Meadows, Dennis, Meadows, Donella, Randers, Jorgen and Behrens, William, (1972) *Rapport van de club van Rome*. Uitgeverij Het Spectrum N.V., (originally published as: *The Limits to growth – A Report for the Club of Rome Project on The Predicament of Mankind*, Universe Books, New York)
91. Mello, Anthony De, (1992) *Awareness, The Perils and Opportunities of Reality*. Doubleday, Image Books.
92. Michalko, Michael, (2001) *Cracking Creativity, the secret of creative genius*. Ten Speed Press.
93. Minski, Marvin, (1988) *The Society of Mind*. Touchstone.
94. Moravec, Hans, (2000) *Robot*. Oxford University Press.
95. Morrison, Reg. (1999) *The Spirit in the Gene*. Cornell University Press.
96. Mulhall, Douglas (2002) *Our Molecular Future, how nanotechnology, robotics, genetics, and artificial intelligence will transform our world*. Prometheus Books.
97. Naisbitt, John, Aburdene, Patricia, (1990) *Megatrends 2000*. Megatrends Ltd.
98. Naisbitt, John, with Naisbitt, Nana and Philips, Douglas, (1999) *High tech High touch*. Broadway Books.
99. National academy of Sciences, Steering committee on Science and Creationism (1999) *Science and Creationism: A View from the National Academy of Sciences, Second Edition*. The National Academy Press. Electronic version free at: http://books.nap.edu/books/0309064066/html/index.html
100. Negroponte, Nicholas, (1995) *Being Digital*. First vintage books edition.
101. Orwell, George, (1949) *1984*. Signet Classic.
102. Pal, Nirmal, Ray, Judith M., (2001) *Pushing the Digital Frontier*. Amacom.
103. Paul, Gregory S., Cox, Earl D. (1996) *Beyond Humanity, Cyber Evolution and Future Minds*. Charles River Media, Inc.
104. Pascale, Richard T., Millemann, Mark and Gioja, Linda, (2000) *Surfing the edge of Chaos*. Three rivers press.
105. Penrose, Roger, (2000) *The Large, the small and the Human mind*. Cambridge University Press, Canto Edition.
106. Penrose, Roger, (1994) *Shadows of the Mind*. Oxford University Press.
107. Pico, Richard M. (2002) *Consciousness in Four Dimensions, biological relativity and the origins of thought*. McGraw Hill.
108. Pink, Daniel H. (2001) *Free agent Nation*. Warner Books.
109. Pinker, Steven. (2002) *The blank slate, The modern Denial of Human Nature*. Penguin Group.
110. Popper, Karl. (2002, first published 1935) *The logic of Scientific Discovery*. Routledge Classics.
111. Pottruck, David S. and Piece, Terry, (2000) *Clicks And Mortar*. Jossey-Bass Inc.
112. Quinn, Daniel, (1999) *Beyond Civilization, Humanity's next great adventures*. Three rivers press.
113. Rees, Martin, (2001) *Our cosmic habitat*. Princeton University Press.
114. Redfield, James, Murphy, Michael, Timbers, Sylvia, (2002) *God and the evolving universe*. Jeremy T. Tarcher/Putnam.

115. Regis, Edward, (1995) *Nano: The Emerging Science of Nanotechnology: Remaking the World-Molecule by Molecule.* Little Brown & Company.
116. Russell, Bertrand. (1945) *The History of Western Philosophy.* Simon & Schuster.
117. Sagan, Carl, (1977) *The Dragons of Eden.* Ballantine Books.
118. Savage, Marshall T. (1994) *The Millennial Project.* Little Brown & Company.
119. Schopenhauer, Arthur, *Aphorismen zur Lebensweisheit.* Eduard Kaiser Verlag.
120. Schopenhauer, Arthur, (1851, translation 1970) *Essays And Aphorisms.* Translation 1970 by R. J. Holling Dale, Penguin books.
121. Schrödinger, Erwin, (1944, 1967) *What is Life? & Mind and Matter.* Cambridge University Press.
122. Searle, John R., (1997) *The Mystery of Consciousness.* The New York Review of Books.
123. Searle, John R., (2001) *Rationality in Action.* A Bradford Book.
124. Searle, John R. (reissue 2002) *The Rediscovery of the Mind.* MIT Press.
125. Shannon, Claude, E. (1948), *A Mathematical Theory of Communication.* The Bell System Technical Journal, Vol. 27, pp. 379-423, 623-656, July, October 1948.
126. Shlain, Leonard, (1998) *The Alphabet versus the Goddess.* Penguin Books, Ltd.
127. Smith, Huston, (1990) *The world's Religions.* HarperCollins.
128. Strauss, William and Howe, Neil, (1997) *The Fourth Turning.* Bantam Doubleday Dell Publishing Group, Inc.
129. Tapscott, Don, Ticoll, David and Lowy, Alex, (2000) *Digital Capital.* Harvard Business School Press.
130. Tipler, Frank, (1995) *The Physics of Immortality: Modern Cosmology, God and the Resurrection of the Dead.* Anchor Books.
131. Toffler, Alvin, (1990) *Powershift, Knowledge, wealth, and violence a the edge of the 21st century.* Bantam Books.
132. Toffler, Alvin, (1985) *The Adaptive Corporation.* Pan Books.s
133. United Nations, (2000) *We the Peoples: The role of the United Nations in the 21st Century.* United Nations. (Available at: www.un.org).
134. United Nations Development Programme, (2002) *Human Development Report 2002, Deepening democracy in a fragmented world.* Oxford University Press. Report available at: http://www.undp.org/hdr2002/complete.pdf
135. Weelen, Paul, (1995) *Limburg bevrijd,* (the liberation of Limburg). Van Geyt, Ljubljana.
136. Williams, Reg. (2002) *Arguing A.I.; The battle for Twenty-first-Century Science.* Atrandom.com, Random House Inc.
137. Williamson, Marianne, (2000) *Imagining what America could be in the 21st century.* Highbridge Company.
138. Wright, Robert, (2001) *Nonzero: The logic of human destiny.* Vintage Books.
139. Zey, Michael G., (2000) *The Future Factor. The Five Forces Transforming our Lives and Shaping our Human Destiny.* McGraw-Hill.
140. Zey, Michael G., (1998) *Seizing the future: The Dawn of the Macroindustrial Era.* Second edition, Transaction Publishers.

Index

Acknowledgements

Many people shape the landscapes that we call our lives. They cross our paths, sometimes joining us, sometimes just visiting. They always leave their imprints that change forever the way we think. I owe thanks to many people, especially those that simply let me *be*; they are the landmarks in my journey of discovery that, finally, led to the writing of this book.

As a young boy, my father lit the flame of curiosity about everything unknown when he gave me one of his favorite books. It was a book about the universe, its title *Song from the Heavens*, written by J. N. Lenz and published in 1936. *Songs from the Universe* is the title of part two of my book. Now you know why. The flame never dimmed. Thanks dad, wherever you might be. My mother can create sunshine out of rain and a song out of tears. She never surrendered to the present, but always looked ahead. I guess that's in the genes she gave to me. They made me go on, try again, and again. Thanks mam. I am blessed to be part of a family that forged unconditional trust between us from the sparks of sibling rivalry. Thanks to my six brothers and my one sister that had the impossible task to keep a semblance of a gender balance.

Three of them were actively involved in the progress of this book. I am indebted to Guido Amkreutz for the countless hours he spent on the manuscript, and for the discussions that sharpened my focus, and to Hub Amkreutz, who supported the project in various ways. My deepest gratitude goes to Joachim, colleague of ten years, youngest brother, mentor, and friend. He supported me, asked critical questions, provided new insights, and pointed out the light in the dark days of doubt.

Thanks to my buddy Jan Last, for a lifelong friendship and his unwavering confidence in my ventures, especially this one. This project started to take shape in the weekly discussions with my German American friend William Brooks, and accelerated after continued encouragement from Dr. Linda Groff. I thank them both, not in the least for their constant efforts to make me look beyond the observable. However, I did stay within: the observable is hard enough to understand as it is and that we can understand anything at all is a gift that we have hardly begun to use.

Thanks to Ruby Cheung for her extensive help with the manuscript, and for so many years of support and friendship, and to Sallie Reisner, Ivan Fuchs and Scott Boyd for their help on language. They did what they could, but the flaws are in my blood, or better: in my memes.

Finally, this book is dedicated to my wife Marlies, my children and their partners, and my grandchildren. They are beyond the reach of thanks, because they are the home of my mind.

About the Author

Photographed by sciluxstudios.com

Jan Amkreutz spent his professional life in the trenches of the digital technology frontier, as a programmer, scientist, University professor, corporate executive and entrepreneur. He lived and worked in Europe, the Far East and North America. He currently lives with his wife Marlies in Montana, USA.

Jan Amkreutz is a professional member of the World Futurist Society (www.wfs.org), and a founding member of the Association of Professional Futurists (www.profuturists.com). Through his new venture, *Digital Crossroads*, he offers research and consulting services to organizations that anticipate the long-term influence of digital technology, and is currently working on a new book that discusses *how* digital technology will change our social, professional and personal lives in the 21st Century. He can be contacted at jan@digeality.com or through his website at www.digeality.com.

Digital Spirit can be ordered at:

- 1stbooks.com
- amazon.com
- barnesandnoble.com
- borders.com
- digeality.com
- at your favorite bookstore

Printed in the United States
1140700003B/37-279

A

Humanity is going digital. In a fascinating journey through time and spa bold new perspective on this new reality, a perspective that will change y present and the future. A perspective that will change your mind.

B

"What in the world is going on, Novare?"
"Humankind is crossing a threshold, Alvas. The third threshold in the history of evolution."
"*Third threshold,* Novare?"
"That is where your digital twin comes in."
"My digital twin?"
"Digital technology, Alvas. It is about to transform the future."
"Technology is?"
"Digital technology is. A new reality is emerging, Alvas. It's digital. We call it digeality."
"Digeality?"
"It is the third reality produced by nature. This time, by human nature."
"Third reality?"
"This new reality will transform the way we think. It makes us co-choreographers of the future."
"I hate it when you talk in riddles, Novare. *Co-choreographers*?"
"It's all in the book, Alvas, as you know very well. After all you and I are its co-authors."
"Just teasing, Novare. Let's introduce ourselves."
"I am sorry, dear reader, for not introducing ourselves. Alvas and I control the author's mind. Which, of course, he doesn't admit."

D

Humanity approaches a pivotal point in the history of evolution, a point of no return. We are leaving the Darwinian world of defenseless mutation and entering a world of global knowledge, where choice transcends randomness and purpose redefines destiny. A new beginning waits beyond the threshold, not a lingering Armageddon; a new dance, not the end of the music. Kurzweil's *intelligent machines* are invited to that dance, but the human mind will be the choreographer. Dawkins' *selfish genes* will continue their fight for dominance, but they are loosing the battle against our knowledge. The universe will continue to expand, and our existence will expand with it. Jan Amkreutz and his mind-companions Novare and Alvas explain *why* the emergence of the digital reality lifts human nature above the swamp of biological evolution, and marks our entry into a new domain of human *being.*

E

Jan Amkreutz spent his professional life in the trenches of the digital technology frontier, as a programmer, scientist, University professor, corporate executive and entrepreneur. He lived and worked in Europe, the Far East and North America. He currently lives with his wife Marlies in Montana, USA.